140,-

Technik

'90

Conference Proceedings of the Society for Experimental Mechanics Series

Series Editor

Kristin B. Zimmerman, Ph.D.
Society for Experimental Mechanics, Inc.,
Bethel, CT, USA

More information about this series at http://www.springer.com/series/8922

Andreas Linderholt • Matthew S. Allen • Randall L. Mayes • Daniel Rixen
Editors

Dynamic Substructures, Volume 4

Proceedings of the 37th IMAC, A Conference and Exposition on Structural Dynamics 2019

Springer

Editors
Andreas Linderholt
Mechanical Engineering
Linnaeus University
Växjö, Sweden

Matthew S. Allen
University of Wisconsin–Madison
Madison, WI, USA

Randall L. Mayes
Bldg 860 Room 201D, MS 0557
Sandia National Laboratory
Albuquerque, NM, USA

Daniel Rixen
Lehrstuhl für Angewandte Mechanik
Technische Universitat Munchen
Garching, Germany

ISSN 2191-5644 ISSN 2191-5652 (electronic)
Conference Proceedings of the Society for Experimental Mechanics Series
ISBN 978-3-030-12183-9 ISBN 978-3-030-12184-6 (eBook)
https://doi.org/10.1007/978-3-030-12184-6

This Springer imprint is published by the registered company Springer Nature Switzerland AG.
The registered company address is: Gewerbestrasse 11, 6330 Cham, Switzerland

Preface

Dynamic Substructures represents one of eight volumes of technical papers presented at the 37th IMAC, A Conference and Exposition on Structural Dynamics, organized by the Society for Experimental Mechanics and held in Orlando, Florida, on January 28–31, 2019. The full proceedings also include volumes on *Nonlinear Structures & Systems; Dynamics of Civil Structures*; *Model Validation and Uncertainty Quantification*; *Special Topics in Structural Dynamics & Experimental Techniques*; *Rotating Machinery, Optical Methods & Scanning LDV Methods*; *Sensors and Instrumentation, Aircraft/Aerospace, Energy Harvesting & Dynamic Environments Testing*; and *Topics in Modal Analysis &Testing*.

Each collection presents early findings from experimental and computational investigations on an important area within structural dynamics. Coupled structures or substructuring is one of these areas.

Substructuring is a general paradigm in engineering dynamics where a complicated system is analyzed by considering the dynamic interactions between subcomponents. In numerical simulations, substructuring allows one to reduce the complexity of parts of the system in order to construct a computationally efficient model of the assembled system. A subcomponent model can also be derived experimentally, allowing one to predict the dynamic behavior of an assembly by combining experimentally and/or analytically derived models. This can be advantageous for subcomponents that are expensive or difficult to model analytically. Substructuring can also be used to couple numerical simulation with real-time testing of components. Such approaches are known as hardware-in-the-loop or hybrid testing.

Whether experimental or numerical, all substructuring approaches have a common basis, namely, the equilibrium of the substructures under the action of the applied and interface forces and the compatibility of displacements at the interfaces of the subcomponents. Experimental substructuring requires special care in the way the measurements are obtained and processed in order to assure that measurement inaccuracies and noise do not invalidate the results. In numerical approaches, the fundamental quest is the efficient computation of reduced order models describing the substructure's dynamic motion. For hardware-in-the-loop applications, difficulties include the fast computation of the numerical components and the proper sensing and actuation of the hardware component. Recent advances in experimental techniques, sensor/actuator technologies, novel numerical methods, and parallel computing have rekindled interest in substructuring in recent years leading to new insights and improved experimental and analytical techniques.

The organizers would like to thank the authors, presenters, session organizers, and session chairs for their participation in this track.

Växjö, Sweden — A. Linderholt
Madison, WI, USA — M. Allen
Albuquerque, NM, USA — R. Mayes
Garching, Germany — D. Rixen

Contents

Chapter 1
Using Laser Vibrometry for Precise FRF Measurements in Experimental Substructuring

Francesco Trainotti, Tobias F. C. Berninger, and Daniel J. Rixen

Abstract The acquisition of high quality FRF measurements is a key factor for a successful implementation of coupling/decoupling techniques in Experimental Dynamic Substructuring. Although the use of piezo accelerometers as response transducers is very popular for impact testing due to its easy and fast implementation, the level of accuracy could not be adequate in certain applications. The laser technology provides a non-invasive alternative to standard piezo devices. The choice of a non-contact measurement technique allows to minimize the impact of external dynamic systems on the test component during the measurement process. In this paper, a validation of Lagrange Multiplier—Frequency Based Substructuring coupling by means of a Virtual Point Reduction is performed on a benchmark structure with a non-stiff interface. The necessary FRF data is acquired twice, using accelerometers and a laser Doppler vibrometer respectively. Both coupling results are compared to each other and are shown to match very well simulation data up to a high frequency range. The results underline the potential of high quality, non-intrusive measurements for Frequency Based Substructuring.

Keywords Experimental dynamics · Dynamic Substructuring · Frequency Based Substructuring · Virtual Point Transformation · FRF measurements · Laser vibrometry

1.1 Introduction

In Dynamic Substructuring (DS) the concept of modular design can be reinterpreted from a structural dynamics point of view, making possible the modelling, analysis and optimization of a complex system on a substructure level. Although various methodologies and technical solutions within DS are well documented [1], it still remains challenging to validate theoretical concepts in the framework of an industrial application. Significant difficulties in Experimental Dynamic Substructuring (EDS) concern the coupling/decoupling of the measured components. In this context, a frequency-based formulation of the problem is recommended as it directly includes the measured Frequency Response Functions (FRFs) in the implemented methods. Furthermore, a proper coupling of substructures strongly depends on a complete and accurate modelling of interface dynamics. It is common practice to 'weaken' the interface problem by projecting the dynamics into a subspace composed by Interface Deformation Modes (IDMs), which aren't global vibration modes but rather kinematic assumptions of the local deformation behaviour at the interface. This approach uses so-called virtual points to connect the components as in finite element models, overcoming some of the issues arising from experimental practice [2]. Inaccuracies and sources of error related to the application of experimental Frequency Based Substructuring (FBS) techniques by means of Virtual Point Transformation (VPT) arise from the complexities associated with both reliable and accurate data acquisition and high quality interface modeling. Regarding the errors exclusively generated by FRF measurements, different sources of disturbances can be further distinguished:

- Modification of the signal arising from data acquisition and signal processing [3, 4].
- Influence of external dynamic systems on the measured component (e.g. support mechanisms, attached measurement devices) [5].
- Random and systematic errors (e.g. environmental noise, sensor noise and positioning).

Moreover, the uncertainty on the measurements can be highly amplified by the DS algorithm due to numerical instabilities and induces spurious peaks and inaccuracies in the coupling results [6, 7].

F. Trainotti (✉) · T. F. C. Berninger · D. J. Rixen
Faculty of Mechanical Engineering, Technical University of Munich, Garching, Germany
e-mail: francesco.trainotti@studenti.unitn.it; t.berninger@tum.de; rixen@tum.de

A. Linderholt et al., *Dynamic Substructures, Volume 4*, Conference Proceedings of the Society for Experimental Mechanics Series,
https://doi.org/10.1007/978-3-030-12184-6_1

In particular, the use of standard measurement devices like piezo accelerometers in the FRFs acquisition process may not be suitable for applications requiring a high level of accuracy. Therefore, a valid alternative for motion detection can be identified in the use of laser interferometry, thanks to its non-invasive and high quality measurement technique.

In this contribution, the potential of laser technology in the acquisition of FRFs within FBS is investigated. The reliability of laser measurements compared to those taken with standard accelerometers is evaluated through the analysis of the final coupling results.

In Sect. 1.2, the theoretical background on Experimental Dynamic Substructuring is recalled, with particular focus on frequency-based approaches and interface modelling techniques. The laser vibrometry is shortly discussed in Sect. 1.3 along with its advantages and drawbacks over traditional measurement techniques. In Sect. 1.4, two measurement campaigns are carried out, one with accelerometers and the other with a laser Doppler vibrometer. A DS coupling is performed and the results of both case studies are compared. A brief summary of findings and conclusions is given in Sect. 1.5.

1.2 Experimental Dynamic Substructuring

This section briefly reviews the theoretical concepts underlying the Experimental Dynamic Substructuring. In particular, a general overview of Frequency Based Substructuring and Virtual Point Transformation is provided in Sects. 1.2.1 and 1.2.2 respectively.

1.2.1 Frequency Based Substructuring

The assembly procedure in the frequency domain according to a dual approach is named Lagrange Multiplier—Frequency Based Substructuring (LM-FBS) [1, 8, 9]. This method, which operates with admittance notation, evaluates locally a set of interface DoFs for each single substructure in the system and considers the interface forces as unknown variables. The aim of LM-FBS is to derive the admittance of the assembled system $\boldsymbol{Y}^{AB}$ from the separate admittances of the two subsystems $\boldsymbol{Y}^{A}$ and $\boldsymbol{Y}^{B}$.

Consider the system depicted in Fig. 1.1. The subsystems' admittances are known and the substructures' DoFs are grouped in internal DoFs ($(*)_1^A$ and $(*)_3^B$) and interface DoFs ($(*)_2^A$ and $(*)_2^B$). The vectors of displacements, applied forces and reaction forces are denoted by $\boldsymbol{u}$, $\boldsymbol{f}$ and $\boldsymbol{g}$ respectively. The governing equation of motion for the uncoupled system in the frequency domain is written in a compact form:

$$\boldsymbol{u} = \boldsymbol{Y}(\boldsymbol{f}+\boldsymbol{g}) \implies \begin{bmatrix} \boldsymbol{u}_1^A \\ \boldsymbol{u}_2^A \\ \boldsymbol{u}_2^B \\ \boldsymbol{u}_3^B \end{bmatrix} = \begin{bmatrix} \boldsymbol{Y}_{11}^A & \boldsymbol{Y}_{12}^A & \boldsymbol{0} & \boldsymbol{0} \\ \boldsymbol{Y}_{21}^A & \boldsymbol{Y}_{22}^A & \boldsymbol{0} & \boldsymbol{0} \\ \boldsymbol{0} & \boldsymbol{0} & \boldsymbol{Y}_{22}^B & \boldsymbol{Y}_{23}^B \\ \boldsymbol{0} & \boldsymbol{0} & \boldsymbol{Y}_{32}^B & \boldsymbol{Y}_{33}^B \end{bmatrix} \left(\begin{bmatrix} \boldsymbol{f}_1^A \\ \boldsymbol{f}_2^A \\ \boldsymbol{f}_2^B \\ \boldsymbol{f}_3^B \end{bmatrix} + \begin{bmatrix} \boldsymbol{0} \\ \boldsymbol{g}_2^A \\ \boldsymbol{g}_2^B \\ \boldsymbol{0} \end{bmatrix} \right) \tag{1.1}$$

The matrix $\boldsymbol{Y}$ represents the admittance of the uncoupled system, built in a block diagonal form by the admittances of the subsystems $\boldsymbol{Y}^{A}$ and $\boldsymbol{Y}^{B}$.

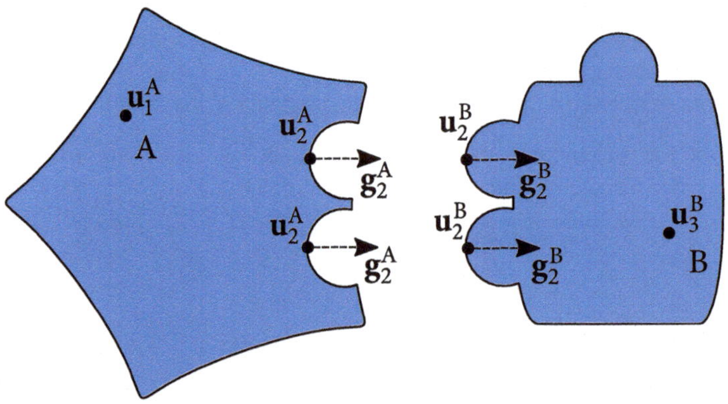

Fig. 1.1 Assembly of subsystems A and B at the interface DoFs $\boldsymbol{u}_2$

The knowledge of the uncoupled system at substructure level and the proper application of boundary conditions on the common interface allow to obtain the dynamical properties of the coupled system. The key step of the assembly procedure is the definition of interface conditions [9]:

- Compatibility condition.
- Equilibrium condition.

The first condition defines the compatibility of displacements at the common boundary:

$$\boldsymbol{B}\boldsymbol{u} = \boldsymbol{0} \implies \boldsymbol{u}_2^B - \boldsymbol{u}_2^A = \boldsymbol{0}, \quad \boldsymbol{B} = \begin{bmatrix} \boldsymbol{0} & -\boldsymbol{I} & \boldsymbol{I} & \boldsymbol{0} \end{bmatrix} \tag{1.2}$$

The $\boldsymbol{B}$ matrix or 'signed Boolean matrix' matches the interface DoFs, enforcing the continuity at the boundary.

The second condition describes the force equilibrium between matching interface DoFs according to Newton's action and reaction principle:

$$\boldsymbol{g} = -\boldsymbol{B}^T\boldsymbol{\lambda} \implies \begin{cases} \boldsymbol{g}_1^A = \boldsymbol{0} \\ \boldsymbol{g}_2^A = \boldsymbol{\lambda} \\ \boldsymbol{g}_2^B = -\boldsymbol{\lambda} \\ \boldsymbol{g}_3^B = \boldsymbol{0} \end{cases} \tag{1.3}$$

Here $\boldsymbol{\lambda}$ is a set of Lagrange multipliers denoting the intensities of interface forces (reaction forces at the common boundary).

The application of the two interface conditions Eqs. (1.2), (1.3) to the equation of motion in Eq. (1.1) leads to the following set of equations:

$$\begin{cases} \boldsymbol{u} = \boldsymbol{Y}(\boldsymbol{f} - \boldsymbol{B}^T\boldsymbol{\lambda}) \\ \boldsymbol{B}\boldsymbol{u} = \boldsymbol{0} \end{cases} \tag{1.4}$$

The continuity of displacements is directly enforced with the compatibility condition and the reaction forces defined via Lagrange multipliers automatically fulfill the equilibrium condition. Solving the system of Eq. (1.4) for $\boldsymbol{\lambda}$:

$$\boldsymbol{\lambda} = \left(\boldsymbol{B}\boldsymbol{Y}\boldsymbol{B}^T\right)^{-1}\boldsymbol{B}\boldsymbol{Y}\boldsymbol{f} \tag{1.5}$$

This result can be interpreted as follows: as a product of the excitation $\boldsymbol{f}$, a gap $\boldsymbol{B}\boldsymbol{Y}\boldsymbol{f}$ is formed between the still uncoupled subsystems' interface. The interface force $\boldsymbol{\lambda}$ is applied in order to close this gap and keep the subsystems together. The stiffness operator between the applied force and the gap is called Interface Dynamic Stiffness, obtained by the inversion of the Interface Flexibility Matrix $\boldsymbol{B}\boldsymbol{Y}\boldsymbol{B}^T$.

The coupled response is then obtained substituting back Eq. (1.5) in Eq. (1.4):

$$\boldsymbol{u} = \boldsymbol{Y}^{AB}\boldsymbol{f}, \quad \boldsymbol{Y}^{AB} = \left[\boldsymbol{I} - \boldsymbol{Y}\boldsymbol{B}^T\left(\boldsymbol{B}\boldsymbol{Y}\boldsymbol{B}^T\right)^{-1}\boldsymbol{B}\right]\boldsymbol{Y} \tag{1.6}$$

Note that, according to the dual formulation of the problem, the assembled admittance $\boldsymbol{Y}^{AB}$ contains twice the interface DoFs and has the same size of the original uncoupled admittance $\boldsymbol{Y}$. Hence, the redundant rows and columns may be removed when deemed necessary.

1.2.2 Virtual Point Transformation

The application of the FBS coupling technique presented in Sect. 1.2.1 requires both compatibility and equilibrium conditions to be satisfied at the interface of the subsystems. In finite element models this two boundary conditions are easily imposed to every coinciding node. In experimental practice, this geometric coincidence can rarely be assured and therefore it is common practice to reduce the interface problem to one or more connecting points [9].

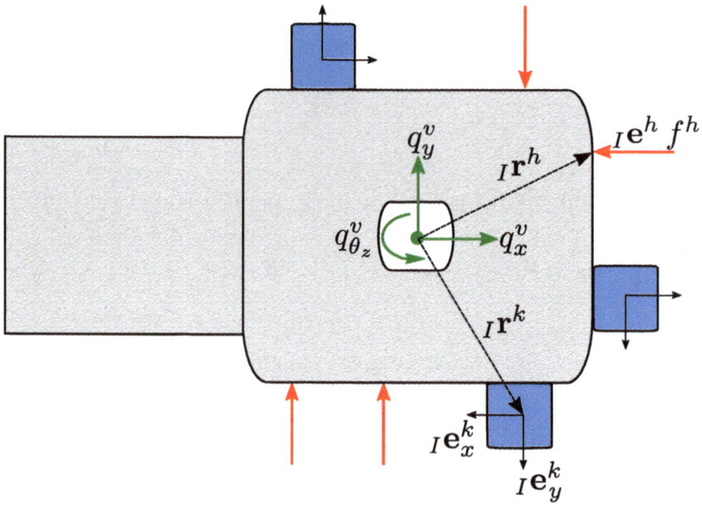

Fig. 1.2 Illustration of a virtual point interface connection. Red: Impacts; Blue: Triaxial sensors; Green: Virtual DoFs

A simple 3 DoFs translational Single-Point Connection (SPC) completely neglects the rotational DoFs. The Equivalent Multi-Point Connection (EMPC) method was introduced to overcome this issue [10]: by coupling translational directions of multiple points in proximity of the interface, which is assumed to be fully rigid locally, rotations are implicitly accounted for. This approach results however in an overdetermination of the coupling problem, leading to unwanted 'stiffening' effects and numerical instabilities. Hence the Virtual Point Transformation method was proposed [2]: translational DoFs are projected into a subspace composed by six rigid Interface Displacement Modes (IDMs), retaining only the dynamics that load the interface in a purely rigid manner. The residual flexibility is left uncoupled and the interface problem is 'weakened'. The interface dynamics is condensed into a single virtual point characterized by a 6 DoFs nodal description.

In a typical experimental measurement setup, measured displacements and forces around the interface are non-collocated, thus making it infeasible to apply the coupling technique presented in Eq. (1.6). A reduction of the measurements in the proximity of the interface, assumed to be fully rigid, into the virtual point displacements $\boldsymbol{q}$ and virtual point forces $\boldsymbol{m}$ is necessary. The transformation of the uncoupled admittance $\boldsymbol{Y}$ from measured to virtual DoFs is performed:

$$\boldsymbol{Y}_{qm} = \boldsymbol{T}_u \boldsymbol{Y} \boldsymbol{T}_f^T \tag{1.7}$$

The matrices $\boldsymbol{T}_u$ and $\boldsymbol{T}_f$ apply the transformation and the Virtual Point Admittance $\boldsymbol{Y}_{qm}$ represents the transformed FRF matrix in the matching generalized DoFs $\boldsymbol{q}$ and $\boldsymbol{m}$. The coupled equation (Eq. 1.6) can so be rewritten in terms of virtual DoFs:

$$\boldsymbol{q} = \left[\boldsymbol{I} - \boldsymbol{Y}_{qm}\boldsymbol{B}^T\left(\boldsymbol{B}\boldsymbol{Y}_{qm}\boldsymbol{B}^T\right)^{-1}\boldsymbol{B}\right]\boldsymbol{Y}_{qm}\boldsymbol{m} \tag{1.8}$$

A back transformation from virtual to measured DoFs is then also possible.

A brief derivation of the transformation of displacements and forces for the subsystem interface of Fig. 1.2 is described in Sects. 1.2.2.1 and 1.2.2.2. More details can be found in [11].

1.2.2.1 Virtual Point Displacements

The measured interface displacements $\boldsymbol{u}_2$ are projected onto the virtual point and rewritten as a function of the generalized coordinates $\boldsymbol{q}$. Let us assume that only rigid modes, sorted in translational $\boldsymbol{q}_t^v = [q_x^v, q_y^v, q_z^v]$ and rotational $\boldsymbol{q}_\theta^v = [q_{\theta_x}^v, q_{\theta_y}^v, q_{\theta_z}^v]$ components, compose the IDM subspace. The relation between $\boldsymbol{u}_2$ and $\boldsymbol{q}$ is geometrically obtained providing the orientation of the sensor axis $_I\mathbf{e}^k$ and the sensors distance from the virtual point $_I\mathbf{r}^k$. The subscript $(*)^v$ denotes the virtual point v, while $(*)^k$ describes the triaxial sensor k, projected on the absolute frame $_I(*)$.

$$\boldsymbol{u}^k = ({}_I\boldsymbol{e}^k)^T(\boldsymbol{q}_t^v + \boldsymbol{q}_\theta^v \times {}_I\,\boldsymbol{r}^k) = ({}_I\boldsymbol{e}^k)^T \begin{bmatrix} 1\,0\,0 & 0 & {}_Ir_z^k & -{}_Ir_y^k \\ 0\,1\,0 & -{}_Ir_z^k & 0 & {}_Ir_x^k \\ 0\,0\,1 & {}_Ir_y^k & -{}_Ir_x^k & 0 \end{bmatrix} \begin{bmatrix} q_x^v \\ q_y^v \\ q_z^v \\ q_{\theta_x}^v \\ q_{\theta_y}^v \\ q_{\theta_z}^v \end{bmatrix} = \boldsymbol{R}_{u^k}\boldsymbol{q}^v \tag{1.9}$$

The reduction operator $\boldsymbol{R}_{u^k}$ is a 3×6 matrix relating the sensor displacement $\boldsymbol{u}^k$ to the generalized coordinates $\boldsymbol{q}^v$. The relation can be extended to all sensor channels $\boldsymbol{u}_2$ and all virtual points $\boldsymbol{q}$:

$$\boldsymbol{u}_2 = \boldsymbol{R}_u\boldsymbol{q} \tag{1.10}$$

This spatial reduction onto the IDM is based on the assumption of an almost rigid behaviour of the interface around the virtual point. Since in reality this condition is not always fully satisfied, a residual term $\boldsymbol{\mu}$, which represents the unprojected flexible motion, must be added:

$$\boldsymbol{u}_2 = \boldsymbol{R}_u\boldsymbol{q} + \boldsymbol{\mu} \tag{1.11}$$

As the number of IDMs is typically lower than the number of interface DoFs, the problem is overdetermined and is handled with a least square procedure. To find the $\boldsymbol{q}$ that best approximates the measured response $\boldsymbol{u}_2$, a residual cost function, namely the squared error $\boldsymbol{\mu}_T\boldsymbol{\mu}$, has to be minimized. The least square projection is performed by applying the Moore-Penrose pseudo-inverse of $\boldsymbol{R}_u$:

$$\boldsymbol{q} = \left(\boldsymbol{R}_u^T\boldsymbol{R}_u\right)^{-1}\boldsymbol{R}_u^T\boldsymbol{u}_2 = \boldsymbol{T}_u\boldsymbol{u}_2 \tag{1.12}$$

In general, a weighting matrix can be used to gain more control over the error minimization by adjusting the importance of certain DoFs in the transformation [2, 11].

1.2.2.2 Virtual Point Forces

The derivation of the force reduction matrix is similar. The full set of input forces $\boldsymbol{f}_2$ has to be related to the generalized forces $\boldsymbol{m}_t^v = [m_x^v, m_y^v, m_z^v]$ and moments $\boldsymbol{m}_\theta^v = [m_{\theta_x}^v, m_{\theta_y}^v, m_{\theta_z}^v]$. For one single impact $\boldsymbol{f}^h$ with orientation ${}_I\boldsymbol{e}^h$ at a distance ${}_I\mathbf{r}^h$ from the virtual point, the relation can be written as follows:

$$\boldsymbol{m}^v = \begin{bmatrix} m_x^v \\ m_y^v \\ m_z^v \\ m_{\theta_x}^v \\ m_{\theta_y}^v \\ m_{\theta_z}^v \end{bmatrix} = \begin{bmatrix} 1 & 0 & 0 \\ 0 & 1 & 0 \\ 0 & 0 & 1 \\ 0 & -{}_Ir_z^h & {}_Ir_y^h \\ {}_Ir_z^h & 0 & -{}_Ir_x^h \\ -{}_Ir_y^h & {}_Ir_x^h & 0 \end{bmatrix} {}_I\boldsymbol{e}^h\boldsymbol{f}^h = \boldsymbol{R}_{f^h}^T\boldsymbol{f}^h \tag{1.13}$$

The operator $\boldsymbol{R}_{f^h}^T$ is the 6×1 matrix describing the reduction of input forces $\boldsymbol{f}^h$ to the generalized forces $\boldsymbol{m}^v$. The extended formulation for all virtual points and the full set of forces can be written:

$$\boldsymbol{m} = \boldsymbol{R}_f^T\boldsymbol{f}_2 \tag{1.14}$$

Note that $\boldsymbol{R}_f$ assumes the same exact formulation as $\boldsymbol{R}_u$ if a 'collocated' setup (sensor channels and forces in the same position and direction) is chosen.

Unlike the displacements transformation, the problem is underdetermined and therefore a standard least square is not applicable. The goal is to find the forces $\tilde{\boldsymbol{f}}_2$ that realize the generalized forces $\boldsymbol{m}$ with a minimal 'effort', or in mathematical language the solution to a minimization problem with Eq. (1.14) as constraint and the quadratic set of forces $\tilde{f}_2^T \tilde{f}_2$ as cost function:

$$\tilde{f}_2 = R_f (R_f^T R_f)^{-1} m = T_f^T m \tag{1.15}$$

Similar to virtual displacements, it is also possible to introduce a weighting matrix, such that preferences in measurement data may be given [2, 11].

1.3 Laser Technology

The acquisition of high quality measurement data is a key step for a successful implementation of the coupling techniques described in Sect. 1.2. It is necessary to ensure maximum reliability and accuracy of the measured data and the results obtained by its subsequent processing. The choice of the measurement equipment is the first step in the delicate process of FRFs acquisition. For motion detection, a distinction can be made between attached and non-attached transducers [12]. The former are widely used in impact testing because of the large displacements in softly supported components induced by the impact. A typical response transducer of this type is the piezo accelerometer, whose mechanism exploits the piezoelectric effect: an internal vibrating mass applies a force, proportional to the acceleration of the mass itself, to a crystal, which generates a voltage. A large variety of sensor types based on different sensitivities, sizes and weights is available on the market. In addition, a broad dynamic range (up to 160 dB), a wide frequency range (from 0.5 Hz to 10 kHz) in common structural dynamics applications, high environmental resistance, easy operation and simple mounting methods are some of the advantages that make piezo sensors so popular in impact testing [5].

An alternative to the standard piezo devices is represented by the laser technology. The laser Doppler vibrometer (LDV) is a scientific instrument that uses light to perform non-invasive displacement and velocity measurements. The physical principle underlying this technology is the Doppler effect. The velocity measurements are based on the Doppler frequency shift of laser light that is reflected by the moving surface of the test object. A vibrometer generally uses a laser interferometer to measure the frequency difference between an internal reference beam and the measurement beam. A comprehensive review of the operating principles and applications of the LDV can be found in [13].

The choice of a non-intrusive measurement technique is mainly driven by the desire to reduce the influence of external dynamic systems on the measured component throughout the data acquisition process. Indeed, the presence of a physical device attached to the structure slightly modifies the overall system dynamics. Loading the structure can affect its mass, stiffness and damping. The dominant effect is mass loading: the spatial distribution and the magnitude of the masses of the sensors can actually lower the eigenfrequencies of the global system. A sensor positioned on an anti-node of a particular mode will highly contribute to the inertial energy of that mode. The effect of additional mass depends on the quantity of inertial energy already present in the mode. For this reason, the relative added mass effect is more significant at high frequencies [5].

In addition, the cables of the sensors as well as a loose sensor attachment and friction phenomena at the mounting interface can introduce damping into the system, altering the dynamic behaviour of the structure [5].

Other benefits of using laser technology compared to traditional triaxial accelerometers are:

- The high level of optical sensitivity allows to obtain a high signal-to-noise ratio even at very low frequencies (0–30 Hz).
- The non-contact technology does not experience the problem of cross-talk between sensors.
- The 'long-distance' operating principle, combined with an extremely small laser target, makes it possible to measure areas that would otherwise be difficult to access.
- In the VPT, for displacements a high degree of overdetermination of the problem can be guaranteed even for very small interfaces.

Most of the disadvantages in using a LDV are associated with the large size and the limited flexibility of the equipment in the measurement setup. In addition, the laser instrumentation is significantly more expensive than standard accelerometers.

With regard to measurements, large displacements of the measuring point may be a problem due to the non-contact nature of the transduction system.

In Table 1.1, the main features of laser vibrometry and piezoelectric sensors are compared.

Table 1.1 Properties of measurement devices in FBS

Type	Laser Doppler vibrometer	Piezo accelerometer
Accuracy	++	+
Structure interaction	++	−−
Ease of use	−	++
Size	−	+
Time-consuming	−	+
Cost	−−	+

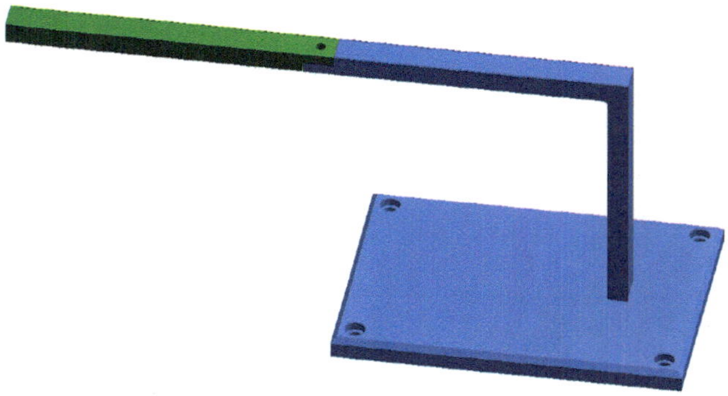

Fig. 1.3 Benchmark structure for DS. Blue: Substructure A; Green: Substructure B

1.4 Measurements

The analysis is performed on the benchmark structure depicted in Fig. 1.3, consisting of two subcomponents A and B. The former is an 'L-shaped' construction welded on a rectangular supporting plate, causing an overall very low damping. The latter is a simple free beam. The structure, made entirely of aluminium, is characterized by a small-sized and highly flexible connection interface, manufactured by means of a CNC milling process. A hex head screw and a locking nut guarantee a proper assembly of the substructures, which are in contact with each other in correspondence of the vertical and upper horizontal flange. A tightening torque of 20 Nm is applied.

The admittance models are developed for each subsystem from both an experimental and a numerical point of view. For the construction of the experimental FRF model, two case studies are carried out on the basis of the adopted response transduction system: one involves the use of triaxial accelerometers, the other exploits the laser technology. A numerical simulation is performed in parallel with the purpose of synthesizing the necessary FRFs starting from the modal properties of the system. Additionally, validation measurements are taken on the assembled structure.

The experimental modelling is briefly described in Sect. 1.4.1. The application of the VPT is addressed in Sect. 1.4.2 and the LM-FBS coupling results are shown and discussed in Sect. 1.4.3.

1.4.1 Measurement Setup

The substructure A is fixed to a vibration-free table, component B is freely suspended in the air by elastic ropes. Particular focus is on the support setup of the free beam for the laser application: on the one hand, it is intended to ensure adequate flexibility and low friction of the support system; on the other hand, the actual movement of the structure has to be minimized in order to guarantee the stability of the laser focal point during the acquisition process. Note that a continuous change in the position of the measurement point during the acquisition may generate distortion around FRF antiresonances.

The experimental FRF model is constructed by impact measurements. The source of excitations is an automatic modal hammer designed by the Chair of Applied Mechanics at the TU Munich equipped with a steel tip [14]. The response transducers are the 10 mV/g triaxial piezo accelerometers by Kistler in one case and a laser Doppler vibrometer, namely the RSV-150 Remote Sensing Vibrometer by Polytec, in the other case.

In laser measurements, an adequate focal distance, a good isolation and the use of reflective tapes are some of the measures taken to maximize the signal-to-noise ratio.

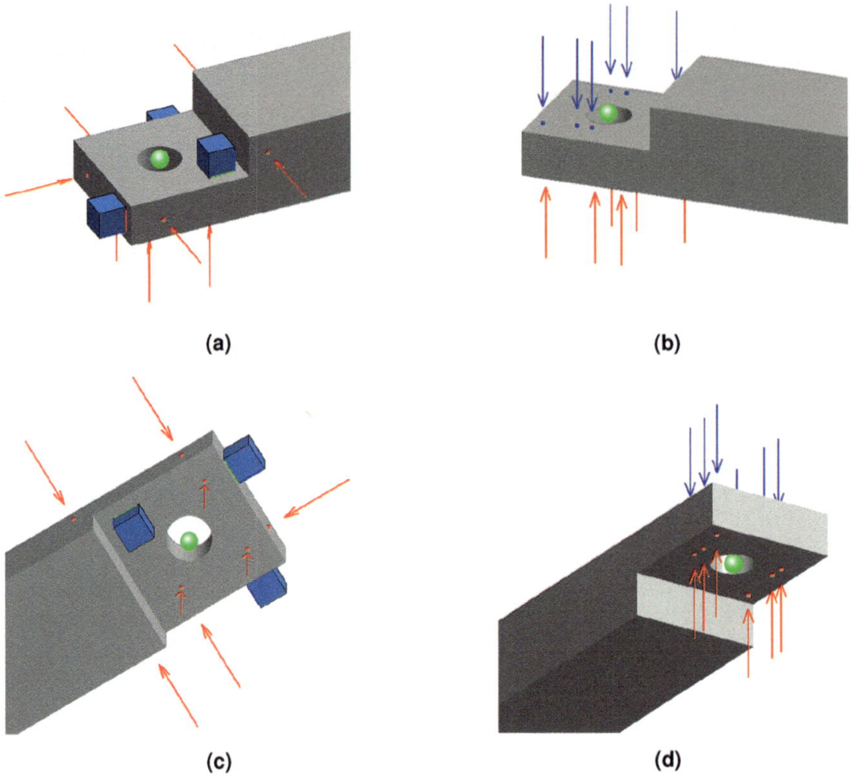

Fig. 1.4 Layout of impacts and responses at the interface for VPT. Red: Impacts; Blue: Responses; Green: Virtual Point. (**a**) Component A—Piezo. (**b**) Component A—Laser. (**c**) Component B—Piezo. (**d**) Component B—Laser

1.4.2 Virtual Point Reduction

The dynamics of the interface is condensed into a virtual point that is located in the center of the hole on the connecting surface of the two substructures. The placement and number of impacts and responses around the virtual point is chosen to ensure the observability of the entire set of selected virtual DoFs and a high quality transformation for both case studies [9, 11].

In the measurement campaign with the triaxial sensors the interface dynamics is reduced into a subspace composed by six rigid IDMs (three translations and three rotations). For this purpose, three sensors and eight excitation points are located on the interface of both substructures (Fig. 1.4a, c). The mono-dimensional nature of the available laser vibrometer, on the other hand, leads to the choice of a single-plane measurement campaign. Indeed, the FRFs are acquired exclusively along the vertical axis and consequently only the dynamics related to the vertical translational DoF and the two out-of-plane rotations is retained in the transformation. A total of six output signals and six impacts are used to describe the virtual point DoFs for each subcomponent (Fig. 1.4b, d).

1.4.3 Coupling Results

The transformed FRF admittances $\boldsymbol{Y}^{A}_{qm}$ and $\boldsymbol{Y}^{B}_{qm}$ are coupled according to LM-FBS (Eq. 1.8). An effective comparison between the two measurement campaigns is achieved by coupling only the dynamics associated with the translational DoF along the vertical axis. The experimental coupled FRF is plotted together with the correspondent assembly validation FRF for both case studies. To assess the accuracy of the experimentally coupled FRF, a cross-validation is performed by comparing the data with simulated results, in which only the vertical translation over the interface area is coupled. The outcome in magnitude and phase is shown in Fig. 1.5a,b. A focus on lower frequencies is depicted in Fig. 1.6a,b.

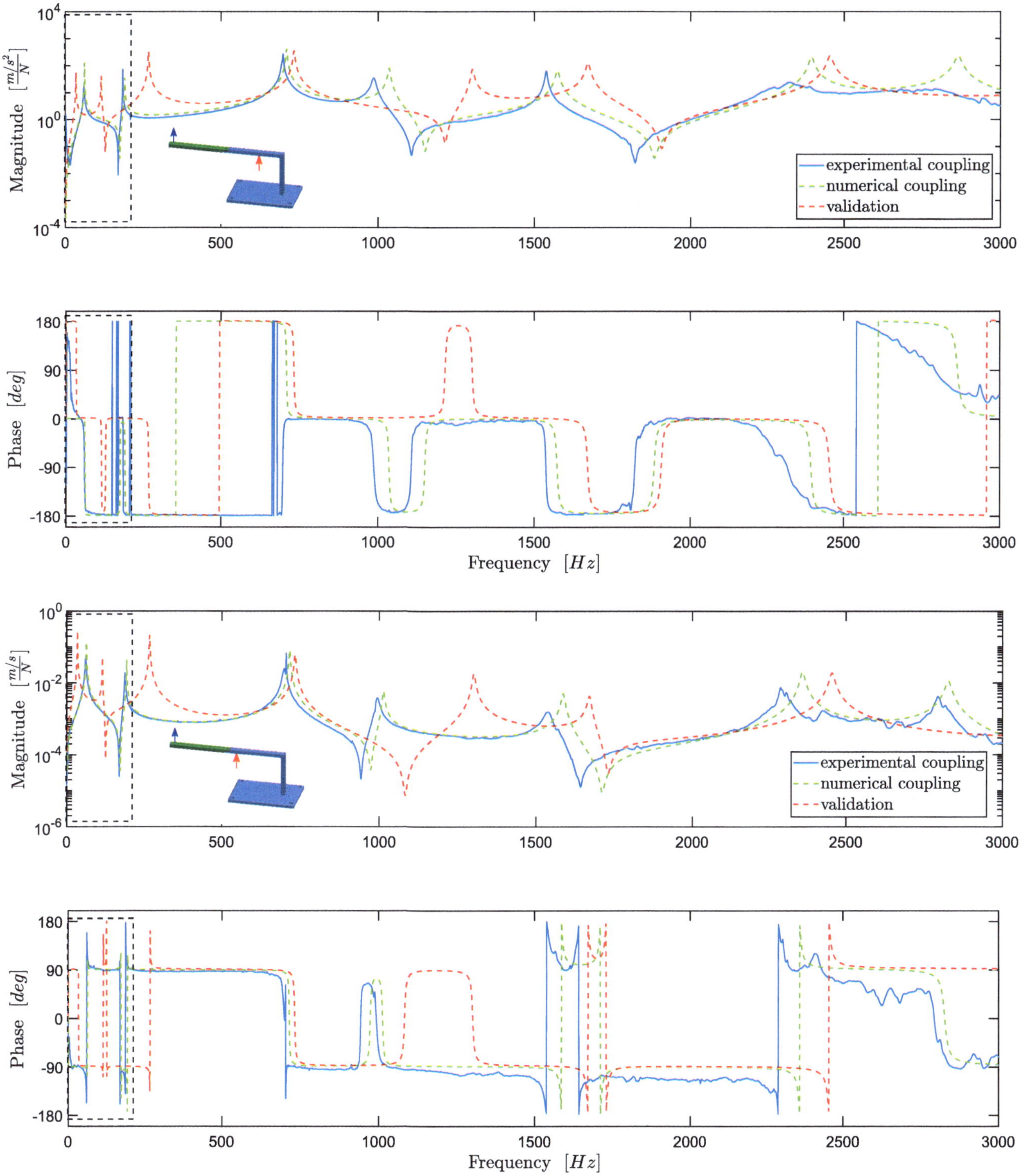

Fig. 1.5 Experimental and numerical coupled FRF with validation. Frequency range [0 – 3000 Hz]. Only the dynamics associated with the vertical translational DoF is coupled for comparative purposes. (**a**) First case—piezo accelerometer. (**b**) Second case—laser Doppler vibrometer

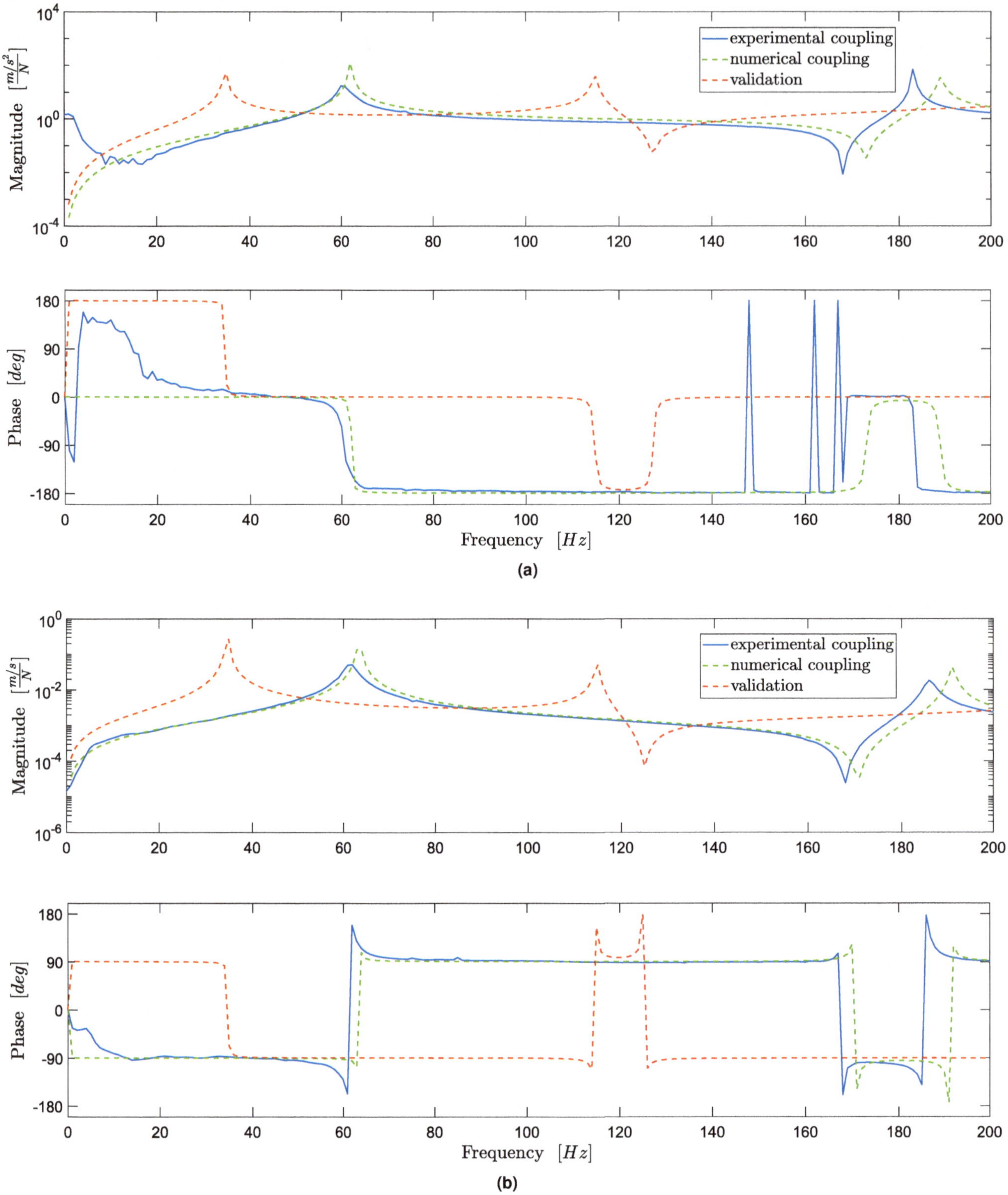

Fig. 1.6 Experimental and numerical coupled FRF with validation. Frequency range [0 – 200 Hz]. Only the dynamics associated with the vertical translational DoF is coupled for comparative purposes. (**a**) First case—piezo accelerometer. (**b**) Second case—laser Doppler vibrometer

The following observations are made:

- Both experimental coupling results match very well numerical data in magnitude and phase up to a high frequency range (Fig. 1.5a,b). Note that the laser seems to be more accurate at very low (0−−30 Hz) and very high (above 2000 Hz) frequencies. In particular, the capability of the laser to measure static or quasi-static responses is highlighted in Fig. 1.6a,b.
- The validation of the coupled FRF with the assembled FRF is highly affected by the rigid IDMs left uncoupled (two translations and three rotations) and the residual flexibility. This result is expected as the uncoupled horizontal DoF is highly correlated with the vertical one and plays an important role in most modes of the assembled structure.
- The discrepancies in the shape of the coupled FRF (in the position of antiresonances) between piezo and laser measurements depend mainly on the different impact location in the observed FRF (Fig. 1.5a,b). In addition, the different dynamics acquired at the interface for the VPT in the two cases may slightly affect the coupling results.

The added mass effect in piezo measurements is minimal due to the small number of lightweight sensors used.

1.5 Conclusions

In this paper, the use of laser vibrometry in the context of experimental LM-FBS coupling via VPT is investigated. The non-contact measurement technique allows to minimize the disturbances in the overall system dynamics related to the presence of a physical device mounted on the measured component. In this sense, the use of a non-intrusive approach in sensing motion is strongly suggested when dealing with small-sized, lightweight structures. A further advantage of laser technology over standard accelerometers is the capability to reach inaccessible areas and measure a large number of points. A comparison between the coupling results of the two different measurement approaches is provided. Although the experimental data acquired with both accelerometers and LDV fit very well with the simulated data, the laser reveals great accuracy over a broader frequency range. Additional analysis can be conducted on more complex applications (e.g. 3D interface coupling) to explore the potential of laser vibrometry in FBS.

References

1. De Klerk, D., Rixen, D.J., Voormeeren, S.: General framework for dynamic substructuring: history, review and classification of techniques. AIAA J. **46.5**, 1169–1181 (2008)
2. Van der Sejis, M., Van der Bosch, D., Rixen, D., De Klerk, D.: An improved methodology for the virtual point transformation of measured frequency response functions in dynamic substructuring. In: COMPDYN (2013)
3. Lyons, R.G.: Understanding Digital Signal Processing, 1st edn. Addison-Wesley Longman, Boston (1997)
4. Brown, D.L., Allemang, R.J., Phillips, A.W.: Forty years of use and abuse of impact testing: a practical guide to making good FRF measurements. In: Experimental Techniques, Rotating Machinery, and Acoustics. Proceeding of the Society for Experimental Mechanics, vol. 8 (2015)
5. Døssing, O., (Firm), B.K.: Structural Testing: Mechanical mobility measurements, pt. 1. Bruël & Kjær, Nærum (1988)
6. Rixen, D.: How measurement inaccuracies induce spurious peaks in frequency based substructuring. In: Proceedings of the Twenty Sixth International Modal Analysis Conference, Orlando, FL. Society for Experimental Mechanics, Bethel (2008)
7. Voormeeren, S.N., De Klerk, D., Rixen, D.J.: Uncertainty quantification in experimental frequency based substructuring. Mech. Syst. Signal Process. **24**, 106–118, (2010)
8. De Klerk, D., Rixen, D.J., De Jong, J.: The frequency based substructuring (FBS) method reformulated according to the dual domain decomposition method. In: 24th International Modal Analysis Conference, St. Louis (2006)
9. Van der Sejis, M.: Experimental Dynamic Substructuring. PhD thesis, Delft University of Technology (2016)
10. De Klerk, D.: Solving the RDoF problem in experimental dynamic substructuring. In: Proceedings of the 26th International Modal Analysis Conference (IMAC) (2008)
11. Häußler, M., Rixen, D.: Optimal transformation of frequency response functions on interface deformation modes. In: Dynamics of Coupled Structures, vol. 4, pp. 25–237. Springer, Cham (2017)
12. Company, H.P.: The Fundamentals of Modal Testing: Application Note 243-3. Hewlett Packard Company, Palo Alto (1986)
13. Rothberg, S., Allen, M., Castellini, P., DiMaio, D., Dirckx, J., Ewins, D., Halkon, B., Muyshondt, P., Paone, N., Ryan, T., Steger, H., Tomasini, E., Vanlanduit, S., Vignola, J.: An international review of laser Doppler vibrometry: making light work of vibration measurement. Opt. Lasers Eng. **99**, 11–22 (2017). https://doi.org/10.1016/j.optlaseng.2016.10.023
14. Maierhofer, J., El Mahmoudi, A., Rixen, D.J.: Development of a lowcost automatic modal hammer for applicaions in substructuring. In: Proceedings of the 37th International Modal Analysis Conference (IMAC) (2019)

Chapter 2
A Priori Interface Reduction for Substructuring of Multistage Bladed Disks

Lukas Schwerdt, Lars Panning-von Scheidt, and Jörg Wallaschek

Abstract When analyzing the dynamics of bladed disks in turbomachinery, most methods focus on a single stage at a time because of the challenges associated with multistage structures. Whereas the cyclic symmetry of individual bladed disks is commonly exploited to yield great savings of computational effort, multistage rotors lack this symmetry due to the differing number of blades in each stage. Substructuring methods can be used to overcome this problem but they still face challenges with non-conforming finite element meshes at the interface between stages. Some state of the art methods expect the nodes at the interface to be arranged in concentric rings and use a truncated Fourier series as basis for the displacement along each ring of nodes. In this paper, a reduction basis for the interface degrees of freedom between adjacent stages is proposed which uses polynomial basis functions in the radial direction in addition to a truncated Fourier series in the circumferential direction. This enables coupling the substructures of multiple stages with arbitrary meshes. Additionally, the resulting reduced order model (ROM) can be smaller while preserving accuracy. The proposed interface reduction is demonstrated in conjunction with a cyclic Craig-Bampton (CB) reduction of each stage. Different ROMs are compared to show the impact of the CB reduction as well as the interface reduction.

Keywords Model order reduction · Component mode synthesis · Multistage · Interface reduction · Mistuning

Nomenclature

n	Number of degrees of freedom
$\boldsymbol{x}$	Physical degrees of freedom
$\boldsymbol{y}$	Degrees of freedom in travelling wave coordinates
$\boldsymbol{F}$	DFT matrix
H	Fourier harmonic basis function
$\boldsymbol{I}$	Identity matrix
$\boldsymbol{K}$	Stiffness matrix
$\boldsymbol{M}$	Mass matrix
N	Number of blades
P	Polynomial basis function
$\boldsymbol{T}$	Transformation/reduction matrix
V	Interface reduction basis function
$\boldsymbol{W}$	Matrix of interface reduction basis
$\boldsymbol{\eta}$	Reduced/modal degrees of freedom
$\boldsymbol{\varphi}$	Mode/eigenvector
$\boldsymbol{\Phi}$	Matrix of fixed interface modes
$\boldsymbol{\Psi}$	Matrix of constraint modes
(r, α)	(Polar) coordinates on the interface
$\square_h$	Substructure harmonic index
$\square_j$	Interface harmonic index
$\square_l$	Mode index
$\square_k$	Stage index

L. Schwerdt (✉) · L. Panning-von Scheidt · J. Wallaschek
Institute of Dynamics and Vibration Research, Faculty of Mechanical Engineering, Leibniz University Hannover, Hannover, Germany
e-mail: schwerdt@ids.uni-hannover.de

A. Linderholt et al., *Dynamic Substructures, Volume 4*, Conference Proceedings of the Society for Experimental Mechanics Series,
https://doi.org/10.1007/978-3-030-12184-6_2

$\square_m$ Master/interface degrees of freedom
$\square_s$ Slave/inner degrees of freedom
$\tilde{\square}$ Reduced model

2.1 Introduction

Finite element models are a vital part of the product development process in turbomachinery to predict the vibratory response, but detailed models of multistage rotors are prohibitively large. Therefore models of single stages are used predominantly although the importance of multistage effects is known [1]. Increasingly more methods are developed to generate reduced order models for multistage assemblies [2–6]. Most use a variant of Component Mode Synthesis (CMS) to split the rotor into substructures, which are reduced separately. This can reduce the amount of memory required for an analysis and allows to perform multiple analyses efficiently using the ROM.

In this paper, a model order reduction method for multistage bladed disks is presented, that uses CMS with the Craig-Bampton reduction (CB-CMS) with a priori interface reduction. As basis functions for the interface motion, a product of Fourier harmonics in circumferential direction and polynomials in radial direction is used. First, the CB-CMS method is presented by itself and then with interface reduction. afterwards, an overview of the existing model reduction methods for multistage bladed disks is given before the proposed method is presented. Finally, the new method is demonstrated on an academic two stage rotor.

2.2 Craig-Bampton Method and Interface Reduction

One method to generate reduced order models is the Component Mode Synthesis. It is widely used in turbomachinery and other applications, by itself and as a basis for more advanced model order reduction methods. In the CMS method, the structure is split up in to multiple components, also known as substructures. Each of the components is then reduced individually and the reduced components are assembled to yield the reduced model of the complete structure. Different methods for the reduction and assembly are available, cf. [7].

2.2.1 *Craig-Bampton Method*

The most popular of these methods is the Craig-Bampton reduction[8]. Here the degrees of freedom (DOF) of each substructure are split into *master* ($\boldsymbol{x}_m$) and *slave* ($\boldsymbol{x}_s$) DOF and the stiffness and mass matrices are partitioned accordingly:

$$\boldsymbol{K} = \begin{bmatrix} \boldsymbol{K}_{mm} & \boldsymbol{K}_{ms} \\ \boldsymbol{K}_{sm} & \boldsymbol{K}_{ss} \end{bmatrix} \qquad \boldsymbol{M} = \begin{bmatrix} \boldsymbol{M}_{mm} & \boldsymbol{M}_{ms} \\ \boldsymbol{M}_{sm} & \boldsymbol{M}_{ss} \end{bmatrix} \tag{2.1}$$

The master DOF are kept in the reduced model while a truncated set of modal DOF ($\boldsymbol{\eta}_s$) represent the slave DOF in the reduced system:

$$\begin{bmatrix} \boldsymbol{x}_m \\ \boldsymbol{x}_s \end{bmatrix} = \begin{bmatrix} \boldsymbol{I} & \boldsymbol{0} \\ \boldsymbol{\Psi} & \boldsymbol{\Phi} \end{bmatrix} \begin{bmatrix} \boldsymbol{x}_m \\ \boldsymbol{\eta}_s \end{bmatrix} = \boldsymbol{T} \begin{bmatrix} \boldsymbol{x}_m \\ \boldsymbol{\eta}_s \end{bmatrix} \tag{2.2}$$

The constraint modes $\boldsymbol{\Psi}$ and fixed-interface normal modes $\boldsymbol{\Phi} = \left[\boldsymbol{\varphi}_1, \boldsymbol{\varphi}_2, \ldots\right]$ are obtained by:

$$\begin{gathered} \boldsymbol{\Psi} = -\boldsymbol{K}_{ss}^{-1}\boldsymbol{K}_{sm} \\ \left[\boldsymbol{K}_{ss} - \omega_i^2 \boldsymbol{M}_{ss}\right]\boldsymbol{\varphi}_i = \boldsymbol{0} \end{gathered} \tag{2.3}$$

The reduced stiffness and mass matrices are

$$\tilde{\boldsymbol{K}} = \boldsymbol{T}^{\mathrm{H}}\boldsymbol{K}\boldsymbol{T} \qquad \tilde{\boldsymbol{M}} = \boldsymbol{T}^{\mathrm{H}}\boldsymbol{M}\boldsymbol{T} \tag{2.4}$$

To facilitate the assembly of the reduced substructures, the DOF on the interface between adjacent substructures are selected as master DOF and kept in the reduced model. For this reason the modes in $\boldsymbol{\Phi}$ are called the fixed-interface normal modes although additional DOF can be selected as master DOF if desired.

The biggest disadvantage of the regular CB-method is, that the interfaces between substructures are not reduced. This leads to two problems: firstly, the number of DOF of the reduced system is limited by the size of the interface. Secondly the reduced matrices and the matrix of constraint modes can contain more non-zero entries than the (sparse) matrices of the unreduced system (cf. [9]). This represents a major problem, making the regular CB-method unattractive for very large models.

2.2.2 *Craig-Bampton Method with Interface Reduction*

To alleviate this problem, various interface reduction methods were developed [10–18]. Interface reduction methods can be classified into methods that reduce the interface prior to the reduction of the slave DOF or after the application of the regular CB-method. Methods that reduce the interface first avoid the problems associated with the partially reduced system altogether, whereas methods that use the reduced model of the regular CB-method generate vastly more computational effort for large-scale systems during the reduction process. Therefore, a prior interface reduction is preferred. One of these methods uses arbitrary assumed displacements of the interface as basis functions for the reduction basis. Carassale and Maurici [12] call this the *Prior Basis Function Method*. Using the Matrix $\boldsymbol{W}$, where each column is one basis function evaluated at all interface DOF, the complete reduction is

$$\begin{bmatrix} \boldsymbol{x}_m \\ \boldsymbol{x}_s \end{bmatrix} = \begin{bmatrix} \boldsymbol{I} & \boldsymbol{0} \\ \boldsymbol{\Psi} & \boldsymbol{\Phi} \end{bmatrix} \begin{bmatrix} \boldsymbol{W} & \boldsymbol{0} \\ \boldsymbol{0} & \boldsymbol{I} \end{bmatrix} \begin{bmatrix} \boldsymbol{\eta}_m \\ \boldsymbol{\eta}_s \end{bmatrix} = \begin{bmatrix} \boldsymbol{W} & \boldsymbol{0} \\ \boldsymbol{\Psi}\boldsymbol{W} & \boldsymbol{\Phi} \end{bmatrix} \begin{bmatrix} \boldsymbol{\eta}_m \\ \boldsymbol{\eta}_s \end{bmatrix} \tag{2.5}$$

with

$$\boldsymbol{\Psi}\boldsymbol{W} = -\boldsymbol{K}_{ss}^{-1}\left(\boldsymbol{K}_{sm}\boldsymbol{W}\right) \tag{2.6}$$

In Eq. (2.6) it is most efficient to perform the interface reduction first as indicated by the parenthesis.

Some important properties that should be considered when selecting the basis functions:

- To yield a reduced model of small size, a small number of basis functions should be able to represent the actual interface displacements in the frequency of interest. Optimally the interface portion of the modeshapes of the full system would be used if they were available.
- The basis functions should be easy to evaluate.
- The basis should be orthogonal to ensure the reduced matrices are not singular. Carassale and Maurici [12] use the Gram-Schmidt procedure to orthogonalize a general basis. It should be noted that perfect orthogonality is not necessary as long as the basis functions are linearly independent with enough margin to avoid numerical problems.

The method allows arbitrary basis functions, but due to their simplicity polynomial bases were used previously[12, 16]. Because of the freedom afforded by the method to choose arbitrary functions, problem specific functions can be superior to multidimensional polynomials. As shown in the next section, Fourier basis functions are useful for systems with cyclic symmetry of the substructures [2]. The newly proposed method uses a combination of polynomials and Fourier basis functions.

2.3 Reduced Models of Multistage Bladed Disks

Rotors consisting of multiple bladed disks, common in turbomachinery applications, represent a special case of mechanical system due to the cyclic symmetry of each bladed disk. This symmetry is commonly exploited to reduce the computational effort when analyzing such systems.

2.3.1 *Single Stage Bladed Disks*

Single bladed disks are cyclic symmetric structures. This results in system matrices that are block circulant. By introducing travelling wave coordinates the matrices become block diagonal. This is only possible for identical disk sectors and blades. When deviations among the blades (mistuning) are considered, the matrices in travelling wave coordinates are fully populated in general. Nonetheless, some reduction methods for mistuned bladed disks make use of this representation as well (cf. [19]). The popular SNM method for example directly uses the modes of the tuned (i.e. cyclic symmetric) structure as a reduction basis[20].

The transformation from the physical coordinates $\boldsymbol{x}$ into travelling wave coordinates $\boldsymbol{y}$ is performed according to

$$\boldsymbol{x} = (\boldsymbol{F} \otimes \boldsymbol{I})\boldsymbol{y} \tag{2.7}$$

where $\boldsymbol{F}$ is the complex DFT matrix with a size of the number of sectors n and $\boldsymbol{I}$ is the identity matrix with a size of the number of DOF per sector. In travelling wave coordinates all analyses can be performed on the sector level i.e. one can do n calculations where the matrices have the size of the number of DOF of a single sector.

2.3.2 *Multistage Bladed Disks*

Due to the variable number of blades among the different stages, a multistage rotor does not in general exhibit cyclic symmetry. Therefore, other methods have to be applied to generate a reduced order model of the rotor. These methods rely on CMS and some exploit the cyclic symmetry of individual stages on the substructure level.

Sternchüss uses individual sectors as components and couples them through intermediate rings which are only one layer of elements thick [21]. Cyclic symmetry is not exploited.

Laxalde presented the multi-stage cyclic symmetry approach [3]. Here the coupling between adjacent stages is reduced to enable exploiting the computational advantages of cyclic symmetry. When representing each stage in travelling wave coordinates, in general all harmonic indices of adjacent stages are coupled together [4]. There can be exceptions depending on the number of blades. The multi-stage cyclic symmetry approach discards most of these connections, but retains the one belonging to the same number of nodal diameters. Thereby the complete rotor model is decomposed into systems with a size of one sector from every stage each. The resulting modes are similar to the actual system modes, but not identical. Nonetheless, they represent a good reduction basis to project the full system into. Disadvantages are the difficulty of connecting adjacent stages with non-matching meshes at the interface and the fact that the substructures are relatively big, as they have a size of multiple sectors, one from each stage.

Song uses a CB-CMS approach where each harmonic index of each stage is a single substructure [2]. A Fourier basis is used to reduce the interfaces. To facilitate this, the interface is assumed to be ring shaped and meshed with concentric rings of nodes. For each of these rings, truncated Fourier series are used to represent the displacement along each direction. Thereby the number of interface DOF is reduced significantly, while retaining the substructure size of a single sector from single stage analyses. Using this prior basis function method adjacent sectors do not need matching meshes at the interface in circumferential direction. This is advantageous because the number of elements in circumferential direction for each stage needed to achieve a matching mesh is the least common multiple of the number of blades of the stage and its neighbors, e.g. 10,100 for a two stage design with 100 and 101 blades. Currently, Song's method for the interface reduction is used with more advanced model order reduction methods for multistage assemblies [6, 22]. The interface reduction was later called Fourier Constraint Modes (FCM).

2.4 Proposed Model Order Reduction Method

The proposed reduction method for multistage bladed disks uses CB-CMS with an a priori interface reduction based on the FCM method. The rotor is split into individual stages. The ring shaped interfaces between the stages are reduced using basis functions defined a priori. These functions consist of Fourier harmonics in the circumferential direction and polynomials in the radial direction. A cyclic Craig-Bampton reduction is then applied to each stage.

In detail, the procedure works as follows: the complete rotor is split into stages. The DOF of each stage k are divided into interior slave DOF $\boldsymbol{x}_{k,s}$ and the sets of DOF of the interfaces to the next and previous stage $\boldsymbol{x}_{k,\Gamma}$ and $\boldsymbol{x}_{k,\gamma}$. Using complex travelling wave coordinates each stage is further divided into N_k substructures with DOF $\boldsymbol{y}_{k,h}$ where the harmonic index h runs from $\lfloor (N_k - 1)/2 \rfloor$ to $\lfloor N_k/2 \rfloor$. Due to the cyclic symmetry each substructure is not directly coupled to any other substructure from the same stage, i.e. the harmonic indices are decoupled. However, each substructure can be coupled to all substructures of the neighboring stages.

This coupling between stages is through the interstage interface, which is assumed to be ring shaped. The nodes from the two adjacent stages corresponding to the DOF $\boldsymbol{y}_{k,\Gamma}$ and $\boldsymbol{y}_{k+1,\gamma}$ lie on this surface but do not have to coincide. For convenience the position on this surface is denoted in polar coordinates by its radius r and angle α. The basis functions for the interface reduction are defined on this coordinate system. In circumferential direction a truncated set of Fourier harmonics $H_{k,j}(\alpha)$ are used with

$$H_{k,j}(\alpha) = \sqrt{N_k}\, e^{ij\alpha} \tag{2.8}$$

and $-j_{k,\max} \leq j_k \leq j_{k,\max}$. In radial direction orthogonal polynomials $P_{k,p}(r)$ are used, where p runs from 0 up to the maximum degree $p_{k,\max}$.

Both parts of the interface basis functions are combined by taking their product to yield

$$V_{k,j,p}(\alpha, r) = H_{k,j}(\alpha) P_{k,p}(r) \tag{2.9}$$

Evaluating V at the location of the nodes on the interfaces, the interface reduction matrices $W_{k,h,\Gamma}$ and $W_{k+1,h,\gamma}$ are generated for each substructure. The basis functions are duplicated for the displacements along each coordinate direction. A reduced interface therefore has $3 \times (2\, j_{\max} + 1) \times (p_{\max} + 1)$ DOF. Note that only some interface harmonics are compatible with each harmonic index. All incompatible basis functions are excluded from the reduction basis for each substructure. The compatibility condition is

$$h = j \pm q\, N_k \qquad \text{with } q \in \mathbb{N}_0 \tag{2.10}$$

The substructures are reduced using Eq. (2.5). All calculations can be done on the sector level, as the cyclic symmetry of each stage is exploited. Fixed interface modes are kept up to a specified maximum frequency. Assembling the substructures is straightforward, as the same basis is used in both neighboring stages to describe the displacement of each interface.

2.5 Application to a Two Stage Academic Rotor

To evaluate the efficiency of the proposed method, it is applied to an academic two stage rotor pictured in Figs. 2.1 and 2.2. The stages have seven and eight blades. They are connected by a hollow shaft, which is clamped at both ends (cf. Fig. 2.2). The model is meshed using quadratic hexaeder elements with 306,684 DOF in total, pictured in Fig. 2.3. Two thousand six hundred and eighty eight of those DOF belong to the interface between the two stages. It is ring shaped and consists of two radial layers of elements i.e. five rings of nodes. To enable the calculation of the reference solution it is meshed with matching nodes, although the reduction method does not require a matching mesh.

First, only the interface is reduced using the proposed basis functions. The maximum harmonic of the Fourier basis is varied from 0 to 14 while the degree of the polynomial basis is varied from 0 to 4. A Polynomial of degree four can represent all displacements of the five rings of nodes independently and therefore exhibits the performance of the FCM method.

To assess the impact of the reduction the first eigenfrequencies of the reduced model $\tilde{\omega}$ are compared to those of the full model ω. The maximum relative error among the first 15 eigenfrequencies as defined by

$$\text{Error} = \max_{l=1..15} \left| \frac{\tilde{\omega}_l - \omega_l}{\omega_l} \right| \tag{2.11}$$

is plotted in Fig. 2.4 against the number of DOF of the reduced model. All 15×5 combinations of polynomial degree and Fourier harmonic are evaluated and grouped according to the polynomial degree. The best combinations are on the Pareto frontier in the bottom left part of the plot with a low number of DOF and a small error of the eigenfrequencies. These are balanced combinations of polynomial degree and maximum harmonic. Increasing one a lot without the other leads to

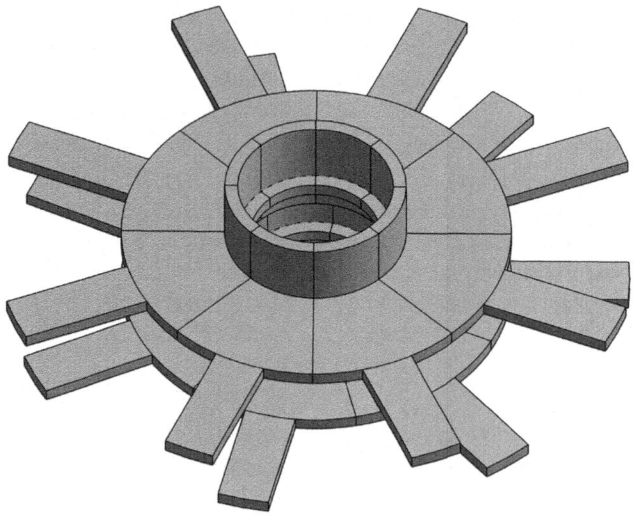

Fig. 2.1 Geometry

Fig. 2.2 Boundary conditions

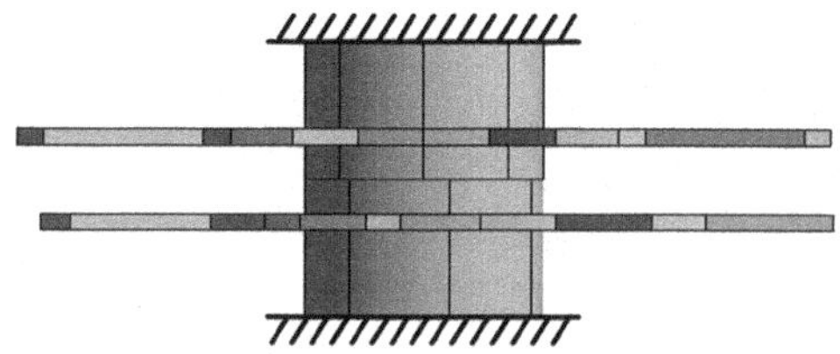

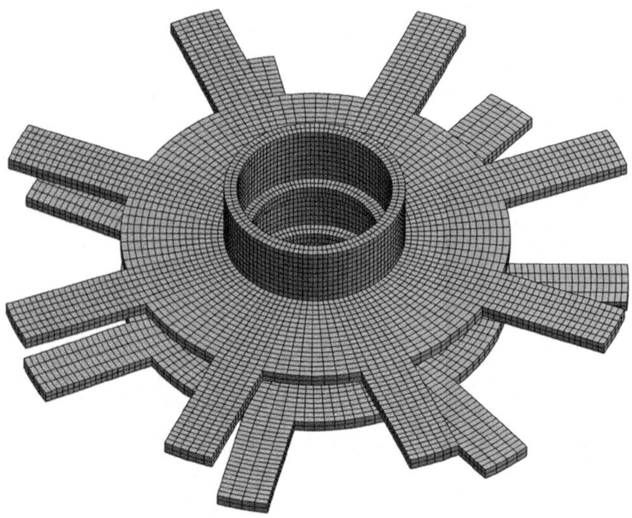

Fig. 2.3 Mesh

inefficient reduction bases. Overall, the number of DOF is still very high because only the interface is reduced. Nonetheless the reduction is quite efficient: reducing the interface by almost 98% from 2688 to 54 DOF results in an error of only 2.2×10^{-5} (304,050 DOF in total, polynomial degree 1). Achieving this accuracy with the FCM method or with polynomial degree 4 requires 135 DOF for the interface—more than twice the amount.

Incorporating the cyclic CB-reduction of the two stages a maximum total number of 120 fixed interface modes were used. They have only 72 distinct eigenfrequencies due to the double modes caused by the cyclic symmetry of the substructures. For each ROM all modes up to a specified frequency were included leading to a total number of $72 \times 15 \times 5 = 5400$ different ROMs. The interface reduction stays the same as the maximum harmonic of the Fourier basis is varied from 0 to 14 while the degree of the polynomial basis is varied from 0 to 4.

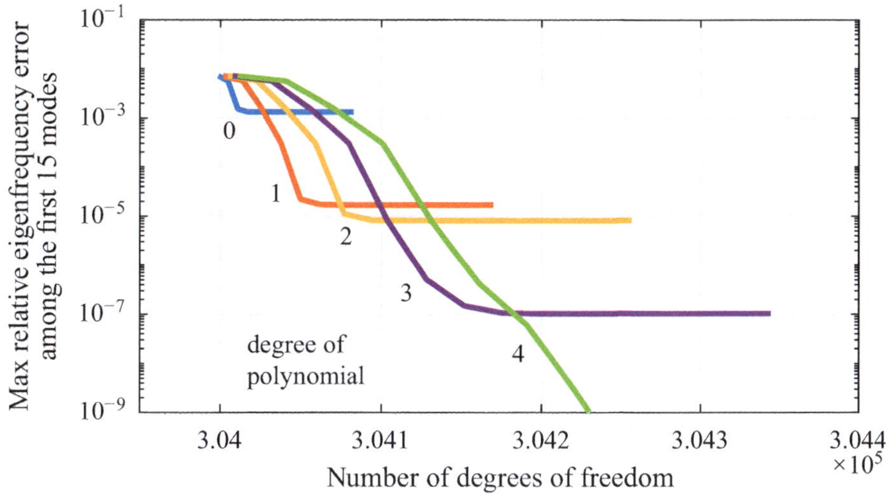

Fig. 2.4 Eigenfrequency error and size of different ROMs with only interface reduction

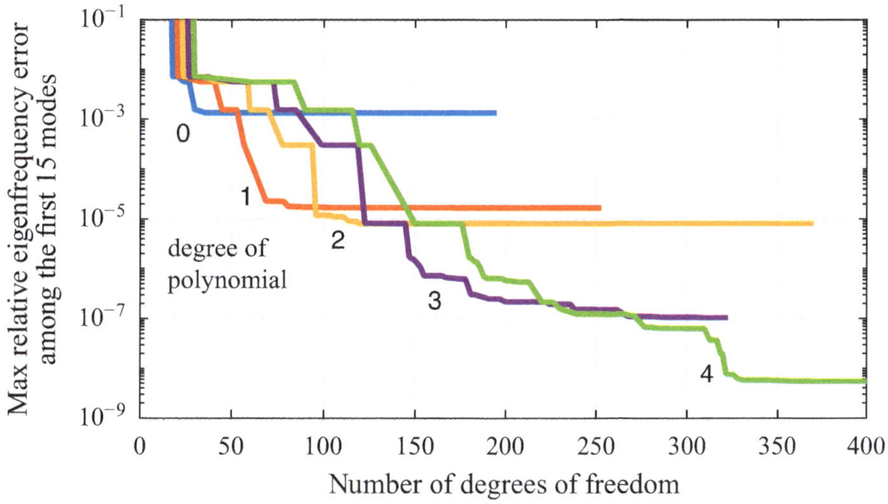

Fig. 2.5 Non-dominated ROMs for a fixed value of the polynomial degree

The maximum relative error among the first 15 eigenfrequencies is plotted for different ROMs with CB reduction in Figs. 2.5 and 2.6. In Fig. 2.5 each line represents the non-dominated combinations of maximum harmonic index and number of CB-modes for a fixed value of the polynomial degree. In Fig. 2.6 the harmonic index is fixed. Obviously, it is best to use a balanced amount of each of the three types of DOF. Similar to the results without fixed interface modes each polynomial degree is optimal for a different accuracy range. For the analyzed case, it is sufficient to include the Fourier harmonics up to harmonic order seven as the higher harmonics bring negligible accuracy improvements. The resulting ROM shows a good efficiency with the proposed use of polynomial basis functions for the interface providing a reduction of the number of DOF up to a factor of three compared to using only the Fourier basis. Simultaneously the requirement for the meshes at the interface are removed.

When evaluating the usefulness of the method in general, some observations can be made:

- Models with finer meshes benefit more from the interface reduction.
- To accurately represent more system modes the number of interface and modal DOF in the ROM must be increased.
- The interface between two stages should be located far away from the blade roots. Near the blades the modeshapes have a higher harmonic content and more interface DOF are needed.

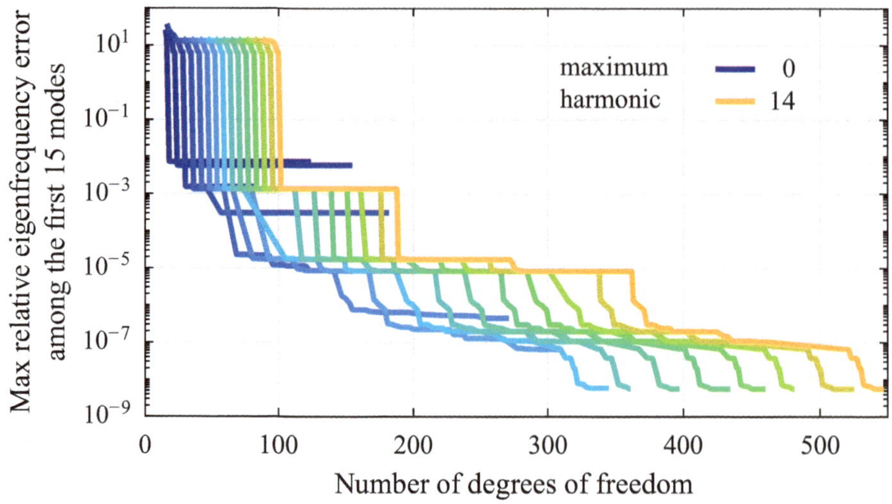

Fig. 2.6 Non-dominated ROMs for a fixed value of the maximum interface harmonic

2.6 Conclusions

In this paper, a model order reduction method for multistage bladed disks is presented. It is based on a cyclic Craig-Bampton reduction of each stage. The interfaces are reduced beforehand using basis functions that are a product of Fourier harmonics in the circumferential direction and polynomials in radial direction. It is a strict improvement over the approach of using only Fourier constraint modes, as the requirements for the interface meshes are removed and ROMs with less DOF are possible. The method is evaluated on an academic two-stage rotor. The possible trade-offs between ROM size and accuracy are demonstrated for different choices of number of interface harmonics and polynomial degrees as well as number of fixed interface modes. The benefits of the proposed method are even greater for larger models with finely meshed interfaces but decrease with higher accuracy requirements for the ROM. All calculations can be performed on single sector sized matrices. Using prior basis functions for the interface ensures that no intermediate reduction step potentially requires extreme amounts of memory.

For mistuning analyses the presented ROM can be extended using an additional modal analysis [23], or by implementing the interface reduction into more advanced methods such as MMDA [5] and PRIME [6].

Acknowledgement The authors kindly thank the German Research Foundation (DFG) for enabling this publication by funding the research project *Influence of Regeneration-induced Mistuning on the Aeroelasticity of Multi-Stage Axial Compressors* as part of the Collaborative Research Center 871 *Regeneration of Complex Capital Goods.*

References

1. Bladh, R., Castanier, M.P., Pierre, C.: Effects of multistage coupling and disk flexibility on mistuned bladed disk dynamics. J. Eng. Gas Turbines Power **125**(1), 121 (2003)
2. Song, S.H., Castanier, M.P., Pierre, C.: Multi-stage modeling of turbine engine rotor vibration. In: Volume 1: 20th Biennial Conference on Mechanical Vibration and Noise, Parts A, B, and C, pp. 1533–1543. ASME, Long Beach (2005)
3. Laxalde, D., Thouverez, F., Lombard, J.P.: Dynamical analysis of multi-stage cyclic structures. Mech. Res. Commun. **34**(4), 379–384 (2007)
4. Sternchüss, A., Balmes, E., Jean, P., Lombard, J.P.: Model reduction applied to multi-stage assemblies of bladed disks. In: ISMA, 15p (2008)
5. Bhartiya, Y., Sinha, A.: Reduced order model of a multistage bladed rotor with geometric mistuning via modal analyses of finite element sectors. J. Turbomach. **134**(4), 041001 (2012)
6. Kurstak, E., D'Souza, K.: Multistage blisk and large mistuning modeling using Fourier Constraint Modes and PRIME. In: Volume 7B: Structures and Dynamics, p. V07BT35A012. ASME, Long Beach (2017)
7. Klerk, D.D., Rixen, D.J., Voormeeren, S.N.: General framework for dynamic substructuring: history, review and classification of techniques. AIAA J. **46**(5), 1169–1181 (2008)
8. Craig, J.R., Roy, R., Bampton, M.C.C.: Coupling of substructures for dynamic analyses. AIAA J. **6**(7), 1313–1319 (1968)
9. Schwerdt, L., Willeke, S., Panning-von Scheidt, L., Wallaschek, J.: Reduced-order modeling of bladed disks considering small mistuning the disk sectors. J. Eng. Gas Turbines Power. **141**(5), 052502 (2018). Advance online publication https://doi.org/10.1115/1.4041071

10. Aoyama, Y., Yagawa, G.: Component mode synthesis for large-scale structural eigenanalysis. Comput. Struct. **79**(6), 605–615 (2001)
11. Balmès, E.: Use of generalized interface degrees of freedom in component mode synthesis. In: Office national d'études et de recherches aérospatiales, Châtillon-sous-Bagneux (1996)
12. Carassale, L., Maurici, M.: Interface reduction in craig–bampton component mode synthesis by orthogonal polynomial series. J. Eng. Gas Turbines Power **140**(5), 052504 (2018)
13. Castanier, M.P., Tan, Y.C., Pierre, C.: Characteristic constraint modes for component mode synthesis. AIAA J. **39**(6), 1182–1187 (2001)
14. Donders, S., Pluymers, B., Ragnarsson, P., Hadjit, R., Desmet, W.: The wave-based substructuring approach for the efficient description of interface dynamics in substructuring. J. Sound Vib. **329**(8), 1062–1080 (2010)
15. Gibanica, M., Abrahamsson, T.J., Rixen, D.J.: A reduced interface component mode synthesis method using coarse meshes. Procedia Eng. **199**, 348–353 (2017)
16. Holzwarth, P., Eberhard, P.: Interface reduction for CMS methods and alternative model order reduction. IFAC-PapersOnLine **48**(1), 254–259 (2015)
17. Tran, D.M.: Component mode synthesis methods using interface modes. Application to structures with cyclic symmetry. Comput. Struct. **79**(2), 209–222 (2001)
18. Zhang, G., Castanier, M.P., Pierre, C.: Efficient component mode synthesis with a new interface reduction method. In: Proceedings of IMAC-XXII: A Conference and Exposition on Structural Dynamics (2004)
19. Castanier, M.P., Pierre, C.: Modeling and analysis of mistuned bladed disk vibration: current status and emerging directions. J. Propuls. Power **22**(2), 384–396 (2006)
20. Yang, M.T., Griffin, J.H.: A reduced-order model of mistuning using a subset of nominal system modes. J. Eng. Gas Turbines Power **123**(4), 893 (2001)
21. Sternchüss, A., Balmès, E., Jean, P., Lombard, J.P.: Reduction of multistage disk models: application to an industrial rotor. J. Eng. Gas Turbines Power **131**(1), 012502 (2009)
22. Battiato, G., Firrone, C.M., Berruti, T.M., Epureanu, B.I.: Reduced order modeling for multistage bladed disks with friction contacts at the flange joint. J. Eng. Gas Turbines Power **140**(5), 052505 (2018)
23. Bladh, R., Castanier, M.P., Pierre, C.: Component-mode-based reduced order modeling techniques for mistuned bladed disks, Part I: theoretical models. J. Eng. Gas Turbines Power **123**(1), 89–99 (2001)

Chapter 3
Using Hybrid Modal Substructuring with a Complex Transmission Simulator to Model an Electrodynamic Shaker

Benjamin Moldenhauer, Matt Allen, Washington J. DeLima, and Eric Dodgen

Abstract When conducting a vibration test on an electrodynamic shaker or shaker table, the layout of fixtures and test components on the shaker adapter changes the vibration modes and anti-resonances of the system. During a test, these dynamics can be excited by the shaker itself, leading to an invalid test if the desired environment is exceeded at some points, and potentially damaging the shaker or test components. A high-fidelity shaker model would allow for accurate pretest planning, in which test component layout and control accelerometer placement can be optimized to mitigate problem areas. However, shaker systems are notoriously difficult to model analytically due to a multitude of joints with indefinite properties, unknown stiffness and damping of the magnetic field, and the scarcity of available technical drawings for the internal components. This work explores the use of hybrid modal substructuring to create a test-based model of a shaker table with a dynamically complex shaker fixture. The transmission simulator method is used with an experimentally derived modal model of the shaker and a fixture to decouple a finite element model of the fixture and replace it with a finite element model of the fixture with an attached test article. Special care is taken to ensure that an optimized test layout and accurate FEMs are created. The resultant modal model for the total system is shown to accurately retain the fixture dynamics while successfully adding the test component dynamics through the usable frequency range of the shaker.

Keywords Experimental substructuring · Component mode synthesis · Transmission simulator · Shaker modeling · Modal testing

3.1 Introduction

Structures that experience extreme vibrational loads during their service life are typically validated by performing life tests on a shaker. A shaker is able to generate arbitrary frequency and amplitude content that can be defined to recreate whatever environment the structure is likely to encounter. However, the shaker will also experience these environments as it produces them, possibly exciting its own resonances. Also, to provide the necessary excitation, the shaker must be directly attached to the structure, creating a coupled system. The nature of this coupling can alter the dynamics of the shaker in such a way that it may respond to its own excitation in unexpected ways. This can lead to invalid test results or possibly damage the test article and/or the shaker.

A model of the coupled test system that captures the cumulative response of the complete assembly would provide insight into how the test article alters the response of the shaker and form a basis for predictive design of the test setup and allow for a more robust test plan. Unfortunately, high fidelity numerical models of shakers, i.e. a finite element model (FEM), are notoriously difficult to create, due to the multitude of unknowns present in the structure, such as joint properties, material parameters, and internal layout [1]. Purely experimentally derived models are also unlikely to be effective as they are time consuming to create, are highly specialized, and require a test setup and plan to have already been created and implemented.

The research presented here within pertains to generating a hybrid experimental-analytical model of a shaker that effectively circumvents the difficulties mentioned above by leveraging the strengths of each method. This is done using the transmission simulator method of modal substructuring, also known as modal constraints for fixture and subsystem (MCFS) [2, 3]. With this technique, an experimental model of the shaker with a test fixture can be used as a basis to which FEMs of the test fixture and any test articles can be coupled, yielding a model of the full system. This experimentally captures the

B. Moldenhauer (✉) · M. Allen
University of Wisconsin—Madison, Madison, WI, USA
e-mail: bmoldenhauer@wisc.edu

W. J. DeLima · E. Dodgen
Honeywell, Kansas City, MO, USA

A. Linderholt et al., *Dynamic Substructures, Volume 4*, Conference Proceedings of the Society for Experimental Mechanics Series,
https://doi.org/10.1007/978-3-030-12184-6_3

dynamics of the internal shaker components and the linear stiffness and damping characteristics of the joints between the test fixture and the shaker, removing the need to accurately model them analytically. Also, by using FEMs of the fixture and potential test article, many different test configurations can be quickly simulated, avoiding costly prototyping and testing.

Previously, this method was successfully applied to a relatively simple setup as a proof of concept [4]. This research explores a more complicated structure that provides valuable insight into how this framework can be applied to a real test scenario with dynamically complex subcomponents.

This paper is divided into several sections, starting with a summary of dynamic substructuring and the theory behind the TSM. Following this, the various subsystems that will be utilized in the substructuring process are introduced and described. The work done to create accurate FEMs of the fixture and test article is then presented. After which, the test procedure implemented to acquire and process experimental data is defined. With all the necessary information to perform the TSM substructuring, the resultant experimental/analytical shaker model is evaluated relative to experimental truth data. Conclusions are then stated pertaining to the accuracy and capability of implementing the TSM to efficiently model electrodynamic shakers with complex fixtures and test articles.

3.2 Theory

3.2.1 Dynamic Substructuring

Modeling and analyzing system dynamics becomes increasingly difficult as the size and complexity of the structure grows. Large FEMs can require an unreasonable amount of time and computational power to directly solve, and as the assembly becomes larger and more complicated it becomes more difficult to design and perform an experiment that can capture all of the modes of interest. However, these difficulties can be somewhat alleviated by dividing the structure into smaller, more manageable subcomponents that can be analyzed individually in whatever domain is most appropriate, e.g. experimental or analytical. When one of the components is experimental, the methods used to combine experimental and numerical models to predict the response of the coupled system are termed experimental dynamic substructuring [5, 6].

Experimental dynamic substructuring methods are divided into two groups: Frequency-Based Substructuring (FBS) and Component Mode Synthesis (CMS) [7]. FBS uses the frequency response functions (FRFs) of the subcomponents to predict the response of the assembled system. CMS, also known as Modal Substructuring, combines the subcomponent equations of motion in modal coordinates to build the equation of motion of the assembled system. The CMS methods differ from each other in which component's modes are chosen to represent the subcomponent. It is imperative that the component modes of the individual subcomponents form an adequate basis to represent the motion of the coupled assembly. In the research presented herein, this is achieved using the Transmission Simulator Method.

3.2.2 Transmission Simulator Method

The Transmission Simulator Method (TSM), as proposed by Allen, Mayes, et al. [2, 3, 8], allows for experimental and analytical models to be coupled together through a common subcomponent that is present in both. This component, referred to as the Transmission Simulator (TS), acts as a distributed interface between the experimental and analytical subsystems, allowing their coupling constraints to be satisfied by the modal dynamics of the TS, as opposed to the more difficult process of enforcing compatibility in all six physical degrees of freedom at each point on the surface of the interface. Also, the TS effectively mass-loads the interface between the other subsystems and simulates the forces that would act there in the assembled structure. In practice, an experimental model is obtained by performing a modal test on some physical system that includes the TS. By decoupling a FEM of the TS from this model, the physical presence of the TS is removed, while the dynamics of a loaded interface are left on the physical component model. If this same process is carried out on a FEM that includes the TS and a third subcomponent, an analytical model of that subcomponent with a mass loaded interface is produced. The experimental and analytical models with mass loaded interfaces can then be coupled together with CMS, yielding a hybrid experimental/analytical model of the assembled structure.

This research applies the TSM to create a model of a floor-mounted electromagnetic shaker that has a three-sided half cube mounted on the armature, and a cantilever beam attached to the half cube. The shaker with the half cube is the experimental subsystem, the half cube is the TS and represents a shaker fixture, and the beam represents a prospective test article and, together with the half cube, is the analytical subsystem. Figure 3.1 displays a schematic of the effective process that will be

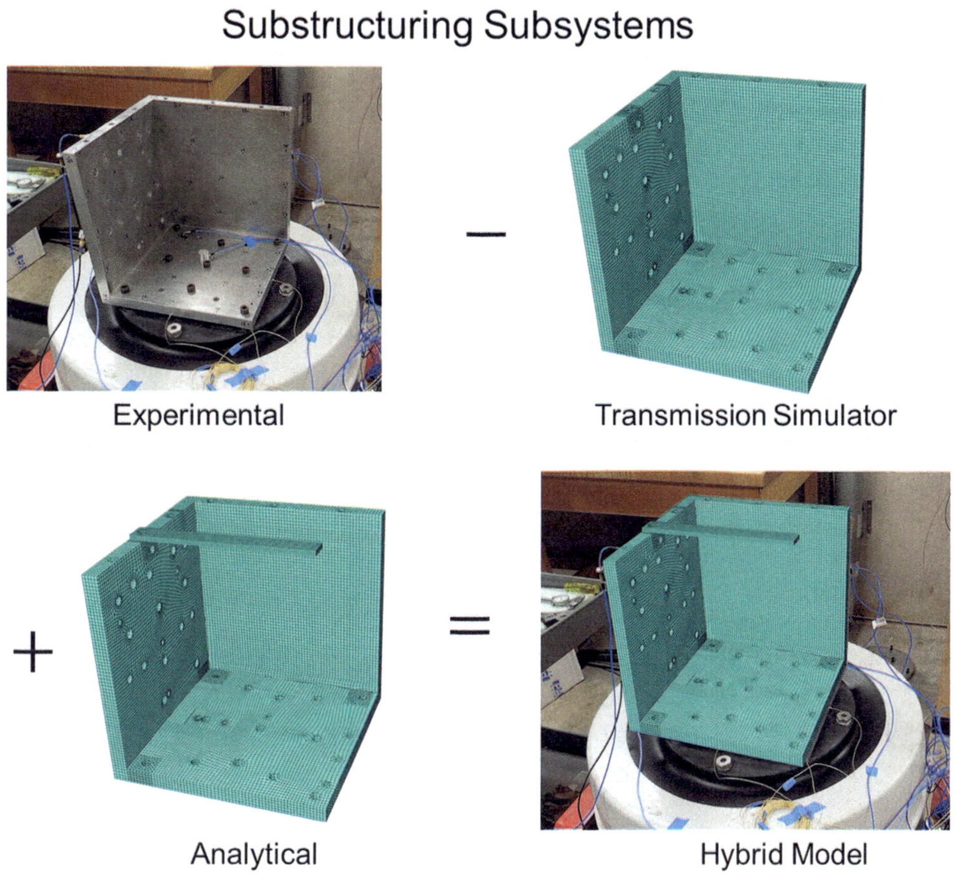

Fig. 3.1 Transmission simulator method schematic and subsystems

completed. Essentially, from an initial experimental modal model of the shaker and half cube, the half cube will be removed and replaced by a FEM of the half cube with the cantilever beam, yielding a hybrid model predicting the response of the complete assembly. While the response of a cantilever beam is not necessarily complex, the dynamics of the half cube are. This work seeks to assess if the half cube response can be preserved through the decoupling and coupling process, and if that will affect how accurately the beam is added to the system. This will be done by comparing the substructuring model to experimental truth data of the half cube and the beam on the shaker.

3.2.3 Mathematical Basis

To combine the subsystems, the equations of motion for each are first formulated in terms of modal coordinates, as given in Eq. (3.1) [7], where the damping is left out for brevity. In this equation, $\boldsymbol{\omega}$ is frequency, $\mathbf{I}$ is an appropriately dimensioned identity matrix, $\boldsymbol{\eta}$ are generalized modal coordinates vectors, $[\boldsymbol{\omega}_{\mathbf{n}}]$ are diagonal matrices of subsystem natural frequencies, $\boldsymbol{\phi}^{\mathbf{T}}$ are transposed subsystem mode shape matrices, and $\mathbf{F}$ are vectors of applied forces. The terms representing the TS are negative to signify that this subsystem is being removed, while the experimental and analytical FEM subsystems are being coupled.

$$-\omega^2 \begin{bmatrix} \mathbf{I}_{\mathbf{EX}} & \mathbf{0} & \mathbf{0} \\ \mathbf{0} & -\mathbf{I}_{\mathbf{TS}} & \mathbf{0} \\ \mathbf{0} & \mathbf{0} & \mathbf{I}_{\mathbf{AN}} \end{bmatrix} \begin{bmatrix} \boldsymbol{\eta}_{\mathbf{EX}} \\ \boldsymbol{\eta}_{\mathbf{TS}} \\ \boldsymbol{\eta}_{\mathbf{AN}} \end{bmatrix} + \begin{bmatrix} \left[\boldsymbol{\omega}^2_{\mathbf{n,EX}}\right] & \mathbf{0} & \mathbf{0} \\ \mathbf{0} & \left[-\boldsymbol{\omega}^2_{\mathbf{n,TS}}\right] & \mathbf{0} \\ \mathbf{0} & \mathbf{0} & \left[\boldsymbol{\omega}^2_{\mathbf{n,AN}}\right] \end{bmatrix} \begin{bmatrix} \boldsymbol{\eta}_{\mathbf{EX}} \\ \boldsymbol{\eta}_{\mathbf{TS}} \\ \boldsymbol{\eta}_{\mathbf{AN}} \end{bmatrix} = \begin{bmatrix} \boldsymbol{\Phi}^{\mathbf{T}}_{\mathbf{EX}} & \mathbf{0} & \mathbf{0} \\ \mathbf{0} & -\boldsymbol{\Phi}^{\mathbf{T}}_{\mathbf{TS}} & \mathbf{0} \\ \mathbf{0} & \mathbf{0} & \boldsymbol{\Phi}^{\mathbf{T}}_{\mathbf{AN}} \end{bmatrix} \begin{bmatrix} \mathbf{F}_{\mathbf{EX}} \\ \mathbf{F}_{\mathbf{TS}} \\ \mathbf{F}_{\mathbf{AN}} \end{bmatrix} \quad (3.1)$$

To enforce the coupling between the subsystems, a set of constraint equations is needed. Unlike more traditional substructuring methods, the TSM does not impose a strong physical constraint between the components in the physical degrees of freedom at the surface of their interface. Instead, constraints are enforced using the modal degrees of freedom of the TS, distributing the effective interface and softening the constraints [2, 3]. These constraint equations are given in modal form in Eq. (3.2), in which $\boldsymbol{\Phi}^{\dagger}$ denotes the Moore–Penrose pseudo-inverse of the mode shape matrix.

$$\begin{bmatrix} \boldsymbol{\Phi}_{\mathbf{TS}}^{\dagger} & \mathbf{0} \\ \mathbf{0} & \boldsymbol{\Phi}_{\mathbf{TS}}^{\dagger} \end{bmatrix} \begin{bmatrix} \boldsymbol{\Phi}_{\mathbf{EX}} & -\boldsymbol{\Phi}_{TS} & \mathbf{0} \\ \mathbf{0} & -\boldsymbol{\Phi}_{TS} & \boldsymbol{\Phi}_{\mathbf{AN}} \end{bmatrix} \begin{bmatrix} \boldsymbol{\eta}_{\mathbf{EX}} \\ \boldsymbol{\eta}_{\mathbf{TS}} \\ \boldsymbol{\eta}_{\mathbf{AN}} \end{bmatrix} = \mathbf{0} \tag{3.2}$$

From the subsystem natural frequencies and modeshape matrices and the constraint conditions, effective modal mass, damping, and stiffness matrices for the assembled structure can be found. The modal parameters of the assembly are found after performing an Eigen solution with the modal mass and stiffness matrices. For a more rigorous and complete derivation, see [2, 7].

3.3 Subsystems

3.3.1 The Transmission Simulator

The TS subsystem is a three sided aluminum half cube, shown below in Fig. 3.2. The sides are 3/4″ thick, the inside dimensions make a 10″ cube, and the fixture weighs approximately 21 lb. The three sections are bolted together with thirteen 3/8″ steel bolts distributed along the three seams, torqued to 50 N m. The half cube provides a plethora of attachment points for test articles and for anchoring to the shaker.

3.3.2 Experimental Subsystem

The basis of the experimental subsystem is an LDS V830 electromagnetic shaker, which is located at the University of Wisconsin—Madison. The shaker, as shown in Fig. 3.3, consists of a top armature measuring 13 in. in diameter, features 21 attachment points and, according to the manufacturer specs, has a usable frequency span of 0–3000 Hz and can produce a maximum sine force of 2200 lb. The experimental subsystem is the shaker with the TS bolted down to the armature with nine 3/8″ steel bolts also torqued to 50 N m.

Fig. 3.2 The transmission simulator. (Left) A three piece aluminum half cube, and (Right) CAD Model

Fig. 3.3 (Left) The shaker system used to evaluate the proposed substructuring approach, and (Right) the experimental subsystem; the TS mounted on the shaker

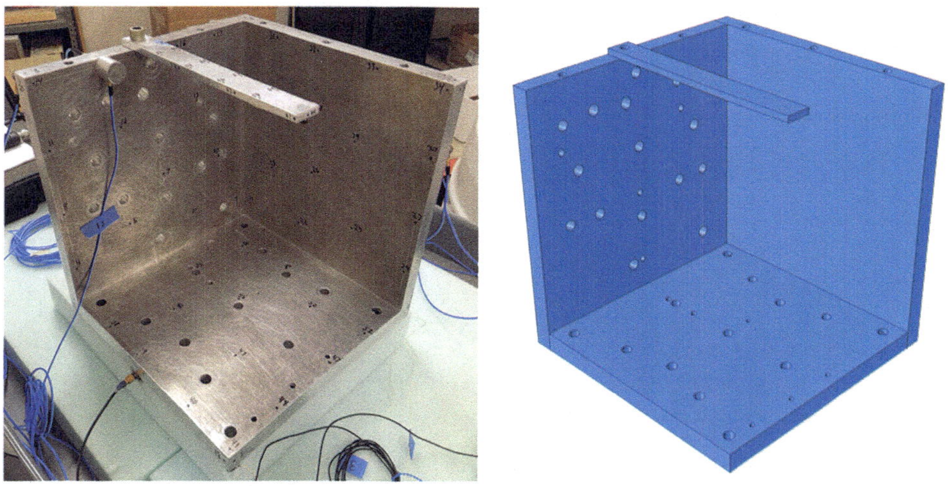

Fig. 3.4 The analytical subsystem. (Left) The half cube with a cantilever beam, and (Right) CAD model

3.3.3 Analytical Subsystem

The analytical subsystem, the half cube with a cantilever aluminum beam, is shown in Fig. 3.4. While the exact test article would generally not be available a priori, in this case the beam was machined in order to produce experimental truth data to validate the substructuring predictions. The beam measures 9″ × 1″ × .25″ and weighs 0.22 lb.

3.4 Finite Element Modeling

For the substructuring procedure, FEMs of the TS and the analytical subsystem were created. In the current work, this was done in Abaqus, resulting in the mesh shown in Fig. 3.5. This mesh contains 350,000 nodes from 67,000 20-node hex elements and 16,500 10-node tet elements. Complex hole geometry made an all hex element mesh unrealistic, but extensive partitioning resulted in a predominantly hex mesh with tet elements filling in around the regions with intersecting holes. Also, it is of note that bolt models in the form of simple cylinder plugs were used in prior versions of this FEM, but were found to not significantly alter the results, thus they have been omitted to reduce model complexity.

To calibrate the baseline FEMs, natural frequencies were taken from an initial test performed on the physical half cube in which a modal hammer was used to strike the half cube once near the free corner on each of the three sides. Material

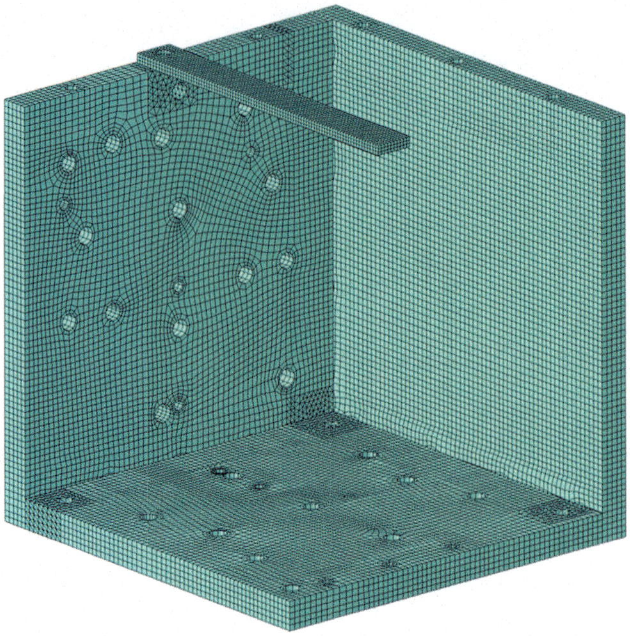

Fig. 3.5 Finite element subsystem mesh

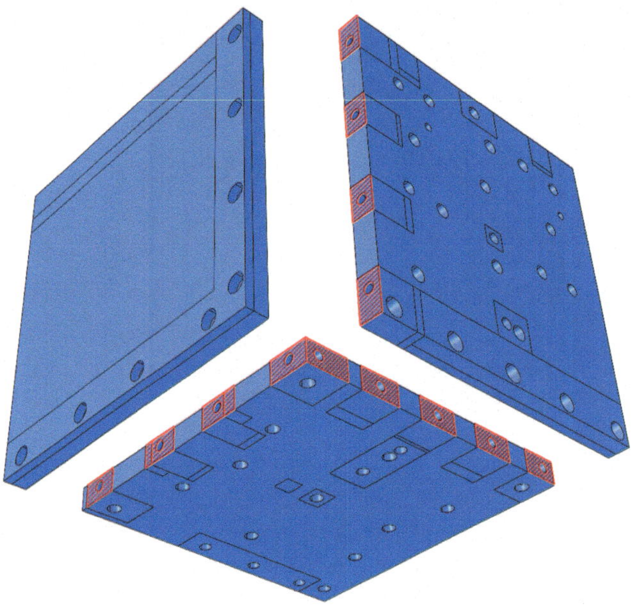

Fig. 3.6 Bonded contact areas defined in the FEM

properties, including elastic modulus and Poisson's ratio, were tuned via a guess and check method until good agreement was found between test and FEM frequencies. Final values of 10,150 kpsi for elastic modulus and 0.33 for Poisson's ratio were found. An average density of 0.1 lb/in^3 was set from mass values measured from the physical components. Also, a point mass of 0.03 lb was added to the beam tip to account for accelerometer weight.

The contact between the aluminum plates was also optimized. This involved varying the size of the bonded contact area around each bolt hole. The final contact areas, shown above in Fig. 3.6, extend approximately 1/2″ to either side of the bolt hole. Setting the open area between bonded zones to various states of contact was found to only degrade results, thus these zones were left free of a defined contact property.

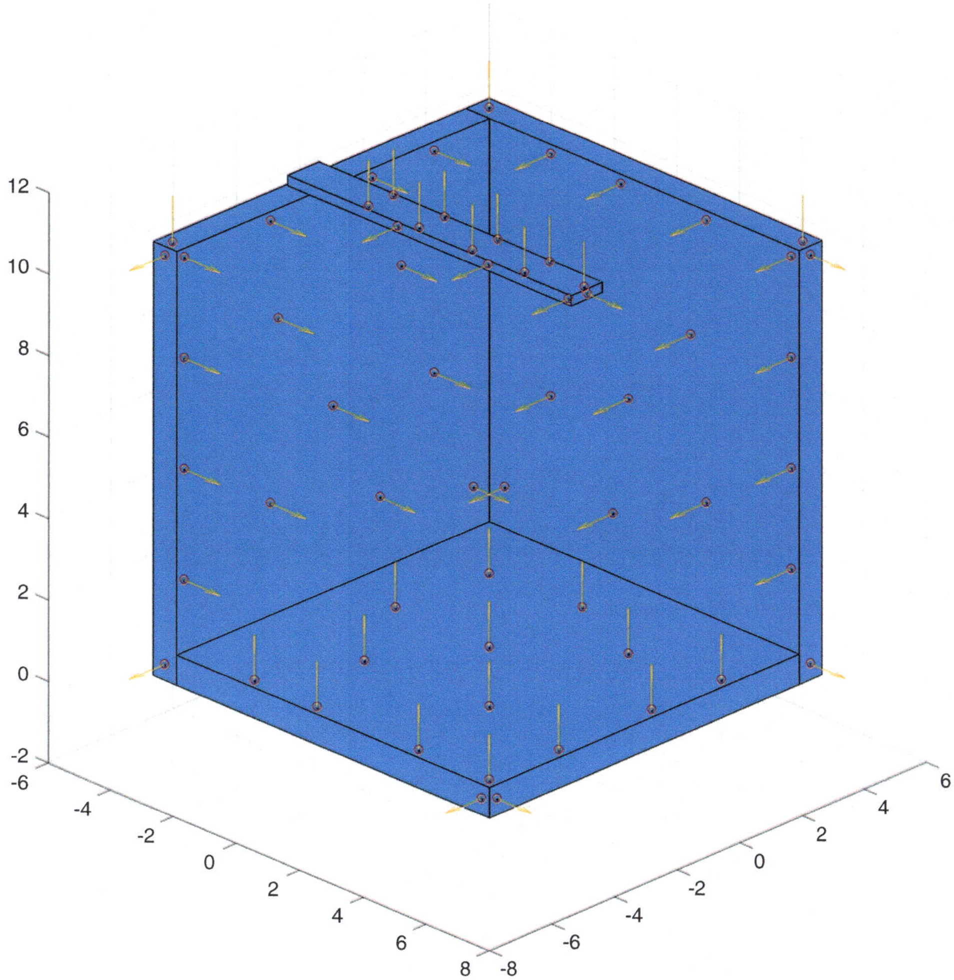

Fig. 3.7 Roving hammer hit locations and directions

3.5 Experimental Setup

An experimental modal model for the experimental subsystem, the TS mounted on top the shaker, was created by conducting a roving modal hammer test on the structure. The location and direction of each hammer hit are shown above in Fig. 3.7. The coordinates and direction of this set of points was chosen with Effective Independence [9]. This is a process through which an initial large set of potential test points are iteratively reduced to the most important set based upon preserving the information contained in the modeshapes. From an initial set of approximately 4000 points, the set of 51 points on the half cube and 13 points on the beam were produced.

The accelerometer responses were recorded in Data Physics SignalCalc® 830 software using the Data Physics Abacus® frontend (Fig. 3.8). A frequency span of 0–10,000 Hz was set to capture enough modes of the experimental subsystem to produce accurate substructuring results for the operable range of the shaker, as TS modes out to twice the target range are typically needed [8].

A roving modal hammer test was performed on each of the following: the half cube isolated on foam, the half cube and the cantilever beam on foam, the half cube attached to the shaker, and the half cube and beam attached to the shaker. The first two tests simulate free boundary conditions, providing data to further validate the FEMs. The third test is the only one that would be strictly necessary in the application of interest and is used to create an experimental model of the shaker and half cube. The final test provides truth data, allowing assessments to be made on how well substructuring has worked in this application. A summary of the results from these test cases is shown in Fig. 3.9. The blue curve provides the average of all of the FRFs from the isolated half cube, or the transmission simulator. To facilitate comparison, the FRFs from the beam plus half cube are divided into two parts. The orange curve shows the average of the measurement points that are on

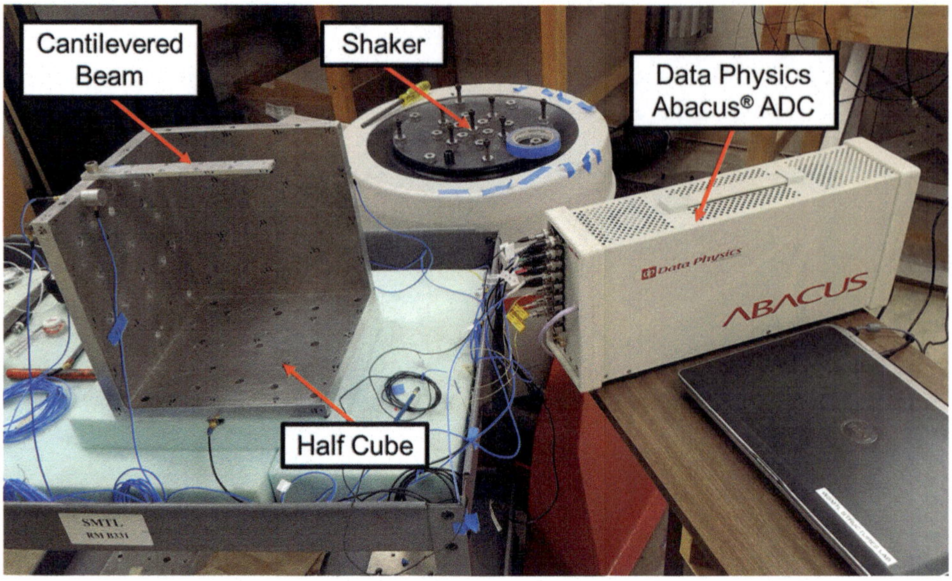

Fig. 3.8 Instrumentation for experimental data acquisition

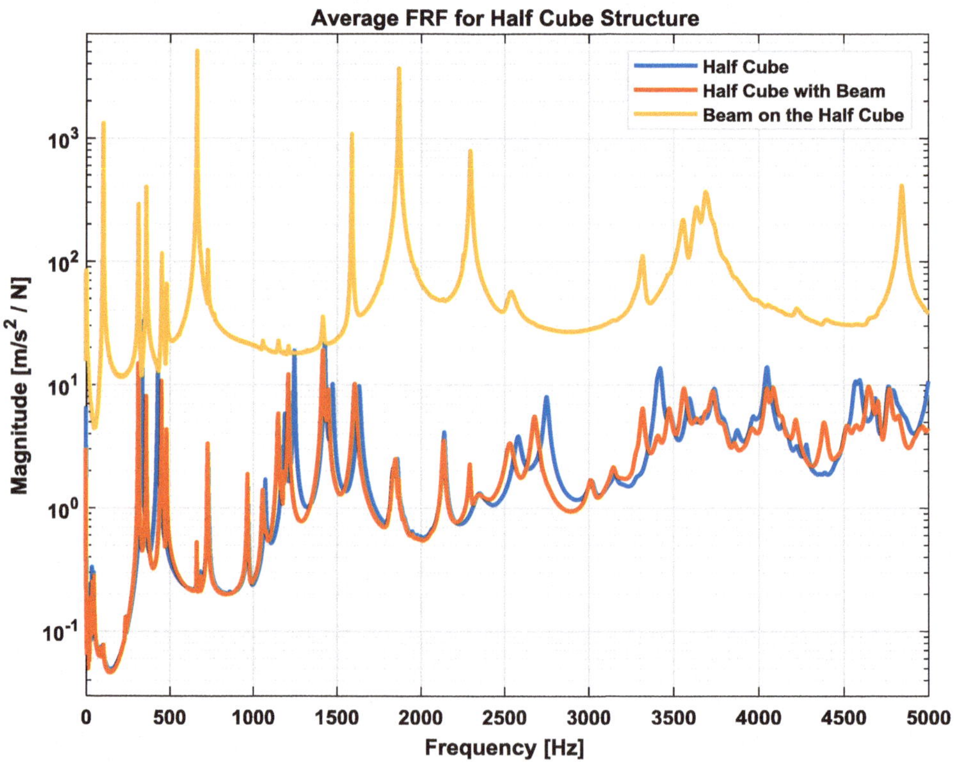

Fig. 3.9 Average FRFs from the free tests of the TS and FE subsystems

the half cube, while the yellow curve shows the measurements from the beam. It can be seen that the response of the cube does not drastically change due to the presence of the beam. This is expected given the size and mass difference between the two components. As a result of this, it is expected that substructuring should predict the slight changes observed on the transmission simulator, but mainly preserve its dynamics. The beam measurements show that several beam modes are within the target frequency range of the shaker and should be captured in the final assembly.

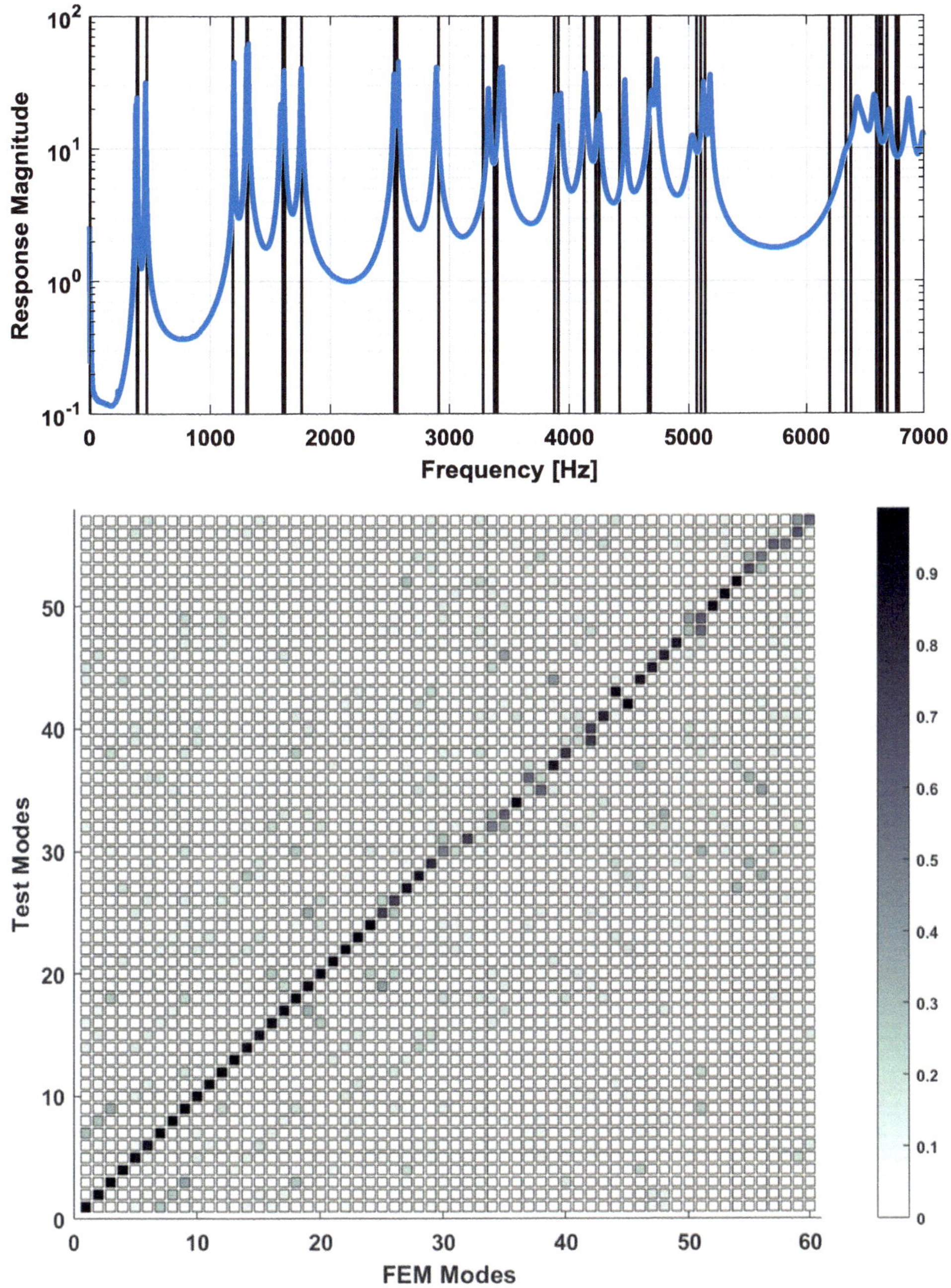

Fig. 3.10 TS test vs FEM—(Top) natural frequency, and (Bottom) MAC plot

Curve fitting for the experimental data was done using the Algorithm of Mode Isolation (AMI) [10, 11]. This program successively fits and subtracts single DOF modal models from the experimental data until all significant peaks have been eliminated. Natural frequencies, damping ratios and mass normalized modeshapes are computed and gathered from the AMI output data.

To validate the FEMs, the natural frequencies and modeshapes of each were compared to their experimental counterparts. For the TS, or the half cube isolated on foam, the comparisons are shown above in Fig. 3.10. In the top portion, the composite FRF of all the test points is given as the blue curve. The black vertical lines represent the FEM natural frequencies. There is very good agreement through 5000 Hz. The lower portion displays a Modal Assurance Criterion (MAC) plot, comparing the test and FEM modeshapes. A value closer to 1, or a color closer to black in the plot, indicates that the modes are similar in shape. This then shows that the first 29 test modes very closely correspond to FEM modes. These are the same modes whose frequencies corresponded closely in the upper plot.

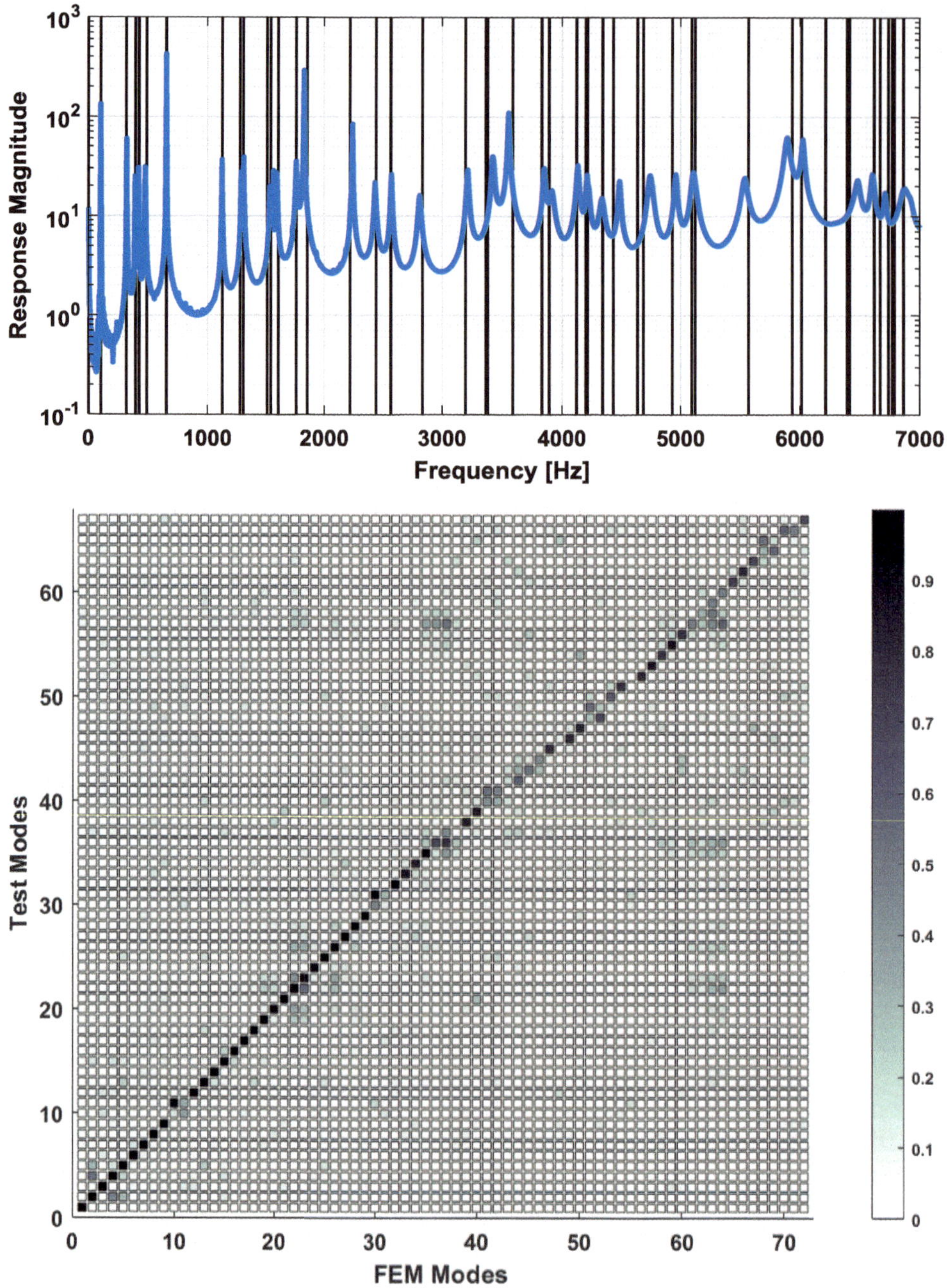

Fig. 3.11 Analytical subsystem test vs FEM—(Top) natural frequency, and (Bottom) MAC plot

The comparison for the half cube with the beam, or the analytical subsystem, is shown in Fig. 3.11. As with the TS, the FEM for the analytical subsystem agrees well in frequency and modeshape within the first ~5000 Hz.

3.6 Substructuring Prediction

With the necessary modal data collected for each subsystem, the TSM may be applied as given in Eqs. (3.1) and (3.2). It is of note that, to define the constraint equations, a suitable basis of linearly independent TS FEM modeshapes must be chosen. If linearly dependent shapes are included, the pseudo-inverse in Eq. (3.2) becomes ill conditioned and an insufficient number of constraints are enforced to satisfy the coupling equations of motion, resulting in wildly inaccurate predictions for the

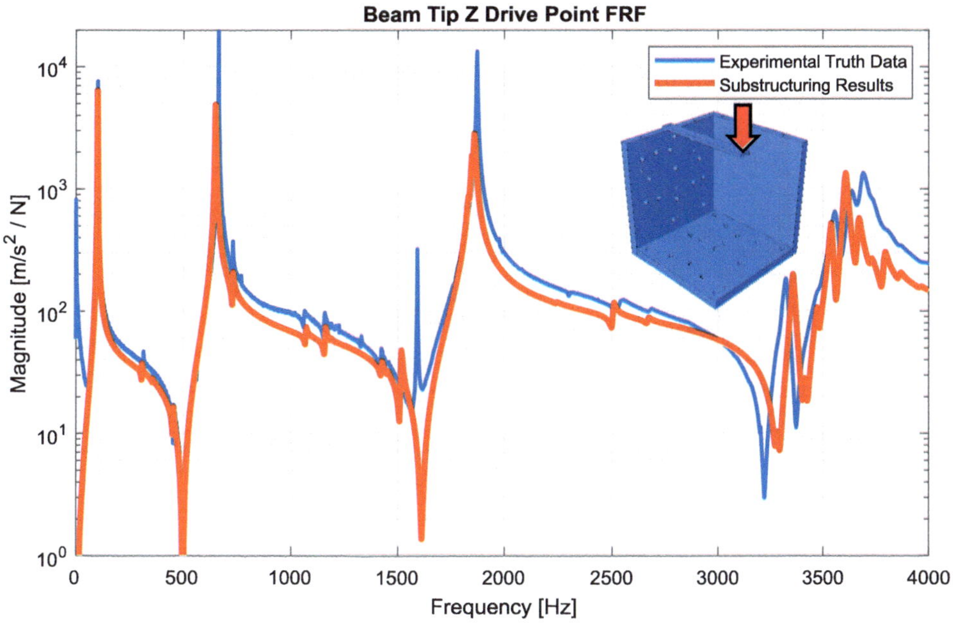

Fig. 3.12 Prediction vs. truth drive point FRF at the beam tip

assembled structure. Reference [12] provides several checks and criterion for determining a viable mode selection. In the current work, the first 54 modes of the experimental subsystem were used, spanning from 0 to 5000 Hz. For the TS, the first 35 modes were used, which corresponds to the number of FEM modes up to 6500 Hz. These were the FEM modes of the TS that were found above to agree very well with experiment. The analytical subsystem (TS plus beam) was represented by its first 48 modes, which spans 0–6700 Hz. This was chosen so that the TS and analytical models were roughly removing and then adding the same range of dynamic information.

The resultant natural frequencies and mode shapes of the hybrid structure predicted by the TSM were used to reconstruct drive point FRFs for easy visual comparison with the truth data. It should be noted that, for this work, damping was not accounted for in the FEMs, and thus zero damping is predicted. For the FRF reconstruction, a damping ratio of .25% was applied to all modes. Figure 3.12 shows a comparison between the predicted and truth drive point FRFs at the tip of the beam in the vertical direction. The transmission simulator method has successfully transferred in the first three cantilever beam modes in this direction, as given by the three large modal peaks in the FRFs. These predicted modes are quite accurate with respect to frequency out to around 3000 Hz, after which the results vaguely resemble the truth data, but are inaccurate.

Figure 3.13 shows the predicted and truth drive point FRFs at the top left corner of the half cube, next to the beam. As with the beam tip, the prediction is quite accurate and is in good agreement with the truth data. The results again become somewhat less accurate past 2500 Hz. This shows that the dynamics of the half cube were successfully preserved through the substructuring process and that complex transmission simulators can be handled quite effectively.

3.7 Conclusion

A framework for implementing the transmission simulator method on an electrodynamic shaker has been applied to a system containing a complex transmission simulator. It was shown that a cantilever beam, representing a mock test article, could be assembled onto a relatively complicated half cube shaker fixture, the TS, while preserving the dynamics of the half cube in the process. This was done by first creating FEMs of each subsystem and utilizing Effective Independence to determine an efficient set of roving hammer points to collect experimental modal data. From the steps laid out in previous work, the experimental and FEM data was passed through CMS to produce a prediction for the built-up system of the shaker, half cube, and the beam. When compared to truth data, the results were shown to be accurate out to near 3000 Hz. Thus, the predictions are quite good in the target range for this shaker, in that the dynamics of the half cube have been preserved while the beam has been successfully coupled onto the half cube.

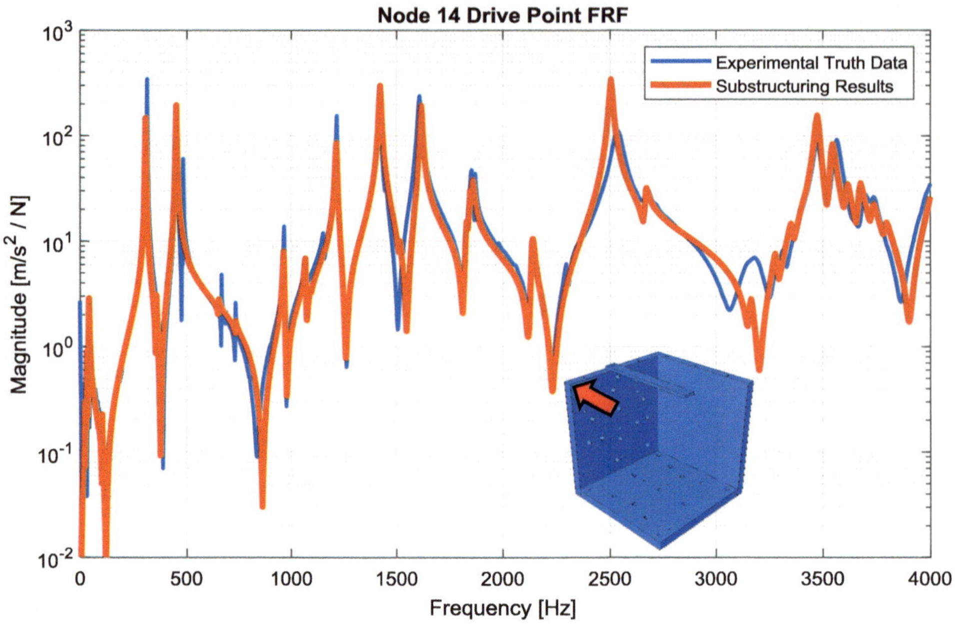

Fig. 3.13 Prediction vs. truth drive point FRF at the top left corner of the half cube

While it was shown that the TSM can be effectively implemented with a complex transmission simulator and modally dense subsystems, the addition of the beam did not significantly alter the system dynamics. A system in which a complex transmission simulator has multiple mock test articles of various masses and sizes would demonstrate the full range of scenarios that this framework for shaker substructuring could be implemented. Currently, such a setup is being tested, in which a large steel block is mounted to the half cube near the cantilever beam. This will sufficiently alter the half cube dynamics that the shaker and the beam should also be affected. A successful substructuring prediction of this setup would offer compelling evidence as to the robustness and effectiveness of modeling shakers with the transmission simulator method. Also, further work should be done to understand how to best incorporate damping in substructure models.

References

1. Delima, W.J., Jones, R., Dodgen, E., Ambrose, M.: A numerical approach to system model identification of random vibration test. In: Proceeding of 35th IMAC, Garden Grove, California (2017)
2. Allen, M.S., Mayes, R.L., Bergman, E.J.: Experimental modal substructuring to couple and uncouple substructures with flexible fixtures and multi-point connections. J. Sound Vib. **329**, 4891–4906 (2010)
3. Mayes, R.L., Arviso, M.: Design studies for the transmission simulator method of experimental dynamic substructuring. In: International Seminar on Modal Analysis, Lueven, Belgium (2010)
4. Moldenhauer, B., Allen, M.S., DeLima, W.J., Dodgen, E.: Modeling an electrodynamic shaker using experimental substructuring. In: Proceedings of the 36th IMAC, Orlando, Florida (2018)
5. Dynamic Substructuring Wiki [Online]. http://substructure.engr.wisc.edu
6. de Klerk, D., et al.: General framework for dynamic substructuring: history, review, and classification of techniques. AIAA J. **46**, 1169–1181 (2008)
7. Allen, M.S., Mayes, R.L.: Comparison of FRF and modal methods for combining experimental and analytical substructures. In: Proceedings of the 25th IMAC, Orlando, Florida (2007)
8. Roettgen, D.R., Allen, M.S., Mayes, R.L.: Ampair 600 wind turbine three-bladed assembly substructuring using the transmission simulator method. In: Sound and Vibration, November 2016
9. Kammer, D.C.: Sensor placement for on-orbit modal identification and correlation of large space structures. J. Guid. Control. Dyn. **14**(2), 251–259 (1991)
10. Allen, M.S., Ginsberg, J.H.: A global, single-input-multi-output (SIMO) implementation of the algorithm of mode isolation and applications to analytical and experimental data. Mech. Syst. Signal Process. **20**, 1090–1111 (2006)
11. Allen, M.S., Ginsberg, J.H.: Global, hybrid, MIMO implementation of the algorithm of mode isolation. In: Proceedings of the 23rd IMAC, Orlando, Florida (2005)
12. Allen, M., Kammer, D., Mayes, R.: Metrics for diagnosing negative mass and stiffness when uncoupling experimental and analytical substructures. J. Sound Vib. **331**, 5435–5448 (2012)

Chapter 4
Hybrid Substructure Assembly Techniques for Efficient and Robust Optimization of Additional Structures in Late Phase NVH Design: A Comparison

Benjamin Kammermeier, Johannes Mayet, and Daniel J. Rixen

Abstract In certain circumstances, not all desired NVH properties of a given mechanical structure, e.g. a vehicle, are satisfied at the end of a development process. In this situation, NVH properties of an existing structure must be improved while extensive changes of this structure are not practicable. Consequently, additional components such as mass dampers are included to improve the NVH properties. The arising task is to determine the optimal configuration of these additional components. If one assumes that no valid or accurate simulation model of the underlying structure exists, a hybrid substructuring approach is essential. The existing structure is measured at the required positions, the additional structures are modeled virtually, subsequently they are combined to a hybrid assembly. The optimization includes the repeated evaluation of such an hybrid assembly. In this contribution two major strategies are regarded: frequency based substructuring (FBS) and state-space substructuring (SSS). The possibly large number of evaluations imposes a greater demand on the computational efficiency compared to onetime assemblies. Furthermore, properties concerning the robustness towards measurement noise of the assembly technique play an important role. Based on a common notation for both assembly techniques, the relevant properties—efficiency and robustness—are compared on a numerical example.

Keywords Hybrid substructuring · Frequency-based substructuring · State-space substructuring · System identification · Frequency response estimation

4.1 Related Work and Outline

The paradigm of hybrid substructuring is straightforward: the assembled substructures are represented by models which are obtained from both—measurement and numerical modeling. In the measurement process, time series of inputs and outputs, e.g. forces and accelerations, are recorded. The process to obtain a model from measured data is called system identification. Besides system identification, numerical methods, e.g. FEM, can be used to derive a model. Therefore, the results of both processes will be referred to as *identified* and *numerical* model. In this contribution, a certain practical example for hybrid substructuring, rooted in vehicle NHV design, is anticipated: determine one (or more) (sub)optimal configuration(s) of additional substructures on a vehicle for a given objective. The vehicle can be measured and identified; the additional structures are available as numerical models. Such a hybrid substructuring task can be achieved by two strategies—Frequency-Based Substructuring (FBS) and State-Space Substructuring (SSS).

The FBS technique was widely used during the last few years. One application case which is often considered in the NVH context is the Transfer Path Analysis. The framework set up in [18] is foremost based on FBS. A general framework for substructuring, including FBS is given in [8]. In case of FBS, the assembly process is achieved in the frequency domain. The

B. Kammermeier (✉)
Faculty of Mechanical Engineering, Technical University of Munich, Garching, Germany

Forschungs- und Innovationszentrum FIZ, BMW Group, München, Germany
e-mail: benjamin.kammermeier@tum.de; Benjamin.Kammermeier@bmw.de

J. Mayet
Forschungs- und Innovationszentrum FIZ, BMW Group, München, Germany
e-mail: Johannes.Mayet@bmw.de

D. J. Rixen
Faculty of Mechanical Engineering, Technical University of Munich, Garching, Germany
e-mail: rixen@tum.de

A. Linderholt et al., *Dynamic Substructures, Volume 4*, Conference Proceedings of the Society for Experimental Mechanics Series,
https://doi.org/10.1007/978-3-030-12184-6_4

required frequency response functions (FRF), synonymously transfer functions (TF), are calculated using non-parametric identification schemes. Since measurements are never free from deviations, error propagation plays an important role in hybrid approaches. This issue is for example investigated in [7, 12, 13] and [6]. One possibility is to condition the acquired measurement results as described in [11]. These efforts emphasize the need for accurate identification procedures, especially when the resulting systems are coupled. An alternative to the FBS approach and the applied non-parametric identification is the use of state-space models for identification.

An early approach to couple state-space models is presented in [16]. The approach is essentially based on partitioning the in- and outputs and using generalized coordinates for both sets. The idea of identifying structures as state-spaces and using them in substructuring was continued by Abrahamsson. In [14], a different coupling procedure is introduced. This strategy is also based on partitioning the state space, yet a transformation is used which inherently reduces the order of the state-space according to the number of coupling constraints. When dealing with state-spaces, several system properties, which are expected from mechanical systems, are directly accessible. In [15] and [14], parameter-based optimization techniques are presented which generate an identified system which exhibits consistent mechanical properties—stability, passivity and reciprocity, among others.

Since state-space models can be converted into a corresponding transfer function, the coupling procedure in the frequency domain is applicable in both cases. In presence of errors, there might be considerable differences between the available coupling procedures. However, this issue is subject of a different investigation. Therefore, the major difference between FBS and SSS which is investigated in this contribution is the identification procedure. The classification into non-parametric and parametric approaches is used in the standard reference for system identification written by Ljung [10]. In [1] a similar classification is used, the identification of systems for mechanical structures is focused. In [17], Tangirala gives a more recent and comprehensive overview.

The objective of this work is to compare the performance of FBS and SSS concerning accuracy and efficiency using numerical examples. The comparison of the accuracy and computational costs of the identification procedure affects each hybrid substructuring scenario. The efficiency of the assembly evaluation is connected to the anticipated optimization problem, which requires many evaluations of these calculations. In order to provide a basis for the comparison, the theoretical background is summarized in Sect. 4.2. At first, the regarded identification schemes are briefly introduced in Sect. 4.2.1. Subsequently, a coupling technique for SSS similar to the FBS case is introduced in Sect. 4.2.2. Basic information on the comparison procedure is given in Sect. 4.3.1, followed by the used numerical examples in Sect. 4.3.2. These scenarios are employed to generate the results presented in Sect. 4.4. Findings concerning the accuracy of the identification and a subsequent assembly can be found in Sect. 4.4.1. Computation times for identification and assembly evaluations are compared in Sect. 4.4.2. Concluding remarks can be found in Sect. 4.5.

4.2 Theoretical Background

4.2.1 Identification Methods

In this section, the identification methods used and compared in Sects. 4.3 and 4.4 are briefly introduced. At first, the so-called non-parametric identification methods are presented. Afterwards, the applied, parametric, state-space methods are considered. The analysis refers to the following identification scenario. A system with output $y(t) = \left(y^1 \dots y^p\right)^\mathsf{T} \in \mathbb{R}^p$ and input $u(t) = \left(u^1 \dots u^o\right)^\mathsf{T} \in \mathbb{R}^o$ is regarded. The measurement takes place at constant sample time T_s. Therefore, the measured data in time domain can be written as $\mathcal{Y} = [y_0, \dots, y_k, \dots, y_{N-1}] \in \mathbb{R}^{p \times N-1}$ and $\mathcal{U} = [u_0, \dots, u_k, \dots, u_{N-1}] \in \mathbb{R}^{o \times N-1}$ where $y_k = y(kT_s)$ and $u_k = u(kT_s)$. For a measurement within $t \in [0, (N-1)T_s]$, N samples are generated. The data $\mathcal{Y}$ and $\mathcal{U}$ is the basis for all subsequently considered identification processes.

4.2.1.1 Non-parametric Identification of Frequency Response Functions

The following approaches to estimate FRF are denoted as non-parametric since measured data in time domain is directly translated into the corresponding data in the frequency domain. No additional parameters of a system representation are estimated [17, p. 484]. Using the impulse response g_k of the system, the underlying time-discrete linear in- and output relation is

$$y_k = \sum_{l=0}^{\infty} g_l u_{k-l} + v_k, \tag{4.1}$$

where v_k represents noise which is correlated with the output. Based on this perception of a disturbed linear in- and output relation in time domain, several non-parametric approaches to estimate the FRF exist. Note that the non-parametric approaches will yield a non-symbolic transfer functions of the form $G \in \mathbb{C}^{p \times o \times n_f}$ for an evaluation at n_f frequencies.

Empirical Transfer Function Estimate (ETFE) The empirical transfer function estimate is based on the Discrete Fourier Transformation $U(\omega_k) \in \mathbb{C}^o$ and $Y(\omega_k) \in \mathbb{C}^p$ of the measured input and output signal, which are evaluated at n_f frequencies ω_k with $k \in \{1, \ldots, n_f\}$. The estimate is given by division of the transformed quantities:

$$G_{ETFE}(\omega_k) = [g_{ij}(\omega_k)], \ \ g_{ij}(\omega) = \frac{Y^i(\omega_k)}{U^j(\omega_k)}, \ \ i \in \{1, .., p\}, \ j \in \{1, \ldots, o\}. \tag{4.2}$$

For arbitrary inputs and in the absence of noise, this expression is still an estimate of the actual FRF. The ETFE is only accurate for periodic inputs. In addition, the ETFE is not consistent—the variance of G_{ETFE} does not decrease as the number of samples rises [17, p. 556]. This leads to approaches based on smoothing.

Smoothing, Non-parametric Approaches Several approaches exist that are based on smoothing operations. Two major ideas can be found: one idea is to smooth the ETFE directly. This can be achieved by weighted averages at each frequency or by Welch's averaging process, which divides the available time data into segments, see [17, p. 561] and [20]. A different approach is based on spectral analysis (SPA), where smoothed estimates of the power spectral density of the input $\bar{S}_{uu}(\omega_k)$ and the cross spectral density of the output $\bar{S}_{yu}(\omega_k)$ are used. This leads to

$$G_{SPA}(\omega_k) = \bar{S}_{yu}(\omega_k)\bar{S}_{uu}^{-1}(\omega_k). \tag{4.3}$$

Like the smoothing of the ETFE, the estimates of the spectral quantities can be calculated by weighted averages or Welch averaging. The latter one leads to the so-called H1 estimator. Another approach is to use the Blackman-Tukey method [2, 3]. This approach will be used for the subsequent comparison.

4.2.1.2 State-Space Identification

Compared to the non-parametric approaches, the objective of state-space identification is to find a state-space model of order n that can be used to explain the observed in- and output samples. In this case, the parameters are the state-space matrices $A \in \mathbb{R}^{n \times n}$, $B \in \mathbb{R}^{n \times o}$, $C \in \mathbb{R}^{p \times n}$, $D \in \mathbb{R}^{p \times o}$ and $K \in \mathbb{R}^{n \times p}$ of the time-discrete state-space innovations form [17, pp. 659]

$$x_{k+1} = Ax_k + Bu_k + Ke_k, \ \ y_k = Cx_k + Du_k + e_k, \tag{4.4}$$

where $x_k \in \mathbb{R}^n$ is an unknown state and $e_k \in \mathbb{R}^p$ is a stationary disturbance. Additional to estimating the matrices A, B, C, D, K, the noise variance $E(e_k e_k^{\mathsf{T}})$ is identified. For the identification of state space models, two main strategies are available. On the one hand, the matrices can be determined via subspace methods, which are based on non-iterative projections, which makes them computationally efficient [4]. On the other hand, predictor-error methods can be applied, which results in the minimization of an objective function. This ultimately leads to the application of iterative solution methods which is likely to increase the computational effort, yet can yield more accurate results. Both approaches are sketched in more detail below.

Subspace Identification (SID) The basic process of SID is summarized in [10, p. 341]. Based on the given in- and output data, the extended observability matrix of the system can be estimated by projections. This matrix leads to A, C, and the noise parameters. With this knowledge, the matrices B and D can be determined. In this framework, several different algorithms, which use different weighting matrices in the estimation process to determine A and C were developed. In this context, the algorithms MOESP [19], CVA [9] and SSARX [5] are mentioned, since these alternatives are available off-the-shelf in MATLAB.

Prediction-Error Methods (PEM) Prediction-error methods are based on determining a set of parameters Θ such that

$$\Theta = \arg\min_{\Theta} \sum_{k=0}^{N} l(y_k - \hat{y}_k(\Theta)), \tag{4.5}$$

where l is a scalar function that measures the prediction error $y - \hat{y}(\Theta)$ [10, pp. 199]. The function l can be chosen such that the result is estimated for example in a least squares or maximum likelihood sense. For state spaces, the number of parameters is reduced by a transformation to a canonical form. Since the resulting optimization problems are in general non-linear, the acquired solution accuracy and required computation time highly depends on the initial guess for the optimization parameters [4]. Such an initial guess can for example be generated by subspace methods.

In this section, several identification strategies have been presented. The non-parametric empirical transfer function estimate and smoothing approaches based on spectral analysis were discussed. For the generation of state-spaces two possibilities, subspace- and prediction-error-based methods were introduced, which can be combined. In Sect. 4.4, the identification schemes ETFE, SPA with Blackman-Tukey, SID and PEM are compared. For all these algorithms, off-the-shelf implementations are available in MATLAB, which are used for the comparison.

4.2.2 Hybrid Substructuring—Coupling Methods

In the previous section, strategies to process measured data to either acquire non-parametric frequency response functions or state-space models were discussed. In hybrid substructuring, such identified models and numerically modeled components are assembled. In this section, the assembling strategies for FBS and SSS are considered. Based on the assembly procedure of the FBS scheme, the corresponding procedure for SSS is derived. Substructuring as well as state-space models have well established, unfortunately overlapping syntax. At some point, accustomed notations must be changed for clearness.

Frequency-Based Substructuring A set of N_S structures can be coupled by stacking the equation of motion $Y^{(s)} = G^{(s)}(U^{(s)} + U_I^{(s)})$ of each structure:

$$Y = G(U + U_I) = \begin{pmatrix} Y^{(1)} \\ \vdots \\ Y^{(N_S)} \end{pmatrix} = \begin{pmatrix} G^{(1)} & \dots & 0 \\ \vdots & \ddots & \vdots \\ 0 & \dots & G^{(N_S)} \end{pmatrix} \left(\begin{pmatrix} U^{(1)} \\ \vdots \\ U^{(N_S)} \end{pmatrix} + \begin{pmatrix} U_I^{(1)} \\ \vdots \\ U_I^{(N_S)} \end{pmatrix} \right). \tag{4.6}$$

In this expression, $U_I^{(s)}$ are internal, as yet undetermined forces due to the coupling process. The assembly operation is based on imposing a set of constraints on the degrees of freedom. This can be expressed by a signed Boolean matrix H such that $HY = 0$. Additionally, a set of generalized interface forces λ is defined by $U_I = -H^{\mathsf{T}}\lambda$. Applying these conditions to (4.6) leads to the assembled system [8]

$$Y = (G - GH^{\mathsf{T}}(HGH^{\mathsf{T}})^{-1}HG)U = \bar{G}U. \tag{4.7}$$

In the above procedure, the transfer functions can either be derived from numerical models or can be estimated.

State-Space Substructuring A similar assembly technique as (4.7) can be stated for a set of state-spaces. A sufficient condition for this assembly procedure to succeed is the existence of an non-zero feedthrough matrix $D^{(s)}$. This should always be the case for physically consistent models with acceleration output. Given displacement or velocity, a feedthrough can be generated by derivation with respect to time. The coupling is stated in continuous-time representation. For a set of N_S structures, the deterministic—K is not regarded—equations $\dot{x}^{(s)} = A^{(s)}x^{(s)} + B^{(s)}(u^{(s)} + u_I^{(s)}),\ \ y^{(s)} = C^{(s)}x^{(s)} + D^{(s)}(u^{(s)} + u_I^{(s)})$ can be stacked similarly to (4.6), yielding $\dot{x} = Ax + B(u + u_I),\ \ y = Cx + D(u + u_I)$. Applying the constraints on the acceleration output and solving for λ then yields the assembled state space and transfer function

$$\dot{x} = (A - BH^{\mathsf{T}}(HDH^{\mathsf{T}})^{-1}HC)x + (B - BH^{\mathsf{T}}(HDH^{\mathsf{T}})^{-1}HD)u = \bar{A}x + \bar{B}u \tag{4.8a}$$

$$y = (C - DH^{\mathsf{T}}(HDH^{\mathsf{T}})^{-1}HC)x + (D - DH^{\mathsf{T}}(HDH^{\mathsf{T}})^{-1}HD)u = \bar{C}x + \bar{D}u \tag{4.8b}$$

$$Y(\omega) = G_{SS}U(\omega),\ \ G_{SS} = \bar{C}(j\omega - \bar{A})^{-1}\bar{B} + \bar{D}. \tag{4.8c}$$

The expression (4.8c) will be used to evaluate the assembled transfer functions for state-space models. Compared to the assembly techniques from [16] and [14], the assembly (4.8a) and (4.8b), is not based on partitioning the single structures in interface and interior. Yet, the possible reduction of the model order due to the applied constraints is not achieved. This could be accomplished by a subsequent truncation-based step. The benefit of the present approach is that it can be seamlessly merged with the existing and frequently used FBS assembly procedure since neither additional information nor steps are required.

4.3 Test Scenarios

4.3.1 Testing Procedure

The general test procedure is based on a given numerical model, which is regarded as an exact reference. Used models for the analysis are a two-mass-oscillator and an acoustic testbench, which are presented in Sect. 4.3.2. These models are used to generate exact time series $\mathcal{Y}$ from a given input $\mathcal{U}$. The sample time is selected such that all the frequencies of the simulated model are captured. Corresponding to real structural measurements, the excitations are assumed to be applied sequentially. The time series for all excitations are then stacked and used as input for the identification process. Additionally, since the data is generated by a model, the order is known a priori. Otherwise, the order would have to be estimated, see for example [10, pp. 494] or [17, pp. 646].

Accuracy The testing of the accuracy is based on perturbing the exact in- and output series by superposing a zero-mean Gaussian white noise. The noise intensity is measured by the signal-to-noise ratio (SNR), which is defined as the ratio of signal to noise power. Different values of SNR at the in- and output can be used to generate deviations. Subsequently, the identification process is conducted with the procedures presented in Sect. 4.2.1. The FRF of the identified model can be compared to the FRF of the initial numerical model. Since the identified models will not perfectly match the numerical model anymore, the influence of this error on an assembly can be investigated: the identified models and the numerical model can be equally coupled with additional structures, the resulting assemblies can again be compared.

In order to compare these results for different SNRs at in- and output, deviation of the resulting transfer functions must be quantified. In this contribution, the comparison of two transfer functions $G_1, G_2 \in \mathbb{C}^{p \times o \times n_f}$ is based on their magnitude $|G|$ and phase $\angle G \in [-\pi, \pi[$. For both quantities, the error e as well as the correlation coefficient ρ are used. The correlation coefficient is defined in the standard way for magnitude and phase. A high correlation is indicated by values close to one. Pure offsets of two functions do not influence ρ. The phase error is measured and logarithmized such that it is bounded between $]-\infty, 0]$. The magnitude error is defined by the logarithmized squared error. The findings on accuracy are presented in Sect. 4.4.1. The term "accurate" is used with respect to these comparison criteria.

Efficiency Recalling the anticipated optimization scenario described in Sect. 4.1, two more quantities are of interest: the computational costs of the system identification—denoted as offline costs—and the costs of the evaluation of the assembled FRF—referred to as online costs. For the identification, off-the-shelf functions implemented in MATLAB are used. The offline costs are the computation time of these functions. In case of the assembly costs, the expressions (4.7) and (4.8c) have to be evaluated. For the latter one, a comparatively fast MATLAB implementation is available and therefore used.

4.3.2 Test Structures

The testing procedure from Sect. 4.3.1 is based on numerical models, which are used to generate the required time series. Two different models are used in this contribution and presented subsequently.

Two Mass Oscillator (TMO) A two mass oscillator is used as a low dimensional example. This allows an efficient investigation of a wide range of the impact of different SNRs at in- and output. The hybrid problem is shown in Fig. 4.1. The TMO on the left-hand side is used as identified model, the counterpart on the right-hand side is coupled by enforcing $\ddot{y}_2 = \ddot{y}_3$. This yields a coupling matrix $H = \begin{pmatrix} 0 & 1 & -1 & 0 \end{pmatrix}$. Based on this identification and assembly, the test scheme described above is conducted. This test setup is used to investigate the influence of different combinations of in- and output SNRs on the accuracy of the identification and the subsequent assembly; see Sect. 4.4.1.

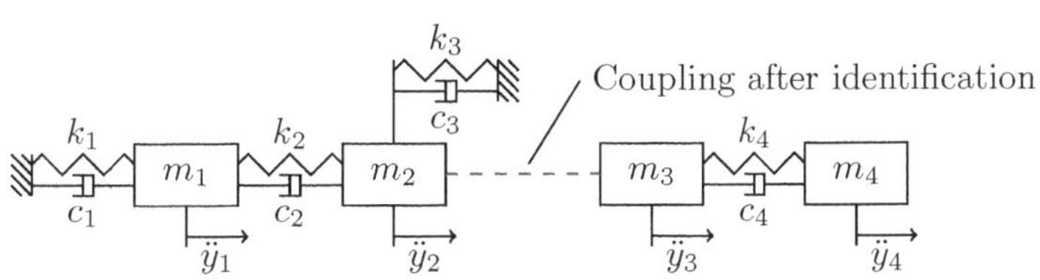

Fig. 4.1 Sketch of the identified two mass oscillator (*left*) and the subsequently coupled two mass oscillator (*right*). The parameters which are used to generate the test results are: $m_1 = m_3 = 1$ kg, $m_2 = m_4 = 0.5$ kg, $c_1 = c_2 = c_3 = c_4 = 1\frac{Ns}{m}$, $k_1 = k_2 = k_3 = k_4 = 10{,}000$ N/m

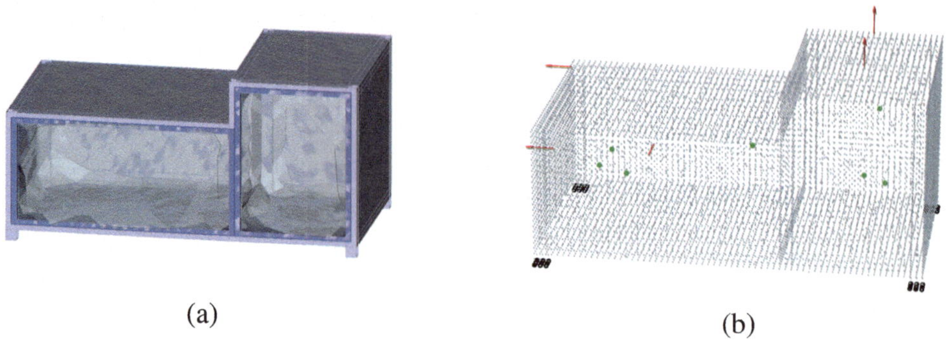

Fig. 4.2 CAD model of the acoustic testbench in (a) and FE grid points including excited and measured degrees of freedom in (b). Length, width and height are equal to 0.66 m, 0.36 m and 0.32 m/0.24 m, respectively

Acoustic Testbench (ATB) In order to provide some insight in a more complex example, the model of an acoustic testbench is used. The ATB is illustrated in Fig. 4.2. Transparent parts in Fig. 4.2 are modeled with acrylic glass. The acrylic glass and steel sheets are mounted on a frame which is built from rectangular hollow and L-sections. This construction encloses an acoustic cavity, which is included in the model. For the subsequent investigations, structural damping is assumed. Additional damping between fluid and structure or in the fluid is not included. The ATB is fixed to the ground at the bottom of the four legs. The inputs are marked with red arrows on the right-hand side of Fig. 4.2. The output is measured at all these input locations. Green dots show the locations where the sound pressure is additionally measured. In total, 10 outputs and 6 inputs are included.

The finite element model of the ATB contains roughly 10^5 degrees of freedom. For the subsequent analysis, a simplified modal model of the ATB, which contains the first 25 modes, is used as reference. This model is used to analyze the impact of a specific noise level with an SNR of 100 at the in- and output. The computational effort for the identification of a larger system can be compared. Additionally, this example is suited to study the online performance of the assembly strategies. The findings of the mentioned investigations are gathered in the next section.

4.4 Test Results

4.4.1 Accuracy

4.4.1.1 Two Mass Oscillator

The two mass oscillator model was used to investigate the impact of different SNRs at in- and output for the identification and the subsequent assembly. An analysis on FRF level is not provided in this section. The results are compared using the criteria from Sect. 4.3.1.

Identification The results of the identification process of the TMO example are shown in Fig. 4.3. It includes the comparison of the identification methods ETFE, SPA, SID and PEM. For all methods, the four criteria—error of phase and magnitude, correlation coefficient of phase and magnitude—described in Sect. 4.3.1 are investigated. The color of a certain pixel indicates the value of a particular criterion for a certain combination of input and output SNRs. All the plots are structured in the same way: the results for high SNRs are located in the center. Moving along the horizontal axis from the center, the SNR of the measured output signal drops. The same principle is used for the input SNR and the vertical axis. All criteria are evaluated for the same noise levels, the results in Fig. 4.3 are generated by using values between 10^{10} and $5 \cdot 10^{-1}$, representing the range between almost perfect signals and signals dominated by noise.

For all methods, the basic expectation that higher SNR come along with more accurate results is met. The non-parametric approaches exhibit similar results—slightly more accurate results can be observed for the SPA approach, especially concerning the error of the phase. The comparison of state-space and non-parametric approaches results is a bipartite picture: for high SNRs, the state-space approaches can be considered as more accurate. For dropping SNRs this observation only extends to the PEM. The subspace identification becomes less accurate and—regarding the correlation of the phase—more

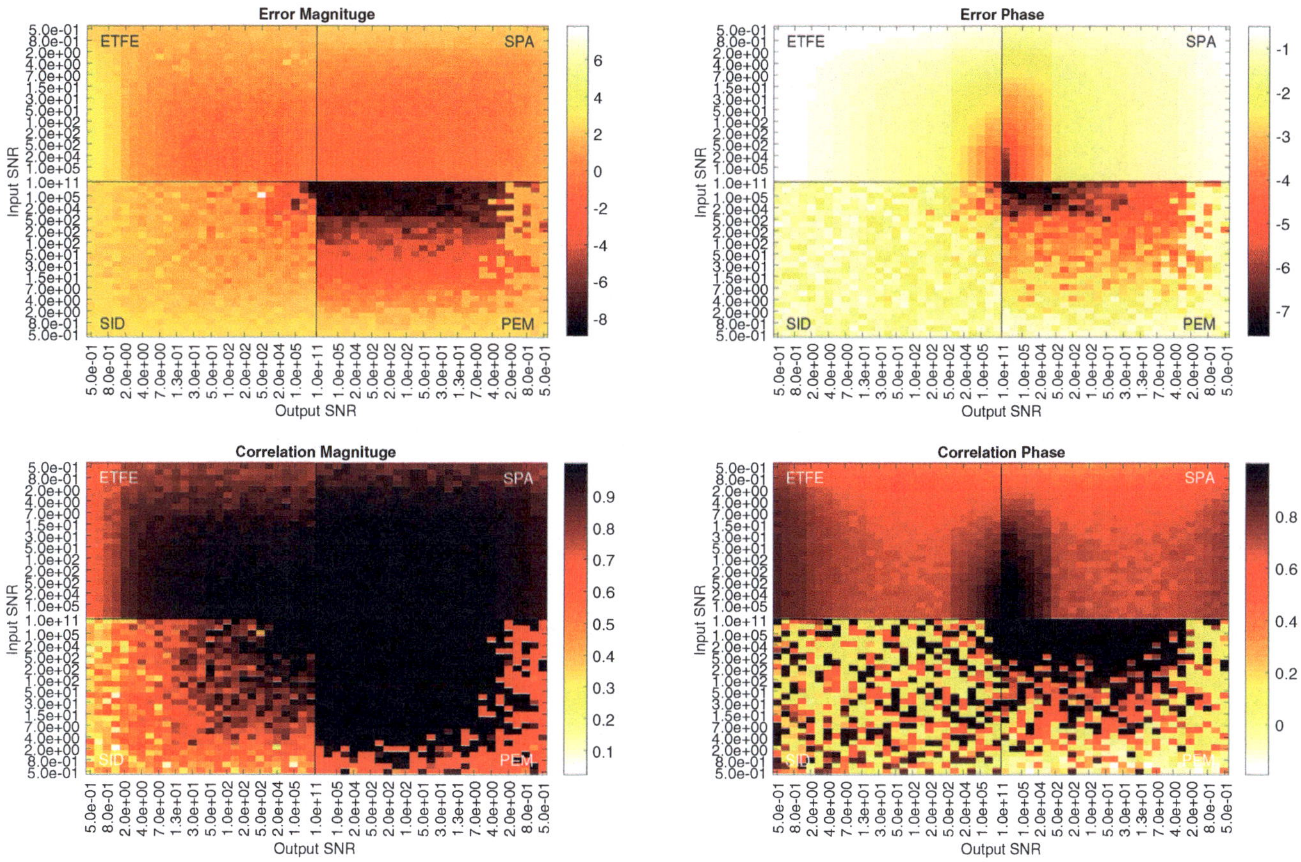

Fig. 4.3 Evaluation results of the identification process of a two mass oscillator for different signal-to-noise ratios at in- and output, evaluated for ETFE, SPA, SID and PEM identification

unreliable, since the evaluation shows no clear tendency for dropping SNRs. In contrast to the SID, more accurate results are generated with the PEM for a larger range of SNR values.

Assembly The results shown in Fig. 4.4 are generated in the same way as in Fig. 4.3. In this case, the assembled system of the identified models and the exact assembly is compared. Except for the phase error, all other criteria show that the inaccuracies made in the identification are increased by the assembly process. The limits of in- and output SNRs in which accurate results are achieved decrease significantly or are not present at all. Additionally, the decrease in the achieved accuracy exhibits a sharp drop with decreasing SNRs in case of the PEM. This underlines that hybrid substructuring demands a high accuracy of the identification process in order to acquire useful results in the assembled system. For the present example, the PEM methods show by far the highest accuracy. Comparing Figs. 4.3 and 4.4 indicates that only the accurate results obtained by PEM deliver accurate results for the assembly. The assessment by an error measure does include a qualitative comparison. At this point, it is necessary to compare the acquired results on FRF level, which is done in the next section.

4.4.1.2 Acoustic Testbench

In this section, the results of the identification and subsequent assembly process of the acoustic testbench are discussed. Additionally, only the methods SPA and PEM are considered, in order to provide clear plots. As shown in the previous section, these methods are more promising than ETFE or SID. The identification was conducted as presented in Sect. 4.3.2. The SNR on the input and output is set to 100. The evaluation of all FRF of the identified and subsequently assembled 10×6 system in terms of the used error measures is gathered in Tables 4.1 and 4.2.

These findings correspond to the expectation that was indicated by the results presented in Sect. 4.4.1. For the chosen SNR values, the PEM shows better results compared to the SPA in terms of the used error measures, for both identification and assembly. The comparison criteria that have been applied can be used to detect changes for entire systems when the

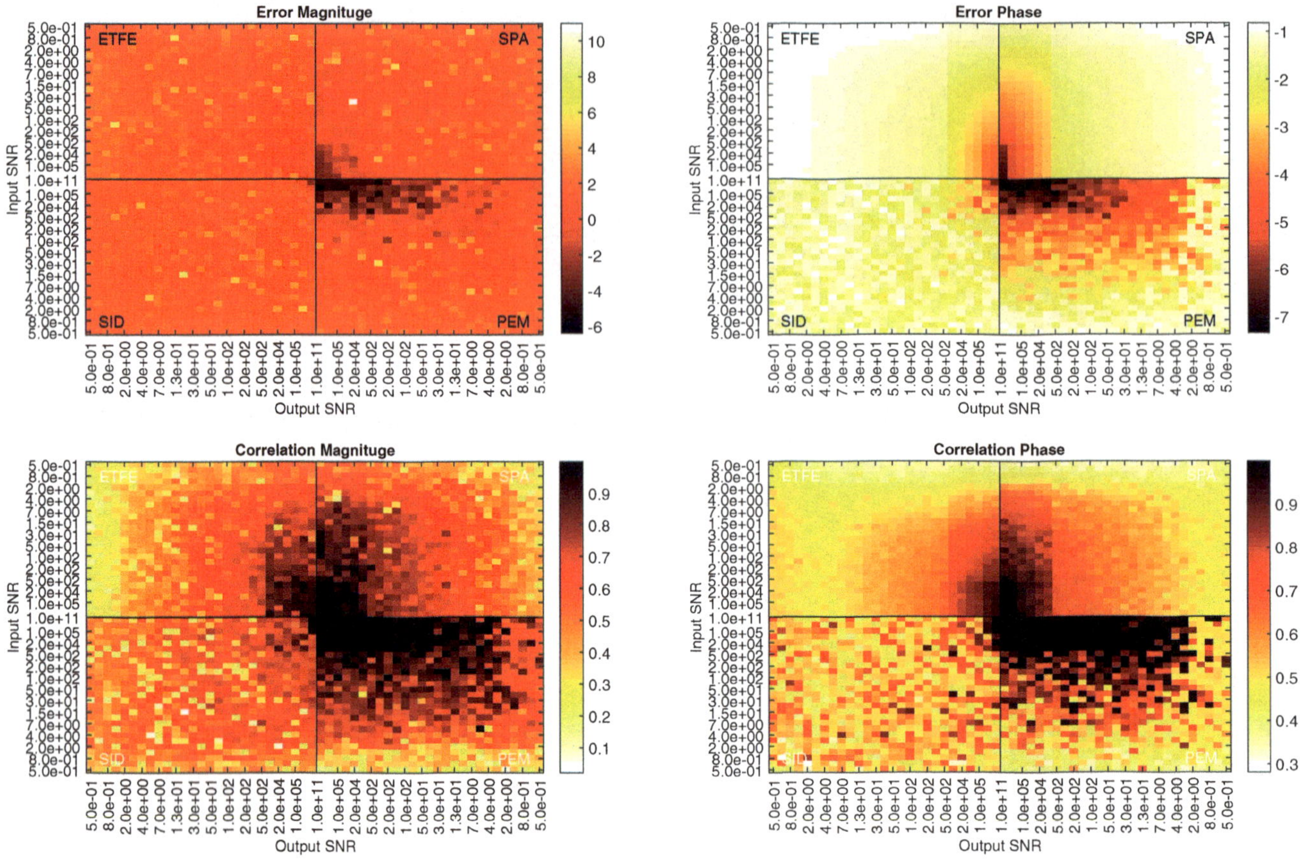

Fig. 4.4 Evaluation results of the hybrid assembly process of an identified and a numerical two mass oscillator for different signal-to-noise ratios at in- and output, evaluated for ETFE, SPA, SID and PEM identification

Table 4.1 Result of the identification process using **SPA** and *PEM*

	Magnitude	Phase
Error	**0.199**, *0.090*	**0.264**, *0.203*
Correlation	**0.994**, *0.996*	**0.291**, *0.327*

Table 4.2 Result of the assembly process using **SPA** and *PEM*

	Magnitude	Phase
Error	**17.42**, *1.93*	**0.306**, *0.138*
Correlation	**0.658**, *0.961*	**0.265**, *0.589*

effects of different SNR are investigated. This purely quantitative tool gives no information on the qualitative differences. In order to have more insight, an FRF of the identified and assembled system is compared in Fig. 4.5.

On the left hand side, the exact and identified FRF without additional structures are shown. In general, the accuracy of the identified results is comparable. Both approaches are able to depict resonance frequencies and amplitudes rather accurately. The foremost difference that can be observed are oscillations that occur in the SPA estimate. These oscillations foremost occur in regions with a small magnitude where noise effects dominate the underlying time series. For the present example, this can mainly be observed in the low-frequency range.

On the right hand side of Fig. 4.5, the resulting FRF with an added two mass oscillator to the identified model is shown. Compared to the identification results, both cases become less accurate. This is in line with the general expectation and the results in Sect. 4.4.1. However, in the assembled case, the results obtained using PEM are more accurate than those based on the SPA estimate. This can foremost be observed at approximately 90 Hz, where two resonance frequencies are not captured by the SPA approach.

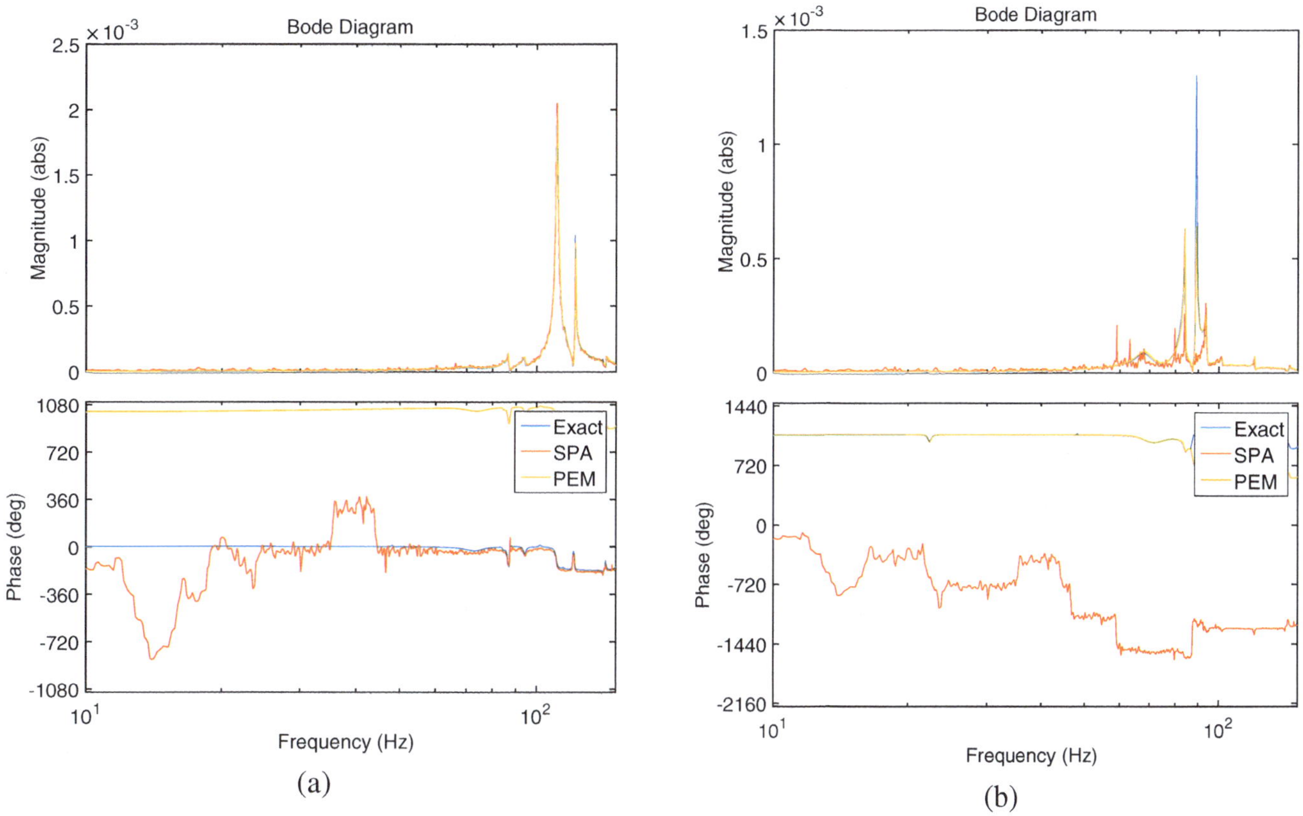

Fig. 4.5 An FRF describing the sound pressure due to structural excitation of ATB for the identification in (**a**) and the hybrid assembly in (**b**), comparing SPA and PEM

Table 4.3 Required offline costs a different number of in- and outputs n_{IO} and different model orders n

$n_{IO}\|n$	2|4	2|50	30|4
ETFE	0.2 s	0.2 s	10.5 s
SPA	5.7 s	5.6 s	1220.7
SID	5.1 s	857.6 s	40,000 s
PEM	5.8 s	1090.8 s	86,520 s

4.4.2 Efficiency

Identification—Offline Costs In the context of the initial optimization problem, the time required to perform the identification is denoted as offline costs. These are briefly mentioned here, since an important observation can be made (Table 4.3).

The effect of a higher order has no impact on the required time for non-parametric approaches. For both, SID and PEM the step from order 4 to order 50 increases the required computation time by a factor of roughly 20. It can be observed that the number of in- and outputs has a dramatic impact on the required identification time for the state-space based methods. This is foremost due to the way the data is handled. As described in Sect. 4.3.1, the results from independent excitations are stacked to form the input time series $\mathcal{U}$ and $\mathcal{Y}$. This lengthens the input time series and results in an impractical computation time.

Assembly: Online Costs The evaluation of the assembly equation requires different mathematical operations for SSS and FBS. Table 4.4 shows the required time to evaluate an assembly at n_f frequencies with n_c coupled oscillators.

The basis for this evaluation was a model of the ATB with 100 arbitrarily chosen in- and outputs of order 50. The results show that based on the chosen example, no method is preferable since the required computation time is in general similar.

Table 4.4 Required computation time for the evaluation of an assembly for a different number of coupled degrees of freedom n_c and evaluated frequencies n_f for *FBS* and **SSS**

$n_c \backslash n_f$	100	1000	5000	10,000
1	*0.03 s* **0.39 s**	*0.27 s* **0.25 s**	*1.38 s* **1.23 s**	*2.71 s* **2.67 s**
10	*0.04 s* **0.21 s**	*0.48 s* **0.43 s**	*2.30 s* **2.01 s**	*4.58 s* **4.00 s**
50	*0.19 s* **0.44 s**	*1.73 s* **1.63 s**	*8.61 s* **8.11 s**	*18.1 s* **16.3 s**
100	*0.60 s* **1.07 s**	*6.16 s* **5.40 s**	*28.5 s* **26.0 s**	*57.7 s* **54.5 s**

4.5 Conclusion and Outlook

In this contribution, two strategies for hybrid substructuring—frequency-based and state-space substructuring—have been compared. The major difference that was investigated was the underlying identification methods, namely non-parametric and state-space-based approaches.

The investigation of the achieved accuracy shows that, over a wide range of noise levels on the in- and output, the state-space methods using prediction-error optimization generate the most accurate results. Additionally, the investigation of a subsequent assembly operation shows that accurate identification results are required to obtain accurate results for the assembly. In detail, these findings are gathered in Sect. 4.4.1. In terms of accuracy, these findings clearly suggest to use a combination of subspace and prediction-error methods.

However, recalling the sequential nature of structural measurements, which leads to the way the identification time series were constructed in the presented approach, the major drawback of the state-space-based schemes is an impracticable computational effort with a rising number of in- and outputs. For this issue, further investigation is required. Additionally, the present contribution did not consider the determination of a suitable model order. This was bypassed since the real order was known a priori. In practice, this is also an additional step to consider. Finally, the nonlinear optimization underlying the PEM methods predisposes them to run into local minima—which can lead to qualitatively wrong solutions. This strongly depends on the used initial guess. For rising noise levels, this risk increases. At the same time, non-parametric approaches tend to generate oscillating FRF, which will not produce accurate results for an assembly but give a reasonable notion of the real solution's order of magnitude.

In conclusion, the present investigation finds that the application of state-space-based approaches for hybrid substructuring can outperform non-parametric alternatives. In order to acquire trustworthy results, a combination of the results of both approaches should be cross-checked. The aforementioned computation time issues concerning PEM approaches can be mitigated by using FRF as objective functions instead of stacked time series. These FRF have been found to not exactly depict the correct solution. The addressing of this issue and the comparison of the alternatives requires further investigations.

References

1. Alvin, K., Robertson, A., Reich, G., Park, K.: Structural system identification: from reality to models. Comput. Struct. **81**(12), 1149–1176 (2003). ISSN: 00457949. https://doi.org/10.1016/S0045-7949(03)00034-8. http://linkinghub.elsevier.com/retrieve/pii/S0045794903000348
2. Blackman, R.B., Tukey, J.W.: The measurement of power spectra from the point of view of communications engineering — Part I. Bell Syst. Tech. J. **37**(1), 185–282 (1958). ISSN: 0005-8580. https://doi.org/10.1002/j.1538-7305.1958.tb03874.x. http://dx.doi.org/10.1002/j.1538-7305.1958.tb01530.x%20http://ieeexplore.ieee.org/lpdocs/epic03/wrapper.htm?arnumber=6773415%20http://ieeexplore.ieee.org/document/6768513/
3. Blackman, R.B., Tukey, J.W.: The measurement of power spectra from the point of view of communications engineering - part II. Bell Syst. Tech. J. **37**(2), 485–569 (1958). ISSN: 00058580. https://doi.org/10.1002/j.1538-7305.1958.tb01530.x. http://ieeexplore.ieee.org/lpdocs/epic03/wrapper.htm?arnumber=6773415
4. Favoreel, W., De Moor, B., Van Overschee, P.: Subspace state space system identification for industrial processes. J. Process Control **10**(2–3), 149–155 (2000). ISSN: 09591524. https://doi.org/10.1016/S0959-1524(99)00030-X. http://linkinghub.elsevier.com/retrieve/pii/S095915249900030X
5. Jansson, M.: Subspace identification and ARX modeling. In: IFAC Proceedings Volumes, Sept. 2003, vol. 36(16), pp. 1585–1590 (2003). ISSN: 14746670. https://doi.org/10.1016/S1474-6670(17)34986-8. https://linkinghub.elsevier.com/retrieve/pii/S1474667017349868
6. Kammer, D.C., Krattiger, D.: Propagation of uncertainty in substructured spacecraft using frequency response. AIAA J. **51**(2), 353–361 (2013). ISSN: 0001-1452. https://doi.org/10.2514/1.J051771. http://arc.aiaa.org/doi/10.2514/1.J051771
7. Kammer, D.C., Nimityongskul, S.: Propagation of uncertainty in test-analysis correlation of substructured spacecraft. J. Sound Vib. (2011). ISSN: 0022460X. https://doi.org/10.1016/j.jsv.2010.09.029
8. Klerk, D.D., Rixen, D.J., Voormeeren, S.N.: General framework for dynamic substructuring: history, review and classification of techniques. AIAA J. **46**(5), 1169–1181 (2008). ISSN:0001-1452. https://doi.org/10.2514/1.33274. http://arc.aiaa.org/doi/10.2514/1.33274

9. Larimore, W.: Canonical variate analysis in identification, filtering, and adaptive control. In: 29th IEEE Conference on Decision and Control, vol. 2, pp. 596–604. IEEE, New York (1990). https://doi.org/10.1109/CDC.1990.203665. http://ieeexplore.ieee.org/document/203665/
10. Ljung, L.: System Identification: Theory for the User. Prentice Hall, Upper Saddle River (1998). ISBN: 9780132441933
11. Nicgorski, D., Avitabile, P.: Conditioning of FRF measurements for use with frequency based substructuring. In: Mechanical Systems and Signal Processing (2010). ISSN: 08883270. https://doi.org/10.1016/j.ymssp.2009.07.013
12. Nicgorski, D., Avitabile, P.: Experimental issues related to frequency response function measurements for frequencybased substructuring. Mech. Syst. Signal Process. **24**(5), 1324–1337 (2010). ISSN: 08883270. https://doi.org/10.1016/j.ymssp.2009.09.006. http://linkinghub.elsevier.com/retrieve/pii/S0888327009002660
13. Rixen, D.J.: How measurement inaccuracies induce spurious peaks in Frequency Based Substructuring. In: Proceedings of the XXVI International Modal Analysis Conference. Society for Experimental Mechanics, Orlando (2008)
14. Sjövall, P., Abrahamsson, T.: Component system identification and state-space model synthesis. Mech. Syst. Signal Process. **21**(7), 2697–2714 (2007). ISSN: 08883270. https://doi.org/10.1016/j.ymssp.2007.03.002. http://linkinghub.elsevier.com/retrieve/pii/S088832700700043X
15. Sjövall, P., McKelvey, T., Abrahamsson, T.: Constrained state–space system identification with application to structural dynamics. Automatica **42**(9), 1539–1546 (2006). ISSN: 00051098. https://doi.org/10.1016/j.automatica.2006.04.021. http://linkinghub.elsevier.com/retrieve/pii/S0005109806001725
16. Su, T.-J., Juang, J.-N.: Substructure system identification and synthesis. J. Guid. Control Dynam. **17**(5), 1087–1095 (1994). ISSN: 0731-5090. https://doi.org/10.2514/3.21314. http://arc.aiaa.org/doi/10.2514/3.21314
17. Tangirala, A.K.: Principles of System Identification: Theory and Practice, 1st edn. CRC Press, Boca Raton (2014). ISBN: 9781439895993
18. van der Seijs, M.V., de Klerk, D., Rixen, D.J.: General framework for transfer path analysis: history, theory and classification of techniques. Mech. Syst. Signal Process. **68–69**, 217–244 (2016). ISSN: 08883270. https://doi.org/10.1016/j.ymssp.2015.08.004. http://dx.doi.org/10.1016/j.ymssp.2015.08.004%20https://linkinghub.elsevier.com/retrieve/pii/S0888327015003647
19. Verhaegen, M.: Identification of the deterministic part of MIMO state space models given in innovations form from input-output data. Automatica **30**(1), 61–74 (1994). ISSN: 00051098. https://doi.org/10.1016/0005-1098(94)90229-1. http://linkinghub.elsevier.com/retrieve/pii/0005109894902291
20. Welch, P. The use of fast Fourier transform for the estimation of power spectra: a method based on time averaging over short, modified periodograms. IEEE Trans. Audio Electroacoust. **15**(2), 70–73 (1967). ISSN: 0018-9278. https://doi.org/10.1109/TAU.1967.1161901. http://ieeexplore.ieee.org/document/1161901/

Chapter 5
Workpiece Coupling in Machine Tools Using Experimental-Analytical Dynamic Substructuring

Prateek Chavan, Christian Brecher, Marcel Fey, and Matthäus Loba

Abstract The process stability of a metal cutting operation is determined by the dynamic compliance behavior of the machine components lying within the force flux. This not only includes the cutting tool and the machine tool structure but also the workpiece. The problem of predicting the tool tip frequency response function (FRF) of a tool in a clamped state has been addressed in the literature using experimental dynamic substructuring. However, the problem of predicting the compliance behavior of a point on a workpiece, which is clamped on a machine table, has not yet been addressed comprehensively.

In case of a milling machine, workpieces are often clamped using bolted joints or fixtures. In this paper, an efficient method for coupling a milling workpiece to a machining table is proposed using experimental-analytical substructure coupling techniques such that, an FRF at any point on the clamped workpiece can be predicted. This FRF can then be utilized for predicting process stability.

In the proposed method, the dynamic behavior of the workpiece is simulated using a discretized model and that of the table is measured experimentally. The accurate measurement of the experimental model of the table is essential for successful substructure coupling. This includes the measurement of translational as well as rotational compliances at the coupling points. For overcoming the challenge of the measurement of rotational compliances, an alternative method using decoupling is proposed and compared with several relevant experimental techniques from the literature. Additionally, a sensitivity analysis is carried out in order to identify the most relevant compliances necessary for an accurate prediction of the dynamics in clamped state. Based on these results, a guideline for successful coupling of workpieces or fixtures with a machining table are proposed. The proposed method is implemented and validated for a workpiece clamped to a milling machine table using bolted joints.

Keywords Experimental substructuring · Frequency domain · Workpiece · Rotational compliance

5.1 Introduction

Vibration amplitudes during metal cutting operations like milling can increase abruptly due to the phenomenon of self-excitation from the cutting process. These chatter vibrations often result in tool and machine damage, apart from poor workpiece quality. The tendency of a process to become unstable depends on the one hand on the cutting parameters (depth of cut, spindle speed, etc.) and on the other hand on the dynamic compliance behavior relative between the workpiece and tool [1]. This compliance behavior, often called as the oriented frequency response function, is influenced by the dynamics of all structural components of the machine lying within the force flux and defines the stability behavior [1, 2].

Since an overhanging, long milling tool often represents one of the most compliant parts of the machine structure, the mounted tool tip frequency response functions (FRFs) are used for determining oriented FRF and predicting the stability behavior. Especially the approach of experimental-analytical dynamic substructure coupling has been widely utilized for the prediction of tool tip FRFs [3–6]. Generally, a dummy tool clamped to the tool holder is decoupled using frequency based substructure decoupling such that the interface properties (tool holder) can be identified. Subsequently, an analytical model of the new tool is coupled to the identified interface and the compliance behavior at the free-end of the tool in the coupled state is calculated. Because of their simple geometry, the dummy and the simplified new tool can easily be modelled analytically using beam theories like the Euler-Bernoulli [4, 5, 7] or Timoshenko beam theory [8]. In this case of compliant tools, the workpiece and workpiece-side dynamics (table, fixtures, work-holding devices, etc.) is ignored. However, the analysis of

P. Chavan (✉) · C. Brecher · M. Fey · M. Loba
Department of Machine Tools and Production Engineering, RWTH Aachen University, Aachen, Germany
e-mail: p.chavan@wzl.rwth-aachen.de

A. Linderholt et al., *Dynamic Substructures, Volume 4*, Conference Proceedings of the Society for Experimental Mechanics Series,
https://doi.org/10.1007/978-3-030-12184-6_5

stability behavior is incomplete without the consideration of workpiece and workpiece-side dynamics, especially in cases where the tool and workpiece-side compliances are comparable. A recent research work deals with predicting the dynamics of turning workpieces which have inherently simple cross-sections and geometries and can be analytically modelled using beam theories similar to the receptance coupling of tools [9]. A similar approach can be seen in [10] for receptance coupling of overhanging thin walled workpiece. Apart from this, Altintas et al. recently published a paper where time varying workpiece dynamics were predicted by frequency domain decoupling of the response model obtained from a rigidly fixed thin-walled workpiece [11]. The influence of work holding devices and workpiece-side dynamics could be ignored because of the highly compliant workpiece. However, in cases where the workpiece and tool are not significantly compliant, the dynamics of the complete workpiece-side must also be considered for the prediction of stability behavior. This has not yet been addressed in the literature.

An important aspect, which limits the implementation of substructuring methods for complex structures like machine tools, is the difficulty in the measurement or estimation of rotational compliances. The research pertaining to receptance coupling of tools utilize substructure decoupling for estimating rotational compliance as mentioned above. However, the experimental estimation of rotational compliances of other interface of the machine tool has not been studied extensively. In the works of [12, 13], only driving point and indirect translational FRFs are used for decoupling and coupling. In this paper, experimental approaches for estimating rotational compliance of the machine table are implemented, compared and used for subsequent coupling calculations.

The goal of this paper is the efficient prediction of workpiece-side dynamics by coupling an experimentally obtained response model of the machine table with a discretized model of the workpiece. For brevity, the term 'workpiece-side' dynamics will be used in this paper to represent the dynamics of a point on the workpiece in the clamped state and hence includes the compliance behavior of the workpiece, work holding as well as the supporting machine tool structure (table).

5.2 Frequency Domain Based Workpiece Coupling

5.2.1 Proposed Method

The starting point of the proposed method is a Computer Aided Design (CAD) model of the workpiece (Fig. 5.1). The focus of this paper are workpieces with geometries, which cannot be modelled analytically using beam or plate theories and which are joint to the table at multiple locations. The time varying dynamics due to material removal is not the focus of this paper but will be addressed in future work. Subsequently, a discretized Finite Element (FE) model of the workpiece is created based on the CAD data. Assuming that the workpiece material is homogeneous and does not have non-linear properties or joints, the FE model does not need to be updated based on FRF measurements. This makes the proposed method efficient. Furthermore, the FE model is then reduced using a model order reduction (such as the Dual Craig-Bampton method) for a quick calculation of FRFs. The coupling degrees of freedom (DOFs) and the DOF at which FRF prediction is required are retained in the reduction process.

Each coupling node has six DOFs (x, y, z, θ_z, θ_y, θ_z). In order to minimize the time and effort for creating the simulated and measured response models, it is necessary to know which DOFs should be coupled for achieving an accurate prediction

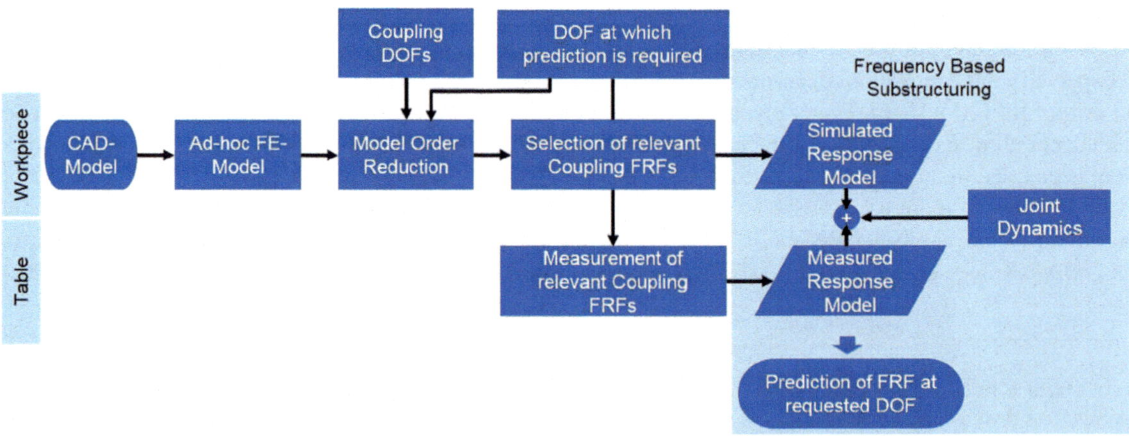

Fig. 5.1 Approach for efficient prediction of workpiece-side dynamics

of FRF at the DOF of the workpiece. Guidelines for selection of relevant DOFs are derived based on the Sect. 5.3 (sensitivity analysis). Once the necessary DOFs are selected, the simulated response model can be created and the FRFs at the corresponding DOFs of the machine table can be measured. However, the measurement of rotational compliances is associated with several practical difficulties. Section 5.4.2 provides a comparison of three methods for experimental determination of rotational compliances for efficient measurement of the response model. Finally, based on these models and the information regarding the joint dynamics, coupling equations of Dynamic Substructuring (DS) can be solved to predict the FRF at the required node and DOF of the workpiece.

5.2.2 Mathematical Background and Formulation

In this section, a brief description of the implementation of frequency based substructuring for experimental-analytical coupling of workpiece and machine table is provided. The mathematical formulation of the problem is derived from the available literature for DS [14–16]. The formulation starts with the Newtonian equations of motion of the uncoupled structures in the physical domain (Eq. 5.1). Here the time dependency is left out for clarity.

$$\begin{bmatrix} \boldsymbol{M}_w & 0 \\ 0 & \boldsymbol{M}_t \end{bmatrix} \begin{Bmatrix} \ddot{\boldsymbol{u}}_w \\ \ddot{\boldsymbol{u}}_t \end{Bmatrix} + \begin{bmatrix} \boldsymbol{C}_w & 0 \\ 0 & \boldsymbol{C}_t \end{bmatrix} \begin{Bmatrix} \dot{\boldsymbol{u}}_w \\ \dot{\boldsymbol{u}}_t \end{Bmatrix} + \begin{bmatrix} \boldsymbol{K}_w & 0 \\ 0 & \boldsymbol{K}_t \end{bmatrix} \begin{Bmatrix} \boldsymbol{u}_w \\ \boldsymbol{u}_t \end{Bmatrix} = \begin{Bmatrix} \boldsymbol{f}_w \\ \boldsymbol{f}_t \end{Bmatrix} + \begin{Bmatrix} 0 \\ \boldsymbol{g} \end{Bmatrix} \tag{5.1}$$

The subscript '*w*' and '*t*' refer to the physical properties or forces applied on the workpiece and machine table respectively. '*g*' refers to the vector of forces at coupling or interface nodes between the table and the workpiece. In order to achieve coupling, the conditions of displacement compatibility and force equilibrium must be satisfied. This enforces that the motion of the interface nodes of the table and workpiece is identical (assuming conforming meshes),

$$\boldsymbol{u}_w^c = \boldsymbol{u}_t^c \tag{5.2}$$

where, the superscript 'c' refers to the coupling or interface nodes. Furthermore, the force equilibrium implies that the forces acting on the interface nodes cancel each other out,

$$\boldsymbol{g}_t^c + \boldsymbol{g}_w^c = 0. \tag{5.3}$$

Since the coupling is to be carried out in the frequency domain, Eq. (5.1) is transformed to the frequency domain. Omitting frequency dependency, we get for the uncoupled structures,

$$\begin{bmatrix} \boldsymbol{Z}_w & 0 \\ 0 & \boldsymbol{Z}_t^c \end{bmatrix} \begin{Bmatrix} \boldsymbol{u}_w^r \\ \boldsymbol{u}_t^c \end{Bmatrix} = \begin{Bmatrix} \boldsymbol{f}_w \\ \boldsymbol{f}_t \end{Bmatrix} + \begin{Bmatrix} 0 \\ \boldsymbol{g} \end{Bmatrix} \tag{5.4}$$

where $\boldsymbol{Z}_w$ represents the dynamic stiffness matrix of the reduced FE model of the workpiece and is given by,

$$\boldsymbol{Z}_w = -\omega^2 \boldsymbol{M}_w^R + j\omega \boldsymbol{C}_w^R + \boldsymbol{K}_w^R \tag{5.5}$$

with,

$\boldsymbol{M}_w^R = \boldsymbol{R}_{DCB}^T \boldsymbol{M}_w \boldsymbol{R}_{DCB}$, $\boldsymbol{C}_w^R = \boldsymbol{R}_{DCB}^T \boldsymbol{C}_w \boldsymbol{R}_{DCB}$ and $\boldsymbol{K}_w^R = \boldsymbol{R}_{DCB}^T \boldsymbol{K}_w \boldsymbol{R}_{DCB}$.
$\boldsymbol{R}_{DCB}$ = basis of Reduction from dual Craig-Bampton method.
$\boldsymbol{u}_w^r$ = vector of reduced DOFs (interface and prediction DOFs are retained).
$\boldsymbol{Z}_t^c$ represents the dynamic stiffness matrix at the interface nodes of the table. This value cannot be measured easily compared to its inverse,

$$\left(\boldsymbol{Z}_t^c\right)^{-1} = \boldsymbol{G}_t^c = \begin{bmatrix} \boldsymbol{G}_{t11} & \cdots & \boldsymbol{G}_{tn1} \\ \vdots & \ddots & \vdots \\ \boldsymbol{G}_{t1n} & \cdots & \boldsymbol{G}_{tnn} \end{bmatrix}_{n \times n} \tag{5.6}$$

where, $\boldsymbol{G}_{tij}$ = FRF measured at *j*th DOF and excited *i*th DOF of the table.

n = number of coupling DOFs.

The compatibility condition and force equilibrium from Eqs. (5.2) to (5.3) can be expressed as,

$$\boldsymbol{B}\boldsymbol{u} = 0 \quad \text{and} \quad \boldsymbol{L}^T \boldsymbol{g} = 0 \tag{5.7}$$

where, $\boldsymbol{B}$ and $\boldsymbol{L}^T$ are signed Boolean matrices for isolating the coupling DOFs in $\boldsymbol{u} = \left[\boldsymbol{u}_w^r \boldsymbol{u}_t^c\right]^T$ and for localizing the set of unique DOFs from the set of merged or global DOFs respectively [15]. Furthermore, interface constraint force vector can be represented as,

$$\boldsymbol{g} = -\boldsymbol{B}^T \boldsymbol{\lambda} \tag{5.8}$$

for obtaining a dual assembly of the structures. Where $\boldsymbol{\lambda}$ represent Lagrange multipliers, signifying the connection forces [16]. Substituting Eqs. (5.7) and (5.8) in Eq. (5.4), we get,

$$\begin{bmatrix} \boldsymbol{Z}_w & 0 & \boldsymbol{B}_w^T \\ 0 & \boldsymbol{Z}_t^c & \boldsymbol{B}_t^T \\ \boldsymbol{B}_w & \boldsymbol{B}_t & 0 \end{bmatrix} \begin{Bmatrix} \boldsymbol{u}_w^r \\ \boldsymbol{u}_t^c \\ \boldsymbol{\lambda} \end{Bmatrix} = \begin{Bmatrix} \boldsymbol{f}_w \\ \boldsymbol{f}_t^c \\ 0 \end{Bmatrix}. \tag{5.9}$$

Eliminating $\boldsymbol{\lambda}$ from the above equation and reformulating by replacing dynamic stiffness matrix $\boldsymbol{Z} = \text{diag}\left(\boldsymbol{Z}_w, \boldsymbol{Z}_t^c\right)$ with dynamic compliance matrix $\boldsymbol{Z}^{-1} = \boldsymbol{G} = \text{diag}\left(\boldsymbol{G}_w, \boldsymbol{G}_t^c\right)$ gives [15],

$$\boldsymbol{G}_{w+t} = \boldsymbol{G} - \boldsymbol{G}\boldsymbol{B}^T \left(\boldsymbol{B}\boldsymbol{G}\boldsymbol{B}^T\right)^{-1} \boldsymbol{B}\boldsymbol{G} \tag{5.10}$$

where, $\boldsymbol{G}_{w+t}$ = dynamic compliance matrix of the assembled workpiece and table.

As explained in the previous sub-section, the FRF matrix of the unconstrained workpiece $\boldsymbol{G}_w$ can be obtained from the FE model. This matrix contains direct and indirect compliances for all nodes at all degrees of freedom. For example, a workpiece model with three nodes (one internal and two interface nodes) with six DOFs each (x, y, z, θ_x, θ_y, θ_z) has a compliance matrix $\boldsymbol{G}_w$ of size (18 × 18) with 324 FRFs. In this case, the measured compliance matrix of the table $\boldsymbol{G}_t^c$ with two interface nodes and six DOFs each, has a size of (12 × 12) with 144 FRFs. The measurement of such a matrix is not practical. In order to reduce the measurement and calculation work, it is necessary to first identify which coupling DOFs should be considered and which can be ignored for predicting the dynamic behavior at the target DOF. This is the subject of the next section.

5.3 Sensitivity Analysis

The question of selection of appropriate coupling DOFs does not arise in case of simple two-dimensional structures like overhanging beams. Here the consideration of a translational compliance and a rotational compliance at the free end are enough for an exact coupling with another beam. However, in case of complex structures with three-dimensional mode shapes and multiple connection points, compliances in multiple directions have to be considered and the question of selection of coupling DOFs becomes relevant.

In this section, the above problem is analyzed in a purely virtual environment. For this, a CAD model of the table of a 4-axis machine tool is created along with that of a real workpiece (Fig. 5.2, left). The machine structure under the table is not considered in order to reduce the complexity of vibration modes. The table is fixed to the ground through four spring-damper elements. The workpiece has material properties of C45 Steel and has two slots along the x-direction. One slot on the side facing the table along the complete length of the workpiece for increasing compliance.

The second slot, on the side facing away from the table, is only until half of the workpiece length. This point '5' can be expected to have the maximum compliance during the slot milling operation as it exactly between the points of attachment (1–4). The bolts are considered rigid for simplicity. The assembly of the two structures at the shown position on the table is treated as the reference structure and the translational FRF in z-direction at point 5 (5z5z) is the reference FRF. With the help of an FE model of the assembly, the reference FRF can be obtained along with the vibration modes (Fig. 5.2, right). There are two significant modes of vibration observable in the reference FRF. The first mode at 44 Hz represents a tilting rigid body mode about the x-axis (θ_x) due to the compliance of the spring-damper elements. The second mode at 2250 Hz corresponds to a flexible bending mode of mainly the workpiece about the y-axis (θ_y).

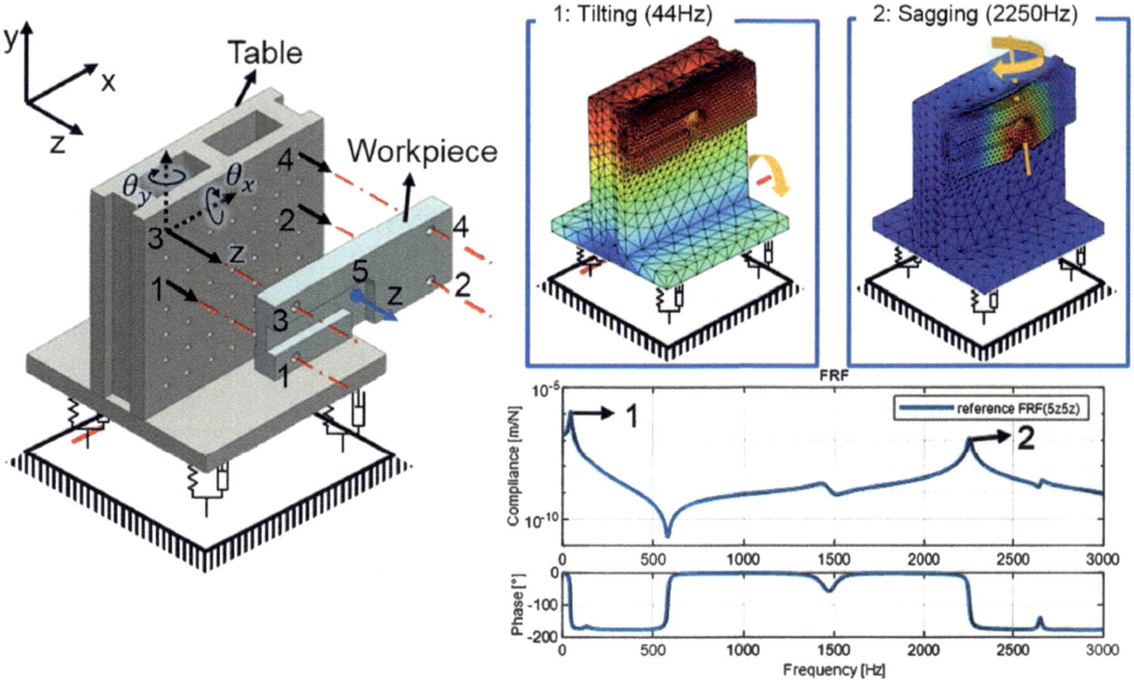

Fig. 5.2 Modelling of workpiece and machine table (*left*) and reference system and its vibration modes in z-direction (*right*)

Aim of the analysis is to implement the proposed method for the simulated structures for predicting the target FRF of 5z5z in the assembled state. For this, firstly, an FE model of the unconstrained (free-free) and reduced workpiece is created. It has four points of attachment (1–4) and one internal point (5). An interface condensation node represents each point of attachment. The condensation node is connected to multiple surface nodes within a circular section with rigid body elements (RBE3 in Siemens NX). The radius of the circular section is calculated based on the radius of the pressure cone of the bolted joint [1]. Subsequently, free-free FRFs are calculated for every combination of excitation and response DOFs at the five nodes which include six DOFs each at the interface nodes and one DOF at the internal node (5z). Thus, a full scale response model of the workpiece is obtained. Similarly, the response model of the discretized table at the four interface points is also achieved.

For analyzing the effect of including and ignoring particular DOFs in the coupling calculations, four cases were created. Since the target FRF is in the translational z-direction, it is obvious that compliance in this direction at interface nodes should be considered in the coupling calculations for all cases. In the first case, only the translational DOFs in z at each interface node of the workpiece and table were chosen for coupling. This includes also the cross compliances between the nodes and results is a $(\boldsymbol{BGB}^T)^{-1}$ matrix of size 4×4 in Eq. (5.10). From cases 2 to 4, the translational DOF z as well as a rotational compliance about each axis were selected (z and θ_x, z and θ_y, z and θ_z). The matrix $(\boldsymbol{BGB}^T)^{-1}$ now has a size of 8×8 for each coupling case. The amplitude and phase plots of the target FRF (5z5z) for the coupling cases can be observed in Fig. 5.3.

As can be seen for most of the cases, an exact match of the reference system could not be achieved. This is mainly because only one or two coupling DOFs are considered and all other DOFs and their cross compliance are ignored. Additionally, the difference in the mesh size of the master nodes of the workpiece and table meshes and the model order reduction could contribute to some of the deviation. In case 1, the first resonance peak at 44 Hz originating form the tilting of the table bout x-axis could be estimated well. A significant error in the predicted frequency of the second resonance peak at 2250 Hz is observable. This is also be seen in cases 2 and 4. The reason is thought to be that the bending mode of the workpiece has a smaller component in the z, θ_x and θ_z. On the other hand, a very good estimation is achieved when z and θ_y are selected as coupling DOFs. The first resonance pole shows some deviation but the zero at ca. 586 Hz and especially the pole at 2250 Hz could be estimated well. This can be attributed to the very good observability of the workpiece bending mode in the rotational compliance about the y-axis, θ_y. Thus, it can be observed that the consideration of merely one translational coupling DOF, as done in previous literature [12, 13], led to insufficiently accurate prediction. The selection of a second, rotational compliance could improve the prediction of the second resonance peak significantly. This indicates, that it is recommendable to evaluate the probable modes of vibration of the assembly and subsequently select one or more coupling DOFs where these modes are observable.

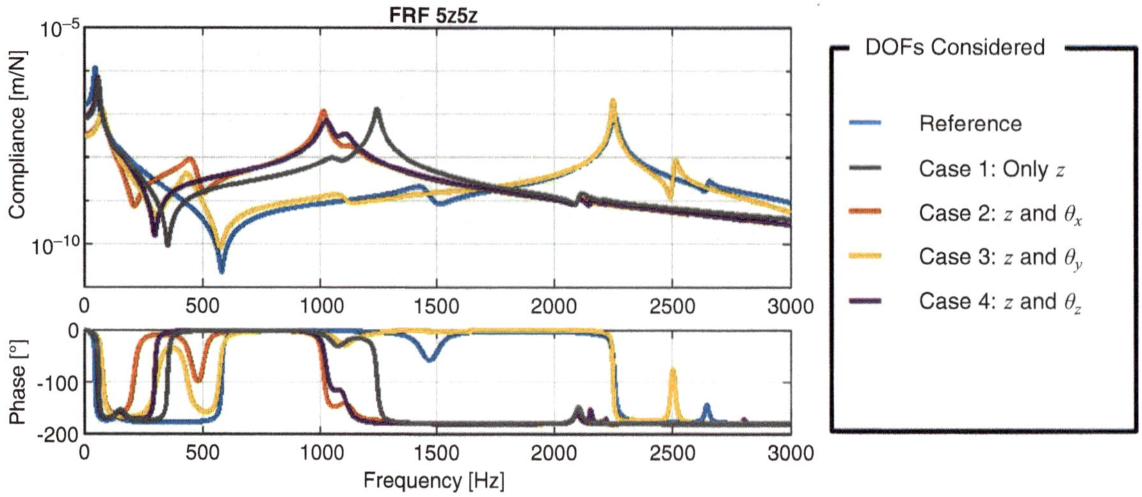

Fig. 5.3 Prediction of target FRF by considering different coupling degrees of freedom

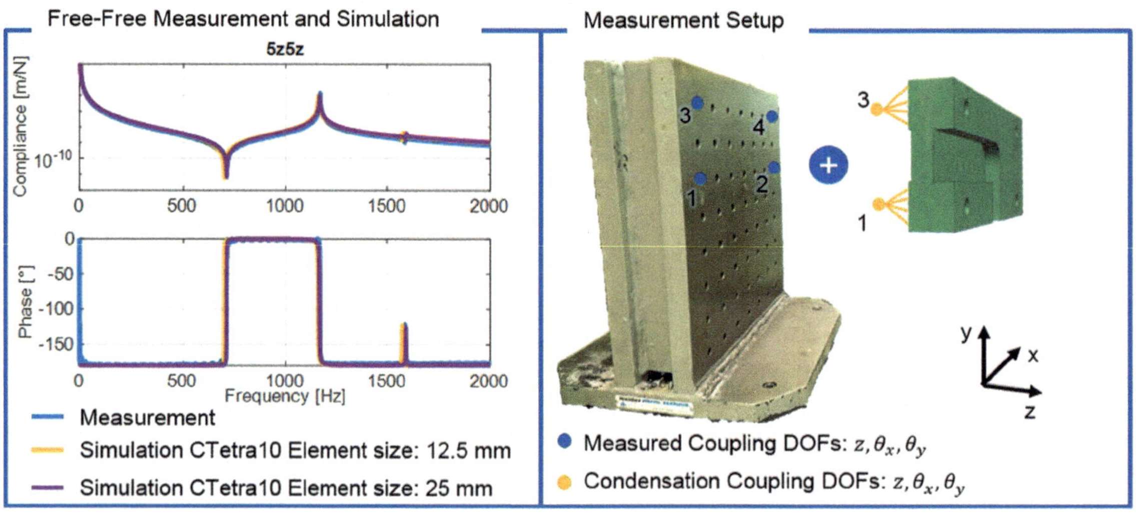

Fig. 5.4 Analysis of workpiece model and measurement setup

5.4 Experimental Validation

5.4.1 Measurement Setup and Workpiece Model Validation

Based on the method proposed in Sect. 5.2.1 and the guidelines derived from the previous section, an experimental-analytical dynamic substructure coupling is implemented in this section. The same combination of machine table and workpiece, which were modelled in the previous section, is chosen. Whereas the complete response model of the workpiece is available from the previous analysis, the response model of the machine table of the 4-axis milling machine is obtained experimentally. The table of the machine and the four interface points are shown in Fig. 5.4. Here, again the aim is to predict the dynamic behavior of the workpiece in the assembled state at DOF 5z.

An assumption in the proposed methodology is that the model of a workpiece with homogenous and linear physical properties does not need to be updated with information from experimental data. This ad-hoc modelling was carried out for the above workpiece as well. In order to validate the approach, a real workpiece with identical dimensions and features was analyzed experimentally. The workpiece was suspended elastically from a bungee chord to simulate nearly unconstrained conditions and several driving point and indirect FRFs were measured on the structure. On the other hand, the virtual model was meshed using standard quadratic tetrahedral elements with ten nodes each. The measured and simulated FRFs at different point were compared with each other to provide an indication of model accuracy. One such comparison for driving point

FRF 5z5z is shown in Fig. 5.4. Two automatically generated meshes were applied. A fine mesh of element size 12.5 mm and another mesh of size 25 mm. Both show a very good correspondence with the measured FRF. This indicates that an ad-hoc model of a linearly behaving workpiece created with a reasonable mesh size and engineering judgement can be used directly for calculating the response model. This is consistent with similar findings in [13].

5.4.2 Measurement of Response Model Including Rotational Compliances

In this sub-section, appropriate coupling DOFs for the response model of the table will be selected and measured based on guidelines derived in Sect. 5.3. Here, again, since the target FRF is the 5z5z translational compliance, it is sensible to include the translational DOFs in the z-direction (1z, 2z, 3z, 4z) in the response model. Additionally, it can be noted that probable modes of vibration of the assembly about the x and y-directions will result in a translational displacement of the point 5 in the z-direction. Therefore, these rotational DOFs should also be included in the response model.

In this section, three experimental strategies for estimating rotational compliances are implemented at the four coupling nodes of a machine tool table. The first approach is the implementation of a second order backwards finite difference method [17] to a point on the machine table. This approach requires the measurement of translational compliance at three points in the same line at fixed distances. Within the framework of creating the response model driving point compliances at four point have to be measured in any case. Hence, only one additional translational compliance has to be measured per node per axis. This corresponds to only four additional FRFs for estimating the rotational compliances at all four points in both θ_x and θ_y and hence makes this approach particularly advantageous and extremely efficient. Figure 5.5 shows a setup for estimating θ_x for point 3. In the second approach, a specialized fixture in the shape of a T-Block along with an arrangement of acceleration sensors is used to extract rotational information [18]. Figure 5.5 shows a setup with T-Block for extracting rotational compliance at point 3. For detailed description of the method, the reader is referred to [18]. The third approach is based on substructure decoupling. In this new approach, a dummy cylindrical beam with a threaded interface is bolted to the table at the coupling point. The idea behind the approach is that through the decoupling of the cylinder from the bolted beam, an accurate estimation of the rotational compliances may be achieved. This is similar to the decoupling of dummy tools to estimate the properties of the tool holder interface [3, 6]. Figure 5.5, right shows the corresponding measurement setup.

Figure 5.5 (below) shows, as a representative example, the comparison of rotational compliances ($2z2\theta_y$ and $2\theta_y 2\theta_y$) obtained from the three approaches. The analysis of these rotational compliances show that similar results are obtained using the T-Block and backwards finite difference approaches. However, much more distinct and pronounced poles and zeros can

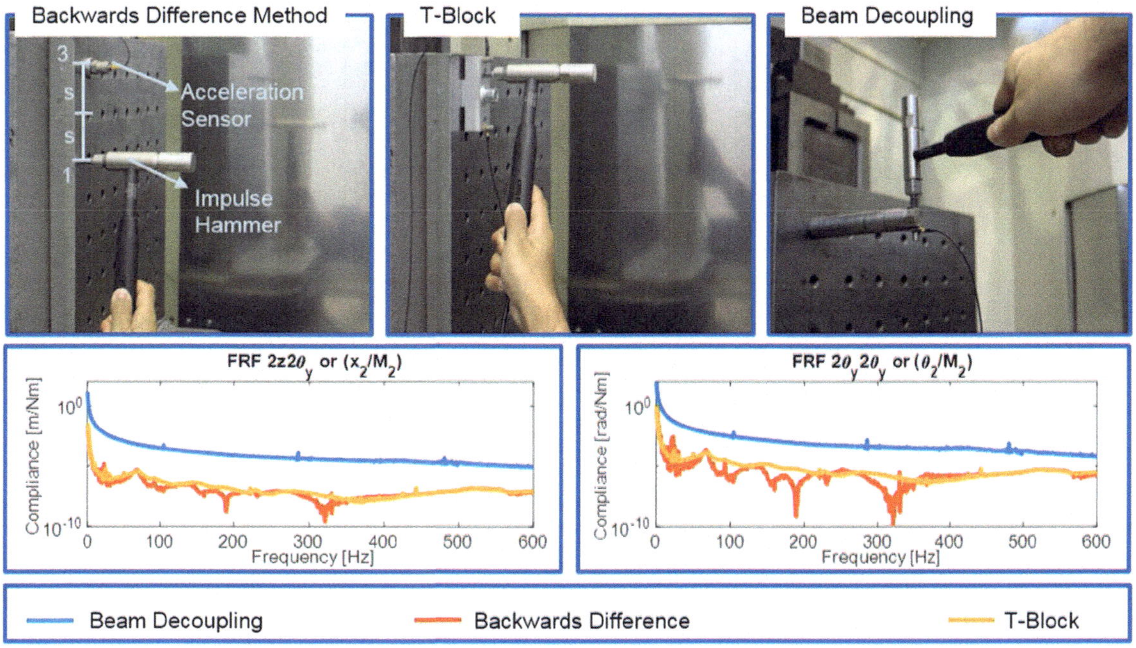

Fig. 5.5 Measurement of rotational compliances at interface nodes of table and representative rotational compliances

be observed in the backwards difference case. The estimation from beam decoupling was unsuccessful. It is thought that the compliance of the cylindrical beam is many magnitudes higher than the table and this influence cannot be effectively decoupled. For this reason, the compliances obtained from beam decoupling will not be used for the subsequent coupling calculations.

5.5 Variations of Coupling Calculations and Corresponding Results

Based on the available compliance measurements, measurement techniques and the consideration of different rotational axis, several variations of the coupling matrices and calculations can be realized. The analysis of these combinations and their corresponding results are presented in this section.

For the first analysis, let us consider and expand the term ($\boldsymbol{BGB}^T$) from Eq. (5.10),

$$\left(\boldsymbol{BGB}^T\right) = \left[\boldsymbol{G}_w^c + \boldsymbol{G}_t^c\right] = \left[\begin{bmatrix} \boldsymbol{G}_w^{11} & \boldsymbol{G}_w^{12} & \boldsymbol{G}_w^{13} & \boldsymbol{G}_w^{14} \\ \boldsymbol{G}_w^{21} & \boldsymbol{G}_w^{22} & \boldsymbol{G}_w^{23} & \boldsymbol{G}_w^{24} \\ \boldsymbol{G}_w^{31} & \boldsymbol{G}_w^{32} & \boldsymbol{G}_w^{33} & \boldsymbol{G}_w^{34} \\ \boldsymbol{G}_w^{41} & \boldsymbol{G}_w^{42} & \boldsymbol{G}_w^{43} & \boldsymbol{G}_w^{44} \end{bmatrix} + \begin{bmatrix} \boldsymbol{G}_t^{11} & \boldsymbol{G}_t^{12} & \boldsymbol{G}_t^{13} & \boldsymbol{G}_t^{14} \\ \boldsymbol{G}_t^{21} & \boldsymbol{G}_t^{22} & \boldsymbol{G}_t^{23} & \boldsymbol{G}_t^{24} \\ \boldsymbol{G}_t^{31} & \boldsymbol{G}_t^{32} & \boldsymbol{G}_t^{33} & \boldsymbol{G}_t^{34} \\ \boldsymbol{G}_t^{41} & \boldsymbol{G}_t^{42} & \boldsymbol{G}_t^{43} & \boldsymbol{G}_t^{44} \end{bmatrix}\right]. \tag{5.11}$$

Here the term $\boldsymbol{G}_w^c$ corresponds to the compliance matrix at the four coupling points obtained from the reduced FE model of the workpiece. $\boldsymbol{G}_t^c$ refers to the corresponding matrix for the coupling points of the table obtained through measurements. If two DOFs are considered for coupling, each term in the above matrices consists of,

$$\boldsymbol{G}^{ij} = \begin{bmatrix} \frac{x_i}{F_j} & \frac{x_i}{M_j} \\ \frac{\theta_i}{F_j} & \frac{\theta_i}{M_j} \end{bmatrix} = \begin{bmatrix} H^{ij} & L^{ij} \\ N^{ij} & P^{ij} \end{bmatrix}. \tag{5.12}$$

Now, in the case of the workpiece compliances, every compliance in the term $\boldsymbol{G}_w^c$ can be easily obtained from the simulation model. The same is not true for the measured compliance matrix $\boldsymbol{G}_t^c$ of the table. In this case, the driving point translational FRFs in the z direction as well as the translational cross compliances in this direction are measurable with conventional measurement systems. Additionally, the direct rotational compliances at each coupling point can also be effectively estimated using the approaches shown in Sect. 5.4.2. However, the cross rotational and cross rotational-translational FRFs cannot be obtained. If the unascertainable compliances are replaced by '0', the matrix $\boldsymbol{G}_t^c$ can be represented as,

$$\boldsymbol{G}_t^{ij} = \left\{ \begin{matrix} for\ \ i = j, & \begin{bmatrix} H_t^{ii} & L_t^{ii} \\ N_t^{ii} & P_t^{ii} \end{bmatrix} \\ else & \begin{bmatrix} H_t^{ij} & 0 \\ 0 & 0 \end{bmatrix} \end{matrix} \right\}. \tag{5.13}$$

Regarding the compliance matrix of the workpiece, there exist two options. The unascertainable DOFs of the table could either also be considered rigid for the workpiece or a fully occupied compliance matrix could be considered. In matrix form,

$$\text{Case A,}\quad \boldsymbol{G}_w^{ij} = \left\{ \begin{matrix} for\ \ i = j, & \begin{bmatrix} H_w^{ii} & L_w^{ii} \\ N_w^{ii} & P_w^{ii} \end{bmatrix} \\ else & \begin{bmatrix} H_t^{ij} & 0 \\ 0 & 0 \end{bmatrix} \end{matrix} \right\} \quad \text{and}\quad \text{Case B,}\quad \boldsymbol{G}_w^{ij} = \begin{bmatrix} H_w^{ij} & L_w^{ij} \\ N_w^{ij} & P_w^{ij} \end{bmatrix} \quad \forall\ i, j.$$

Subsequently, coupling equations were solved for both the cases for DOFs z and θ_x (obtained from T-Block) and the results of the prediction at DOF 5z along with the reference FRF are illustrated in Fig. 5.6. The comparison shows clearly that coupling only the measurable DOFs and considering others as rigid (Case A) led to a poor prediction. On the other hand, coupling with full workpiece compliance matrix results in a comparably better correspondence until about 150 Hz. Further analysis will show that still better results are achievable (Fig. 5.7).

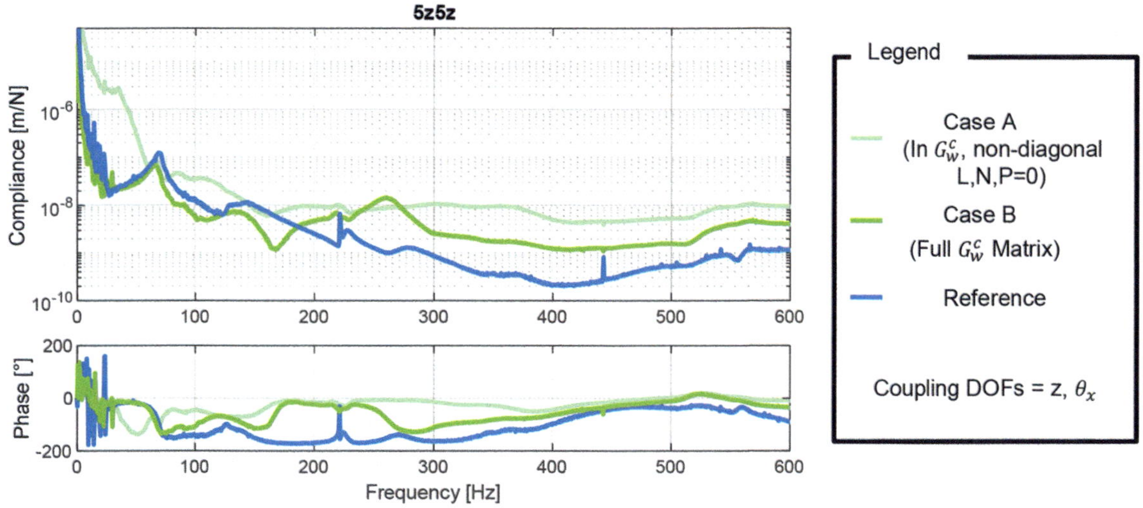

Fig. 5.6 Effect of including a full compliance matrix of the workpiece

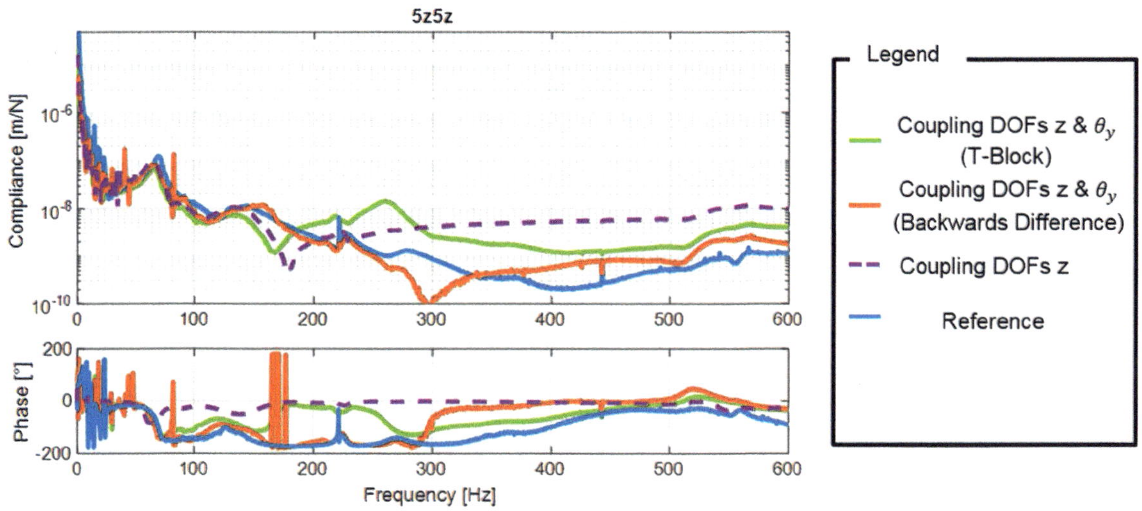

Fig. 5.7 Prediction of target FRF

Another analyzed aspect of the coupling calculations is choice of the method for estimating rotational compliance. As described in the previous section, the backwards difference method derived compliances showed better results as those derived from the T-Block. Although these measurements cannot be directly validated with a reference, the results of coupling using the two approaches can be compared with the reference FRF 5z5z. This comparison in Fig. 5.7 shows that, the coupling with backwards difference method results in a very good correspondence with the target FRF, especially in the low frequency range up till 250 Hz. This indicates that, if accurately estimated, the inclusion of rotational compliances can improve the quality of prediction. The prerequisite is that the vibration modes are observable in the coupling DOFs. The prediction based only on translational FRFs in z-direction shows good correspondence in a very limited frequency range up until around 150 Hz.

5.6 Conclusions

In this paper, an efficient method based on experimental-analytical substructuring is proposed in order to predict the compliance behavior at a point on the workpiece in a clamped state. For achieving this, response models of the workpiece and machine table should be obtained accurately and efficiently. An ad-hoc modelling of workpieces with linear properties reduces the modeling effort significantly. An analysis of selection of coupling DOFs in a virtual environment (as well as in subsequent experimental trials) showed that the choice of coupling DOFs is crucial for accurate prediction. This choice depends on the DOF of the target FRF and whether the modes in the target FRF are also observable in the coupling DOFs. However, further research needs to be conducted to formally develop criteria for selecting coupling DOFs based on response models of the structures to be coupled and the target FRF. Subsequently, the proposed method was implemented for coupling a real machine table and virtual workpiece. For experimentally obtaining the rotational compliances, three measurement approaches were evaluated. The rotational compliance derived from backwards difference method showed defined poles and zeros compared to the other methods. The coupling results with these FRFs were also more accurate for a larger frequency range. This indicates the importance of the accurate estimation of rotational compliances. Finally, it can be concluded that although the FRFs of some DOFs in the experimental response model cannot be measured directly, a full compliance matrix of the simulated workpiece should be considered for the coupling calculations. The proposed method and developed guidelines can be helpful for experimental-analytical substructure coupling of complex structures that cannot be modelled analytically and are fixed at multiple locations.

Acknowledgements The authors wish to gratefully acknowledge the support of the German Research Foundation (DFG). This work was funded as part of the DFG Project "Experimental substructure coupling for vibration analysis in machine tools" (Project Number-BR 2905/55-2).

References

1. Brecher, C., Weck, M.: Werkzeugmaschinen Fertigungssysteme 2, Textbook, vol. 9 (2017)
2. Altintas, Y.: Manufacturing Automation: Principles of Metal Cutting, Machine Tool Vibrations and CNC Design. Cambridge University Press, Cambridge (2000)
3. Schmitz, T.L., Duncan, G.S.: Three-component receptance coupling substructure analysis for tool point dynamics prediction. J. Manuf. Sci. Eng. **127**, 781 (2005)
4. Park, S.S., Altintas, Y., Movahhedy, M.: Receptance coupling for end mills. Int. J. Mach. Tools Manuf. **43**(9), 889–896 (2003)
5. Albertelli, P., Goletti, M., Monno, M.: An improved receptance coupling substructure analysis to predict chatter free high speed cutting conditions. Procedia CIRP. **12**, 19–24 (2013)
6. Brecher, C., Chavan, P., Fey, M., Daniels, M.: A modal parameter approach for receptance coupling of tools. MM Sci. J. **2016**, 1032–1034 (2016)
7. Schmitz, T.L., Burns, T.: Receptance coupling for high-speed dynamics prediction. In: Proceedings of the 21st International Modal Analysis Conference (IMAC-2003), Kissimmee, FL (2003)
8. Ertürk, A., Özgüven, H.N., Budak, E.: Analytical modeling of spindle–tool dynamics on machine tools using Timoshenko beam model and receptance coupling for the prediction of tool point FRF. Int. J. Mach. Tools Manuf. **46**(15), 1901–1912 (2006)
9. Li, H., Xue, G., Zhou, Y., Li, H., Wen, B.: Receptance coupling for frequency response prediction of cylindrical workpiece in CNC lathe. J. Vibroeng. **17**(4), 1731–1747 (2015)
10. Honeycutt, A., Schmitz, T.: Receptance coupling model for variable dynamics in fixed-free thin rib machining. Procedia Manuf. **26**, 173–180 (2018)
11. Tuysuz, O., Altintas, Y.: Frequency domain updating of thin-walled workpiece dynamics using reduced order substructuring method in machining. J. Manuf. Sci. Eng. **139**(7), 071013 (2017)
12. Law, M., Rentzsch, H., Ihlenfeld, S.: Predicting mobile machine tool dynamics by experimental dynamic substructuring. Int. J. Mach. Tools Manuf. **108**, 127–134 (2016)
13. Daniels, M.: Substrukturkopplung zur effizienten Schwingungsanalyse von Werkzeugmaschinen, Dissertation. Apprimus Verlag, Aachen (2017)
14. Klerk, D., Rixen, D., Jong, J.: The frequency based substructuring method reformulated according to the dual domain decomposition method. In: Conference: Proceedings of the XXIV International Modal Analysis Conference (2006)
15. Klerk, D., Rixen, D., Voormeeren, S.: General framework for dynamic substructuring: history, review, and classification of techniques. AIAA J. **46**, 1169–1181 (2008). https://doi.org/10.2514/1.33274
16. Craig, R.R., Kurdila, A.J.: Fundamentals of Structural Dynamics, 2nd edn. Wiley, Hoboken (2006)
17. Duarte, M., Ewins, D.: Rotational degrees of freedom for structural coupling analysis via finite-difference technique with residual compensation. Mech. Syst. Signal Process. **14**(2), 205–227 (2000)
18. Ewins, D.: Modal Testing: Theory, Practice and Application, 2nd edn. Research Studies Press, Baldock and Philadelphia (2000)

Chapter 6
Mechanical Characterization and Numerical Modeling of High Density Polyethylene Pipes

Mehrzad Taherzadehboroujeni and Scott W. Case

Abstract The worldwide plastic pipe industry is predicted to experience a dramatic grow over the next decade. As a group of plastic pipes, high density polyethylene (HDPE) pipes are often employed because of their low-cost production, easy installation, and excellent long-term performance. However, due to their complicated semi-crystalline microstructure and nonlinear time-temperature dependent mechanical behavior, the mechanical characterization of HDPE pipes is very challenging and time consuming. In addition, during the manufacturing of HDPE pipes, the processing conditions (such as molecular orientation, cooling rate, and extrusion injection pressure) can introduce different complex microstructures into the material which yield different material properties. In this study, a robust mechanical characterization approach is developed to support numerical modeling of HDPE pipes. The mechanical tests are performed directly on as-manufactured pipe segments. The simulation results are compared with the experimental data for tensile and internal pressurization (burst) tests and a good agreement is observed.

Keywords Long-term hydrostatic strength · HDPE pipe characterization · Numerical modeling · Accelerated method

6.1 Introduction

Thanks for their excellent features such as low cost of manufacturing, low density and easy installation, polymeric pipes have been widely employed in pressure vessel networks and pipelines. However, because of their complicated time-temperature dependent behavior of polymeric materials, understanding and predicting the long-term performance of the pipes are challenging. In particular, the long-term performance and the lifetime of these pipes are dramatically sensitive to the loading level and the working temperature. The challenge can become even greater when the pipes are used in applications with a variable environmental condition such as temperature and/or in a pipe network with load variations. In terms of safety and required standards, the plastic pipes should be designed to have at least 50 years lifetime under a specified loading level. Therefore, investigations are needed to evaluate the long-term performance of the pipes and insure the quality of new designed products.

Recently, a number of studies have been conducted focusing on the long-term creep behavior of polymeric materials. Hu et al. [1] investigated the impacts of temperature and stress on the short-term and long-term creep behavior of ethylene tetrafluoroethylene (ETFE). They found that the time-temperature superposition could underestimate the creep strains while the time-stress superposition could overestimate creep strains. Zhou et al. [2] experimentally studied the creep behavior and lifetime performance of PMMA immersed in liquid scintillator. The comparison between the proposed model perditions and actual long-term creep results demonstrate excellent agreement for short time data, However, the actual creep data shows a higher creep rate for long-time data. The effects of size and thickness on the creep property of a cross-linked polyethylene were studied by Takahashi et al. [3].

In addition, several studies conducted to understand the long term creep behavior in plastic pipes. Fatima et al. [4] investigated the differences between the burst behavior of chlorinated polyvinyl chloride (CPVC) and high density polyethylene pipes. In a different study, Moon et al. [5] suggested two algorithmic methods to construct the long-term strength estimation line of plastic pipes. The methods eliminate uncertainties and errors caused by the typical trial and error approach

M. Taherzadehboroujeni (✉) · S. W. Case
Department of Biomedical Engineering and Mechanics, Virginia Tech, Blacksburg, VA, USA
e-mail: mehrzadt@vt.edu

S. W. Case
Macromolecules Innovation Institute, Virginia Tech, Blacksburg, VA, USA

A. Linderholt et al., *Dynamic Substructures, Volume 4*, Conference Proceedings of the Society for Experimental Mechanics Series,
https://doi.org/10.1007/978-3-030-12184-6_6

for developing the baseline estimation. Zhang et al. [6] investigated the time-dependent behavior of polyethylene pipes. They suggested a new time-strain rate superposition principal to construct the relaxation master curve of polyethylene pressure pipes using horizontal and vertical shifting. The long term creep behavior of HDPE pipes has been experimentally studied by Vakili-Tahami et al. [7] where a creep constitutive model has been suggested to predict and optimize the creep lifetime of the pipes. Furthermore, Taherzadehboroujeni et al. [8] developed a novel experimental protocol to estimate the long term performance of HDPE pipes only base on short term uniaxial tensile and compression tests. Based on the observations, they concluded that the long term performance of the pipes is mainly depends on the rate of deformation which can be evaluated using a hydrostatic pressure modified Eyring formulation for yield kinetics of the material. Although the recent experimental approaches can successfully predict the overall long-term performance and lifetime of the pipes, they have very limited capability of studying the effects of variation in working conditions, such as temperature, on the lifetime and creep behavior of the pipes.

One can employ numerical simulation to overcome some of the limitations of experimental approaches. Despite many experimental studies on long-term creep behavior of plastic pipes, very few studies of numerical simulation and/or modeling have been reported. Kuhl et al. [9] suggested a numerical modeling to describe the nonlinear viscoelastoplastic creep behavior of HDPE pipes at small strains. Cheng et al. [10] used the finite element method to analyze the primary and secondary creep behavior of HDPE pipes and also to estimate the overall lifetime of the pipes. However, the suggested constitutive model was developed by employing a fitting process of a general formula to the actual data from very time consuming long-term creep tests. However, for accelerated characterization purposes it is desirable to develop a combination of a comprehensive numerical simulation with short-term mechanical testing to model the long-term creep rupture behavior of plastic pipes at different variable working temperatures and loadings. Thus, the aim of this study is to develop such robust and efficient combined approach.

6.2 Methods

The general response of HDPE to a large deformation has two major components: the initial viscoelastic and the plastic responses. HDPE is a semi-crystalline polymer with very complex viscoelastic and plastic behavior. To develop a comprehensive constitutive equation to describe the behavior of the material under different temperature, loadings and deformation rates one can combine the viscoelastic and plastic components of the material behavior.

Several research studies focused on the viscoelastic behavior of the material [11–17]. Colak and Dusunceli [12] studied the viscoelastic and viscoplastic behavior of HDPE under uniaxial compression configuration. Later on, Bilgin [14] studied the viscoelastic behavior of polyethylene pipes by conducting several tests to evaluate the time-dependent behavior of the material under different mechanical and thermal loading levels. They also employed finite element methods to simulate the stress relaxation behavior of the pipes. In this study, we assume that the initial viscoelastic part of the material response does not play a significant role in the overall long-term mechanical response and creep behavior of HDPE pipes. Thus, we proceed the modeling with this assumption, however, the sensitivity of long-term material response to the initial viscoelastic behavior is later evaluated to check this assumption.

On the other hand, the plastic deformation of the material is highly sensitive to the working temperature and the loading level. Eyring [18] suggested a generalized expression to relate the viscosity to the rate of shear and the temperature. Several other investigators (e.g. [19–21]) later adopted and modified the expression to evaluate the stress and temperature dependence of the plastic flow rate, $\dot{\varepsilon}$, as:

$$\dot{\varepsilon} = \dot{\varepsilon}_0 \exp\left(-\Delta E/kT\right) \sinh\left(\frac{S_y V}{kT}\right) \tag{6.1}$$

where S_y is yield stress, $\dot{\varepsilon}_0$ is a constant, T is absolute temperature, and k is Boltzmann's constant. In Eq. (6.1), ΔE is the required energy to activate the plastic flow (the, activation energy). The polymer is permitted to flow when several number of polymer chain segments can move [22]. The required volume of those polymer segments to initiate the movement is termed the activation volume [23], denoted by V. Different research studies [21, 24–26] on yield behavior of polymer materials over wide range of deformation rates and temperatures have indicated that more than one plastic flow process exists. Thus, multiple Eyring processes can be employed in parallel to describe the yielding behavior of the material over a wide range of deformation rates and temperates. The overall yield stress can be calculated by adding the yield stress evaluated based on each Eyring process. In the special case of two processes, we have

$$\frac{S_y}{T} = \frac{S_{y1}}{T} + \frac{S_{y2}}{T} = \frac{k}{V_1}\sinh^{-1}\left(\frac{\dot{\varepsilon}}{\dot{\varepsilon}_{0_1}}\exp\left(\frac{\Delta E_1}{kT}\right)\right)\frac{k}{V_2}\sinh^{-1}\left(\frac{\dot{\varepsilon}}{\dot{\varepsilon}_{0_2}}\exp\left(\frac{\Delta E_2}{kT}\right)\right). \tag{6.2}$$

In Eq. (6.2), the subscripts 1 and 2 are associated with each of the Eyring plastic flow processes. Further studies on uniaxial tensile and compression tests of polymers under the same deformation rate [27–29] indicated a significant difference in the magnitude of yield stress which was attributed to the effect hydrostatic pressure on yield stress. The hydrostatic pressure increases the activation energy of each Eyring processes; therefore, the hydrostatic pressure modified form of Eq. (6.2) is [8, 27];

$$\frac{S_y}{T} = \frac{k}{V_1}\sinh^{-1}\left(\frac{\dot{\varepsilon}}{\dot{\varepsilon}_{0_1}}\exp\left(\frac{\Delta E_1}{kT} + \frac{p\Psi_1}{kT}\right)\right) + \frac{k}{V_2}\sinh^{-1}\left(\frac{\dot{\varepsilon}}{\dot{\varepsilon}_{0_2}}\exp\left(\frac{\Delta E_2}{kT} + \frac{p\Psi_2}{kT}\right)\right) \tag{6.3}$$

where, p is the hydrostatic pressure and Ψ_1 and Ψ_2 are pressure activation volumes for processes 1 and 2, respectively. Eight parameters are required to be experimentally determined to calibrate the model given by Eq. (6.3).

6.3 Experiments

Several uniaxial tensile tests at different temperatures and strain rates are required to evaluate the parameters of Eq. (6.3) to fully develop the hydrostatic pressure modified plastic flow rate model. Since pressure activation volumes, Ψ_1 and Ψ_2, are independent parameters, at least two sets of uniaxial compression tests in different temperatures and different strain rates are required to ensure there are sufficient experimental data points for each Eyring plastic flow processes. In addition to the uniaxial tests, several internal pressurization tests (burst tests) are conducted to measure the actual time-to-failure and long term performance of the pipes. The details of all the tests are provided in the previous work [8] where an accelerated testing protocol presented to evaluate the long-term performance of the pipes and validated using measured long-term experimental data.

The uniaxial tensile tests are conducted utilizing a 50 kN Instron load frame at four different temperatures, 20, 42, 60, and 80 °C. A chamber is mounted on the load frame to maintain the temperature constant during the tests. For each temperature, several uniaxial tensile tests are performed under a wide range of displacement rates to measure the yield stresses over four decades of strain rates, form 10^{-2} to 10^{-6}. All the uniaxial tensile tests are performed on as-manufactured HDPE pipe segments with length of 254 mm, outer diameter of 33 mm, and thickness of 3 mm. The material is provided by LyondellBasell Industries (Cincinnati Technology Center, OH). Digital image correlation (DIC) is employed to measure the axial and hoop strain during the tests. In addition, several uniaxial compression tests are performed at two temperatures, 21 and 60 °C, and a wide range of strain rates. The strain rates and temperatures have been selected to that ensure sufficient experimental data points have been recorded to study the effect of hydrostatic pressure on the yield kinetics and evaluate the pressure activation volumes in Eq. (6.3). The Instron machine along with a small chamber are employed to apply the constant uniaxial compressive displacement rates at constant temperatures on as-manufactured HDPE pipe segments with length of 7 mm. The tensile and the compression tests, which are conducted using a single Instron machine within 10 days, provide the required experimental data to develop the model, Eq. (6.3), and construct the finite element analysis.

Burst tests are conducted using a custom-designed testing setup to measure the time-to-failure of the pipes. Water at ambient temperature is pumped into a 254 mm long pipe segment. In the process, air is purged from the pipe. When entire system is filled with water, the desired pressure is applied using a high pressure nitrogen tank controlled with a gas regulator. Meanwhile, the internal pressure of the pipe is measured using a pressure transducer and the results are processed in a data acquisition software. The software sends on and off signals to an electrically activated valve to maintain the constant internal pressure within the expanding pipe during the tests. A schematic chart of the testing setup is shown in Fig. 6.1. The DIC system is employed to measure the deformation during the internal pressurization tests. Several internal pressurization tests at different constant internal pressures are conducted. The recorded time-to-failure associated with different hoop stresses are used to validate the predictions from the finite element simulations.

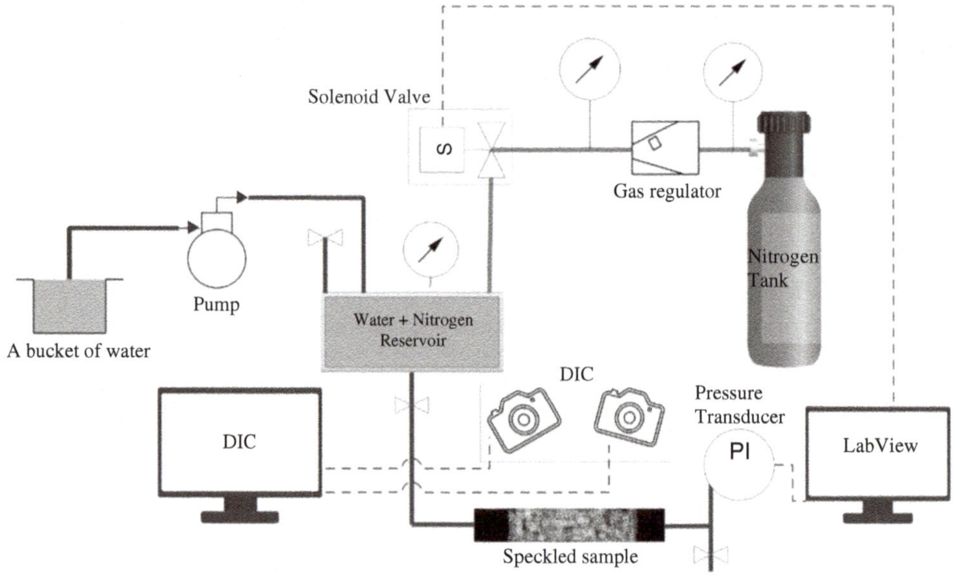

Fig. 6.1 A schematic chart of the internal pressurization tests [8]

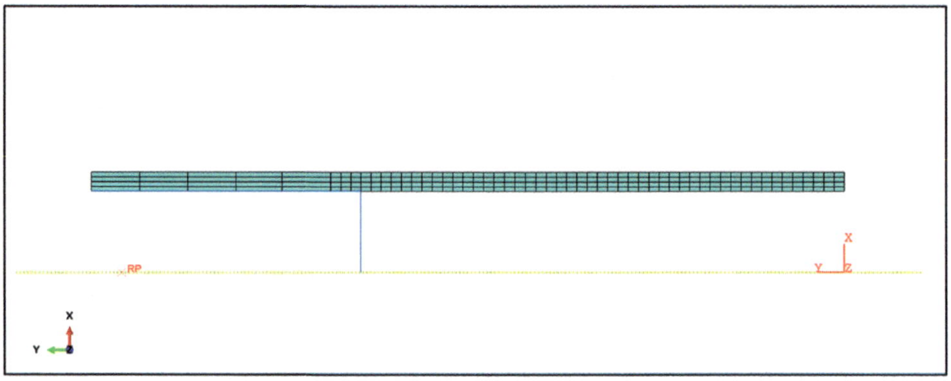

Fig. 6.2 Modeled pipe in the internal pressurization (burst) test

6.4 Numerical Modeling

In this study, the developed plastic flow model along with initial elastic response of the material are utilized to simulate long-term internal pressurization tests and evaluate the lifetime of the pressure pipes under different working conditions. The finite element creep analysis is performed using ABAQUS CAE software, version 6.10.

The pipes are modeled as a deformable axisymmetric shell with the same dimensions as used in the internal pressurization tests. Because of the symmetry in the axial direction of the pipe, only a half of the pipe is modeled and appropriate constrains are applied. Figure 6.2 shows the modeled pipe. In order to use the developed plastic flow model in the simulation, a user creep subroutine is developed which is called at all integration points. The subroutine code is provided in Appendix. We assume that the total deformation has both linear elastic and plastic creep components. Thus,

$$\boldsymbol{\varepsilon} = \boldsymbol{\varepsilon}_e + \boldsymbol{\varepsilon}_{cr} \tag{6.4}$$

For an isotropic material the linear elastic portion of the deformation can easily be calculated using a temperature-dependent elastic modulus and Poisson's ratio. We used the developed plastic flow model to calculate the plastic creep portion of the deformation. The eight parameters used in the plastic flow model and elastic modulus evaluated at different temperatures and the Poisson's ratio are reported in Table 6.1. At each time increment and for each integration point, the equivalent plastic flow rate associated with the local updated equivalent Mises stress, S_y, equivalent pressure stress, p, and the

Table 6.1 Model parameters and elastic properties

	V_i (nm^3)	ΔE_i $\left(\frac{\text{kJ}}{\text{mol}}\right)$	$\dot{\varepsilon}_i$ $\left(\frac{1}{\text{s}}\right)$	Ψ_i (nm^3)
Process 1	17.42	497	5.62E+60	0.943
Process 2	2.63	106	1.48E+14	0.039
Temperature (K)	293	315	333	353
E (MPa)	500	300	150	120
v	0.49	0.49	0.49	0.49

working temperature, T, is calculate by solving the Eq. (6.3). Because of the transcendental nature of Eq. (6.3), the solutions are found numerically using Newton-Raphson method. For tensile test simulation, the updated equivalent plastic strain can be calculated by

$$\overline{\varepsilon}_p^{(i)} = \overline{\varepsilon}_p^{(i-1)} + \dot{\varepsilon}_p \Delta t \tag{6.5}$$

where, $\overline{\varepsilon}_p^{(i-1)}$ is the old plastic strain, $\overline{\varepsilon}_p^{(i)}$ is updated plastic strain, $\dot{\varepsilon}_p$ is the plastic strain rate calculating from solving Eq. (6.3), and Δt is the time increment. The updated elastic portion of the strain, then, can be calculated as

$$\overline{\varepsilon}_e^{(i)} = t\dot{\varepsilon} - \overline{\varepsilon}_p^{(i)} \tag{6.6}$$

where, t and $\dot{\varepsilon}$ are the elapsed time and applied strain rate, respectively. Then, the new stress can be calculated using elastic properties of the material, as

$$\boldsymbol{\sigma}^i = C : \boldsymbol{\varepsilon}_e^i \tag{6.7}$$

where C is the stiffness tensor for isotropic materials. Both the tensile and internal pressurization tests are simulated in the finite element modeling software. To evaluate the long-term hydrostatic strength curves, several simulations conducted using different pressures and temperatures.

To investigate the effect of variations in the working temperature, a burst test was simulated where a constant pressure, 2.275 MPa (330 Psi), with a daily temperature fluctuation was applied on the pipe. The average temperature was set to 25 °C (298 K) with ±°C K daily variation.

6.5 Results and Discussion

To study the effect of different initial elastic modulus on the long-term performance of the pipes, two sets of simulations were performed using the minimum and maximum elastic modulus, 400 and 600 MPa, observed in the tensile tests at room temperature. Based on the results, shown in Fig. 6.3, the simulations using elastic modulus of 400 MPa result in a shorter lifetime estimate than those using elastic modulus of 600 MPa. However, the differences are not significant when compared to the scatter in the experimental data. Thus, it appears that the viscoelastic behavior (in contrast to the plastic behavior) of the HDPE does not play a significant role in the long-term performance of the pipes, and constant elastic modulus for each temperature can be used. The values for the elastic modulus used at each temperature are given in Table 6.1.

The FE model used to regenerate the tensile tests at four different temperatures and different strain rate. The stress-strain results of FE simulation for different configurations are presented in Fig. 6.4, where numerical results, shown with solid lines, are compared with the experimental data, shown with dashed lines. As the results show, the FE simulation reaches to the yield stress at a lower strain than the actual data. However, in general, the simulated stress-strain results for different strain rate and temperatures show very good agreement with the experimental data. The model is less accurate for the highest temperature investigated (353 K) because a secondary yielding occurs at large strain (~0.6) [8]—a behavior which has been not considered in the model.

The model was employed to simulate the pipe burst tests at room temperate. Since the burst tests were conducted at room temperature, 294 ± 2 K, the burst test simulations under different internal pressures and at 292 and 296 K were performed. Figure 6.5 shows the modeled pipe deformation during the pressurization test along with an image of bursting pipe from a burst test, and indicates that the simulation mimics the deformation of the pipe during the test. The model predictions of

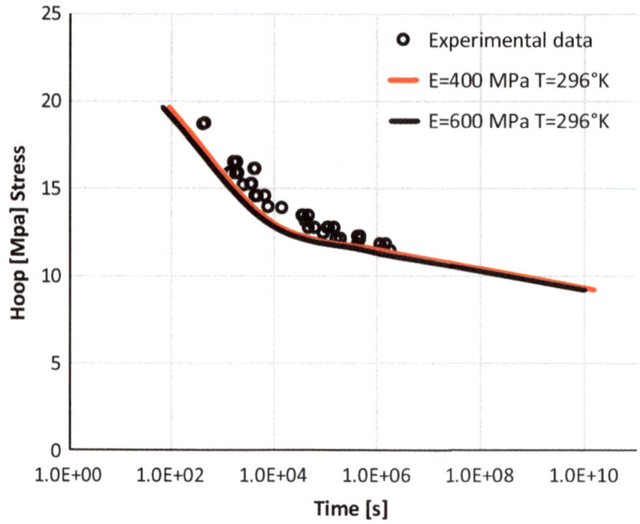

Fig. 6.3 Simulation results of internal pressurization (burst) tests using different elastic modulus

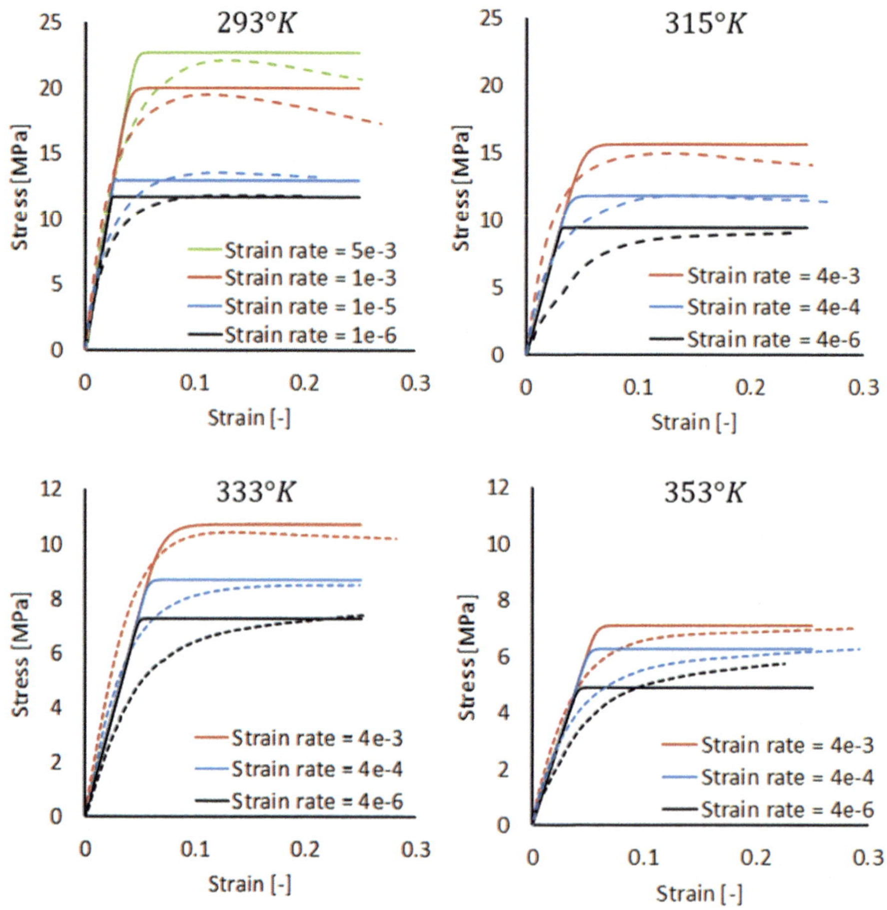

Fig. 6.4 Uniaxial tensile simulation in different temperatures and strain rates, the solid lines. The experimental data shown with dashed lines

time-to-failure for these two temperatures are presented in Fig. 6.6, shown by solid lines, along with experimental data, the solid dots. As it shown in Fig. 6.6, the model can predict the trend of the actual time-to-failure data and the predictions are almost accurate. However, the model looks slightly conservative in predicting the failure time. This discrepancy can be explained by the observation that the yielding occurs at lower strain levels in the model than in the experiments.

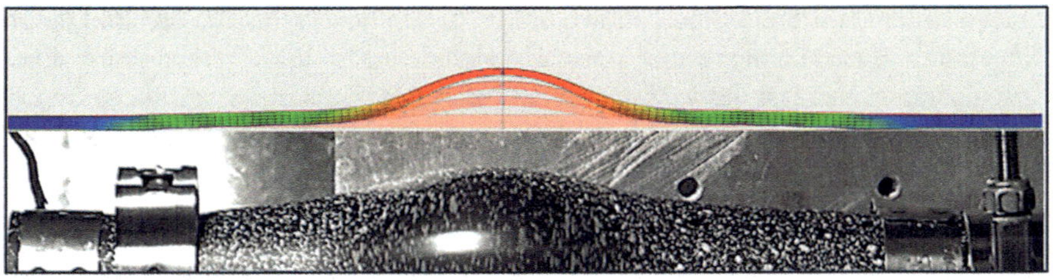

Fig. 6.5 The deformation of the pipe during the internal pressurization (burst) test, top is from the FE simulation, bottom is from the experiments

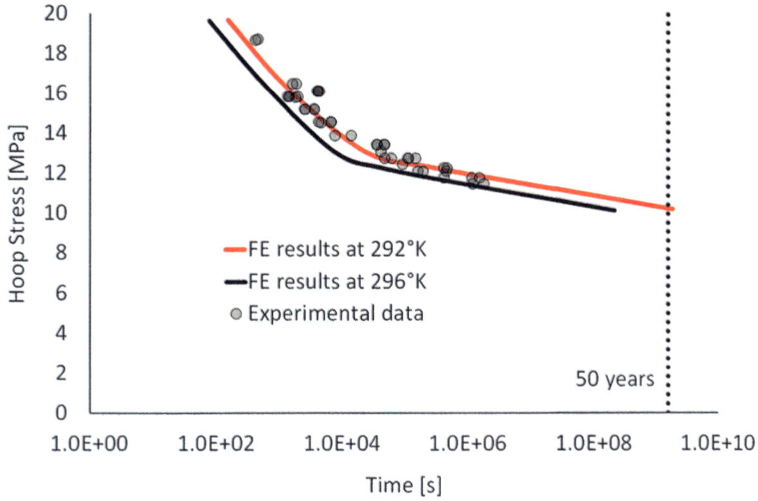

Fig. 6.6 Model prediction of the pipe burst tests at 292 and 296 K and the experimental data

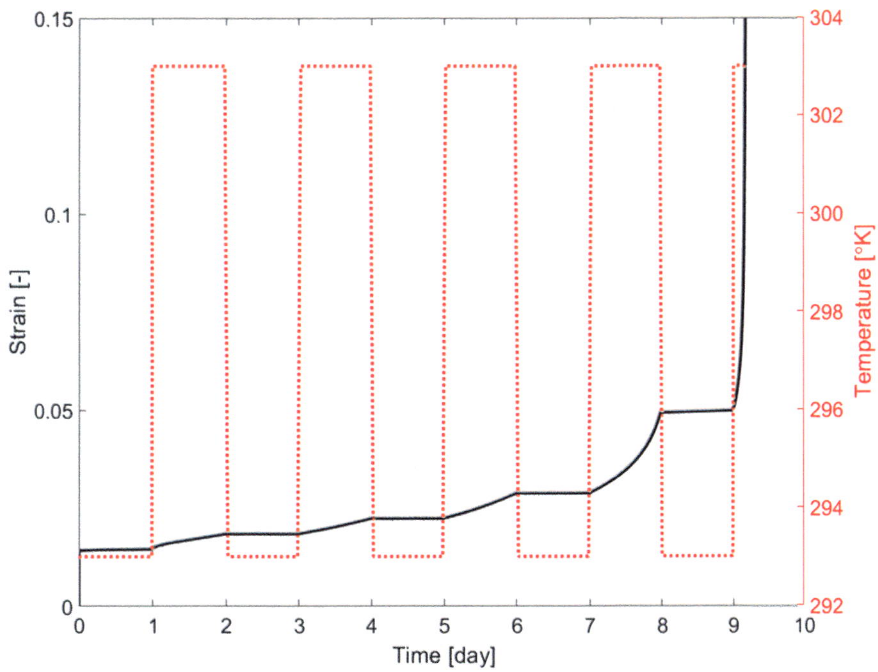

Fig. 6.7 The variable temperature and the corresponding deformation of the pipe under a constant internal pressure

To study the effect of variation in the working temperature on long-term performance of the pipe, a burst test was simulated in which the daily variation of the temperature was included. Figure 6.7 shows the variation of the temperature as the red dashed curve. An internal constant pressure of 2.275 MPa (330 Psi) was applied until the pipe failed due to excessive plastic deformation. The calculated strain at different times, shown in Fig. 6.7, indicates that the material deformation is dominated by the change in the temperature. Based on the results, one can conclude that at lower temperature, does not show significant deformation. However, temperature increase leads to increased creep rate. This observation can be explained based on the very sensitive response of the material to the temperature. In low temperature, 293 K, the rate of deformation is negligible with compare to the rate of deformation of the material when the temperature is only 10° higher, 303 K. Thus, for a pipe working under a variable temperature between T_{low} and T_{high}, the time-to-failure can be estimated as twice of the time-to-failure for the pipe under a constant T_{high} temperature.

6.6 Conclusions

In this study, short-term experimental measurements of plastic strain rate are used to calibrate a new numerical model to describe the time-temperature dependent behavior of high density polyethylene pipes. Several tensile and compression tests were performed on as-manufactured pipe segments to develop the plastic flow Eyring model. Using the developed model and initial elastic properties of the material series of tensile and creep tests were simulated by employing the finite element method. The simulation results showed good agreement with the experimental data from different tensile tests and internal pressurization (burst) tests. In addition, the effect of variation in working temperature was studied. The results of a case study indicated that for a pipe working under a daily variable temperature between 293 K and 303 K and under a constant internal pressure of 2.275 MPa (330 Psi), the time-to-failure can be estimated as twice of the time-to-failure for the pipe working under the same constant internal pressure and at constant temperature of 303 K.

Appendix

```
      SUBROUTINE CREEP(DECRA,DESWA,STATEV,SERD,EC,ESW,P,QTILD,
     1 TEMP,DTEMP,PREDEF,DPRED,TIME,DTIME,CMNAME,LEXIMP,LEND,
     2 COORDS,NSTATV,NOEL,NPT,LAYER,KSPT,KSTEP,KINC)
C
      INCLUDE 'ABA_PARAM.INC'
C
      CHARACTER*80 CMNAME
C
      DIMENSION DECRA(5),DESWA(5),STATEV(*),PREDEF(*),DPRED(*),
     1 TIME(2),COORDS(*),EC(2),ESW(2)
C
C
      REAL*8     :: v1, v2, e2, e1, loge1, h1, h2, g1, g2, K, y0, KK
      REAL*8     :: s, T, p, f, ff, y, x1, x2, sinh1, sinh2, n1, n2, y1
      REAL*8     :: y2, ee1, ee2, yy1, yy2, dly

      v1 = 1.74220376E-3
      loge1 = 139.8808
      h1 = 8.26058E0
      g1 = 9.4290946E-9
      v2 = 2.632239842E-4
      e2 = 1.48346E14
      h2 = 1.75285E0
      g2 = 3.8763635E-10
      K = 1.38064852E0
      KK = 1.38064852E-4

C
      e1=exp(loge1)
      s = QTILD
      T = TEMP
      n1=h1/KK/T+P*g1/KK/T
      n2=h2/KK/T+P*g2/KK/T
      ee1=exp(n1)
      ee2=exp(n2)
      yy1=ee1/(v1*1e-23)
      yy2=ee2/(v2*1e-23)
```

```
      y1=yy1/e1
      y2=yy2/e2
      y0 = s/(y1+y2)
      y = y0

      DO WHILE ( dly < 4. )
              y0 = y
              x1 = EXP((h1+P*g1)/(KK*T))/e1
              x2 = EXP((h2+P*g2)/(KK*T))/e2
              sinh1 = log(y0*x1+sqrt((y0*x1)**2.+1.))
              sinh2 = log(y0*x2+sqrt((y0*x2)**2.+1.))
              f = K*T*(sinh1/v1+sinh2/v2)-s
              ff=K*T*(x1/SQRT((y0*x1)**2.+1.)/v1+x2/SQRT((y0*x2)**2.+1.)/v2)
              y = y0-f/ff
              dly = log10(y0)-log10(abs(f/ff))
      END DO
      DECRA(1) = y*DTIME
C
      RETURN
      END
```

References

1. Hu, J., Li, Y., Chen, W., Zhao, B., Yang, D.: Effects of temperature and stress on creep properties of ethylene tetrafluoroethylene (ETFE) foils for transparent buildings. Polym. Test. **59**, 268–276 (2017)
2. Zhou, F., Hou, S., Qian, X., Chen, Z., Zheng, C., Xu, F.: Creep behavior and lifetime prediction of PMMA immersed in liquid scintillator. Polym. Test. **53**, 323–328 (2016)
3. Takahashi, Y., Tateiwa, T., Shishido, T., Masaoka, T., Kubo, K., Yamamoto, K.: Size and thickness effect on creep behavior in conventional and vitamin E-diffused highly crosslinked polyethylene for total hip arthroplasty. J. Mech. Behav. Biomed. Mater. **62**, 399–406 (2016)
4. Fatima, M., Mohamed, S., Mohamed, E.: Burst behavior of CPVC compared to HDPE thermoplastic polymer under a controlled internal pressure. Proc. Struct. Integrity. **3**, 380–386 (2017)
5. Moon, J., Bae, H., Song, J., Choi, S.: Algorithmic methods of reference-line construction for estimating long-term strength of plastic pipe system. Polym. Test. **56**, 58–64 (2016)
6. Zhang, Y., Jar, P.-Y.B.: Time-strain rate superposition for relaxation behavior of polyethylene pressure pipes. Polym. Test. **50**, 292–296 (2016)
7. Vakili-Tahami, F., Adibeig, M.R.: Using developed creep constitutive model for optimum design of HDPE pipes. Polym. Test. **63**, 392–397 (2017)
8. Taherzadehboroujeni, M., Kalhor, R., Fahs, G., Moore, R., Case, S.: Accelerated testing method to estimate the long-term hydrostatic strength of semi-crystalline plastic pipes. Polym. Eng. Sci. **60**, (2019)
9. Kühl, A., Muñoz-Rojas, P.A., Barbieri, R., Benvenutti, I.J.: A procedure for modeling the nonlinear viscoelastoplastic creep of HDPE at small strains. Polym. Eng. Sci. **57**, 144–152 (2017)
10. Cheng, C., Widera, G.O.: Development of maximum secondary creep strain method for lifetime of HDPE pipes. J. Press. Vessel. Technol. **131**, 021208 (2009)
11. Guedes, R.M.: A viscoelastic model for a biomedical ultra-high molecular weight polyethylene using the time-temperature superposition principle. Polym. Test. **30**, 294–302 (2011)
12. Colak, O.U., Dusunceli, N.: Modeling viscoelastic and viscoplastic behavior of high density polyethylene (HDPE). J. Eng. Mater. Technol. **128**, 572–578 (2006)
13. Reis, J.M.L., Pacheco, L.J., da Costa Mattos, H.S.: Tensile behavior of post-consumer recycled high-density polyethylene at different strain rates. Polym. Test. **32**, 338–342 (2013)
14. Bilgin, O.: Modeling viscoelastic behavior of polyethylene pipe stresses. J. Mater. Civ. Eng. **26**, 676–683 (2014)
15. Taherzadeh, M., Baghani, M., Baniassadi, M., Abrinia, K., Safdari, M.: Modeling and homogenization of shape memory polymer nanocomposites. J. Compos. Part B Eng. **91**, 36–43 (2016)
16. Yang, F., Mousavie, A., Oh, T., Yang, T., Lu, Y., Farley, C., Bodnar, R., Niu, L., Qiao, R., Li, Z.: Sodium–sulfur flow battery for low-cost electrical storage. J. Adv Energy Mater. **8**, 1701991 (2018)
17. Piavis, W., Turn, S., Mousavi, A.: Non-thermal gliding-arc plasma reforming of dodecane and hydroprocessed renewable diesel. Int J Hydrogen Energy. **40**, 13295–13305 (2015)
18. Eyring, H.: Viscosity, plasticity, and diffusion as examples of absolute reaction rates. J. Chem. Phys. **4**, 283–291 (1936)
19. Bauwens, J.: Yield condition and propagation of Lüders' lines in tension–torsion experiments on poly (vinyl chloride). J. Polym. Sci. B Polym. Phys. **8**, 893–901 (1970)
20. Bauwens-Crowet, C., Bauwens, J.-C., Homès, G.: The temperature dependence of yield of polycarbonate in uniaxial compression and tensile tests. J. Mater. Sci. **7**, 176–183 (1972)
21. Bauwens-Crowet, C.: The compression yield behaviour of polymethyl methacrylate over a wide range of temperatures and strain-rates. J. Mater. Sci. **8**, 968–979 (1973)

22. Haward, R., Thackray, G.: The use of a mathematical model to describe isothermal stress-strain curves in glassy thermoplastics. Proc. R. Soc. Lond. A. **302**, 453–472 (1968)
23. Richeton, J., Ahzi, S., Daridon, L., Rémond, Y.: A formulation of the cooperative model for the yield stress of amorphous polymers for a wide range of strain rates and temperatures. Polymer. **46**, 6035–6043 (2005)
24. Roetling, J.: Yield stress behaviour of polymethylmethacrylate. Polymer. **6**, 311–317 (1965)
25. Roetling, J.: Yield stress behaviour of isotactic polypropylene. Polymer. **7**, 303–306 (1966)
26. Roetling, J.: Yield stress behaviour of poly (ethyl methacrylate) in the glass transition region. Polymer. **6**, 615–619 (1965)
27. Truss, R., Clarke, P., Duckett, R., Ward, I.: The dependence of yield behavior on temperature, pressure, and strain rate for linear polyethylenes of different molecular weight and morphology. J. Polym. Sci. Polym. Phys. Ed. **22**, 191–209 (1984)
28. Truss, R., Duckett, R., Ward, I.: Effect of hydrostatic pressure on the yield and fracture of polyethylene in torsion. J. Mater. Sci. **16**, 1689–1699 (1981)
29. Bauwens-Crowet, C., Ots, J.-M., Bauwens, J.-C.: The strain-rate and temperature dependence of yield of polycarbonate in tension, tensile creep and impact tests. J. Mater. Sci. **9**, 1197–1201 (1974)

Chapter 7
Study on Dynamic Stiffness Characteristic of Primary Suspension for High-Speed EMU

Xiugang Wang, Xiaoning Cao, Ai qin Tian, Jian Su, Wei Xue, and Shen Zhan

Abstract This paper made a deep research on the axle-box suspension dynamic stiffness characteristics. And testing models including vertical, transversal and longitudinal and solving method based on stepping swept sine excitation were put forward. Then, experiment was taken on a certain bogie, and curves of frequency-stiffness were obtained based on the evaluation method for the dynamic parameters. The results of this paper play a positive role to parameter matching and dynamic simulation.

Keywords Dynamic · Stiffness · Primary suspension · High-speed EMU

7.1 Instruction

Railway vehicle is a complexity dynamic system with multidimensional and multi-DOF. As the unique running gear of vehicle, dynamic performance of the bogie is the key to the running safety and stability. Suspension system including axle-box suspension and secondary suspension is of the main components, which is closely related to passing ability and passenger comfort. This paper made a deep research on the axle-box suspension dynamic stiffness characteristics; analysis model of axle-box suspension was also established. Meanwhile, testing models including vertical, transversal and longitudinal and solving method based on stepping swept sine excitation were put forward. Then, the experiment was taken on a certain bogie, and curves of frequency-stiffness were obtained based on the evaluation method for the dynamic parameters. The results of this paper play a positive role to parameter matching and dynamic simulation.

7.2 Analysis Model of Axle-Box Suspension

Combined with structure and bearing performance of axle-box suspension, steel spring was simplified as a single degree of freedom model with the vertical stiffness, which ignored the damping characteristics of flexible rubber sheet. Vertical damper was simplified as the ideal linear viscoelasticity model. Considering the viscoelastic characteristics of Guide arm node, the spring-damped parallel model was established. Meanwhile, as Guide arm node endured the 6-dof spatial loading and torque from wheel and framework, it was simplified to a 8-dof spring–damper model. Figure 7.1 was the physical of axle-box suspension, and analysis model was shown as Fig. 7.2.

During the analysis model, it could be set as follows. $O_w - x_w y_w z_w$ was Coordinate system of axle box, $O_t - x_t y_t z_t$, $O_{t1} - x_{t1} y_{t1} z_{t1}$, $O_{t2} - x_{t2} y_{t2} z_{t2}$ were Coordinate system of Guide arm node. Vertical stiffness and damping coefficient was described as K_{zp}, C_{zp}; Stiffness and damping coefficient of vertical, lateral and longitudinal for the Guide arm node were described as K_{tzj}, C_{tzj}, K_{tyj}, C_{tyj}, K_{txj}, C_{txj}; $K_{\theta tj}$ was rotation angle of Guide arm node.

X. Wang (✉) · X. Cao · A. Tian · W. Xue · S. Zhan
CRRC Qingdao Sifang Co., Ltd, Qingdao City, China
e-mail: wangxiugang@cqsf.com

J. Su
College of Transportation, Jilin University, Changchun City, China

A. Linderholt et al., *Dynamic Substructures, Volume 4*, Conference Proceedings of the Society for Experimental Mechanics Series,
https://doi.org/10.1007/978-3-030-12184-6_7

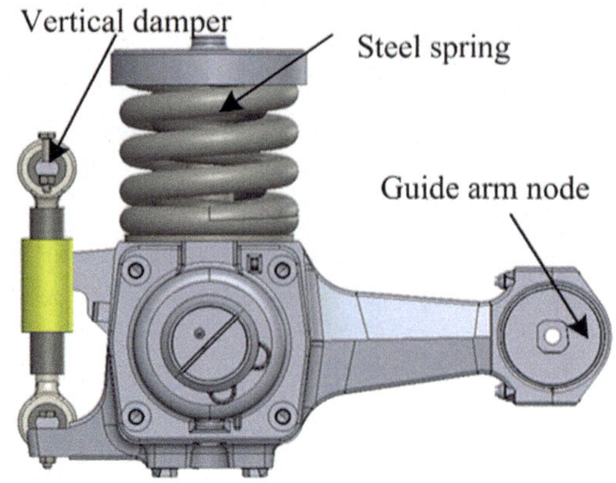

Fig. 7.1 Physical of axle-box suspension

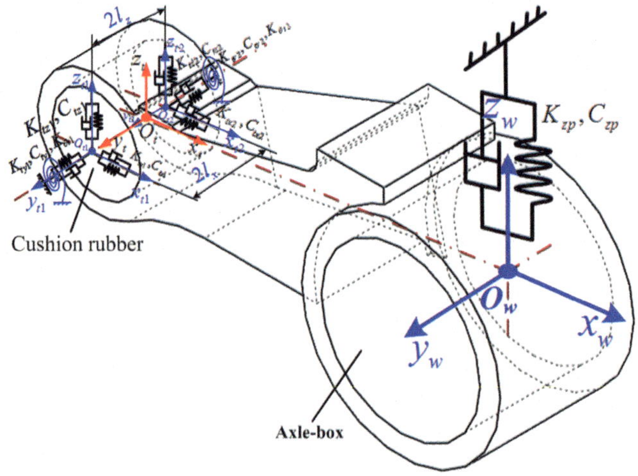

Fig. 7.2 Model of axle-box suspension

7.3 Study on the Dynamic Stiffness Test Model

By taking the operating conditions of axle-box suspension as research object, testing models including vertical, transversal and longitudinal were put forward. The mathematical model was set up by using Darren Bell principle, and equations of equivalent stiffness and single stiffness-frequency characteristic for axle-box suspension were solved, which laid a foundation for axle suspension parameters test.

Vertical stiffness test model was shown as Fig. 7.3. It could be assumed that vertical deformation for spring and Guide arm node was described as $\tilde{\delta}_{zpi}(t)$ and $\tilde{\delta}_{vzi}(t)$, rotation angle of Guide arm node was $\tilde{\theta}_{vpi}(t)$; rotational inertia of wheelset was described as J_{wy}.

It was assumed that load of vertical suspension was $\tilde{F}_{zpi}(t)$, vertical differential equation was established based on Newton second law.

$$m_t\ddot{\tilde{q}}_i(t) = \tilde{F}_{zpi}(t) - \left[\left(\tilde{K}_{zpi}\tilde{\delta}_{zpi}(t) + C_{tzi}\ddot{\tilde{\delta}}_{zpi}(t)\right) + \sum_{j=1}^{2}\left(\tilde{K}_{tzji}\tilde{\delta}_{vzji}(t) + C_{tzji}\dot{\tilde{\delta}}_{vzji}(t)\right)\right] \tag{7.1}$$

Force $\tilde{F}_{ezpi}(t)$ of suspension caused by vertical deformation can be shown as follows.

$$\tilde{F}_{ezpi}(t) = \left[\left(\tilde{K}_{zpi}\tilde{\delta}_{zpi}(t) + C_{zpi}\ddot{\tilde{\delta}}_{zpi}(t)\right) + \sum_{j=1}^{2}\left(\tilde{K}_{tzji}\tilde{\delta}_{vzji}(t) + C_{tzji}\dot{\tilde{\delta}}_{vzji}(t)\right)\right] = \tilde{F}_{zpi}(t) - m_t\ddot{\tilde{q}}_i(t)$$

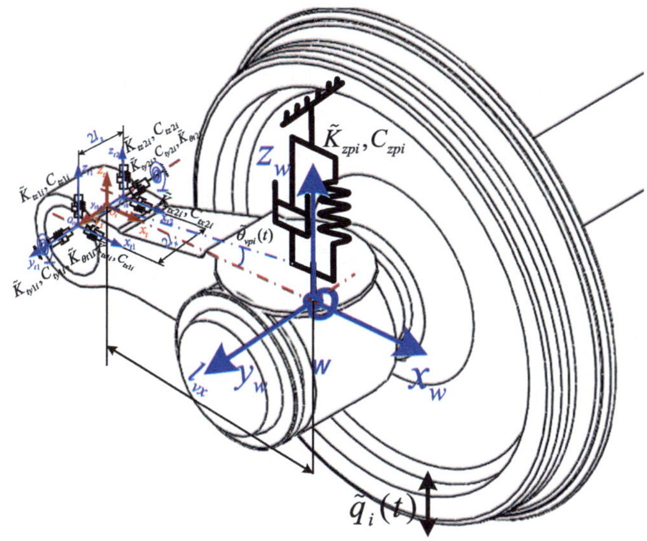

Fig. 7.3 Vertical stiffness test model

So, vertical dynamic stiffness excited by the *i*th frequency can be described as formula (7.2).

$$\tilde{K}_{ezpi} = \left(\tilde{F}_{ezpi}(t)_{\max} - \tilde{F}_{ezpi}(t)_{\min}\right) / \left(\tilde{q}_i(t)_{\max} - \tilde{q}_i(t)_{\min}\right) \tag{7.2}$$

7.4 Test Method and Solving Method

Experiment was taken by the method of exciting the axle-box suspension with sine sweep frequency step by step. The number of exciting points is N from low frequency to high, assuming that when the ith exciting point f_i works, the force $F_i(t)$ is of $F_i \sin(\omega_i t)$, and displacement $X_i(t)$ is of $X_i \sin(\omega_i t + \varphi_i)$, then stiffness value K_i of suspension of the ith exciting point can be obtained. After the exciting, curve of suspension dynamic stiffness base on this frequency range can be acquired as follows:

$$\tilde{k} = G(f_i) \quad (i = 1, 2, \cdots, n, \ n \text{ is a positive integer}) \tag{7.3}$$

Combined with the characteristics of response signals including dynamic load and displacement of suspension, solving method of stiffness $\overline{K}_{di}$ under any excited point based on multi-cycle loading was proposed, which is shown in formula (7.4). Then, suspension dynamic stiffness curve of this frequency range could be acquired.

$$\begin{aligned} \overline{K}_{di} = \frac{\overline{F}_i}{\overline{X}_i} &= \frac{\frac{\omega_i}{4}\sum_{k=0}^{n}\left(\left|\int_{2kt}^{(2k+1)t} F_i(t)dt\right| + \left|\int_{(2k+1)t}^{2(k+1)t} F_i(t)dt\right|\right)}{\frac{\omega_i}{4}\sum_{k=0}^{n}\left(\left|\int_{2kt}^{(2k+1)t} X_i(t)dt\right| + \left|\int_{(2k+1)t}^{2(k+1)t} X_i(t)dt\right|\right)} \\ &= \frac{\sum_{k=0}^{n}\left(\left|\int_{2kt}^{(2k+1)t} F_i(t)dt\right| + \left|\int_{(2k+1)t}^{2(k+1)t} F_i(t)dt\right|\right)}{\sum_{k=0}^{n}\left(\left|\int_{2kt}^{(2k+1)t} X_i(t)dt\right| + \left|\int_{(2k+1)t}^{2(k+1)t} X_i(t)dt\right|\right)} \end{aligned} \tag{7.4}$$

7.5 Test Analysis

Taking a certain bogie shown in Figs. 7.4 and 7.5 as the research object, vertical frequency-stiffness characteristic of axle-box suspension was carried out, which was shown in Fig. 7.3. Result of the four axle suspension stiffness parameters was asymmetric, which would affect the vehicle running stability and security. Also, the result would possessed guidance faction for parameters matching.

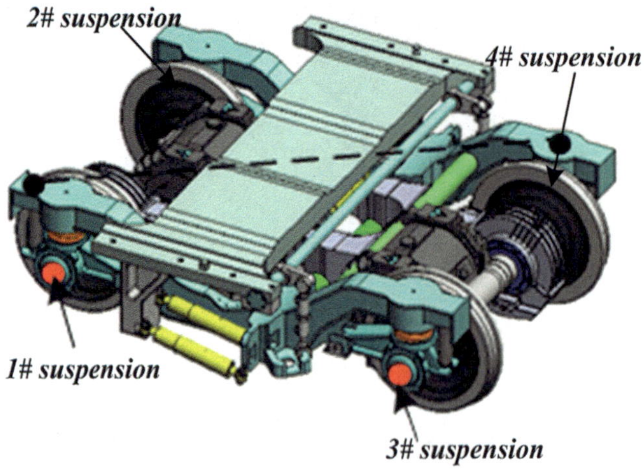

Fig. 7.4 Structure of the bogie

Fig. 7.5 Experimental determination

Time domain signals of deformation and load were shown as Fig. 7.6, and curve of dynamic stiffness for the vertical suspension was described as Fig. 7.7. The result shown that Stiffness parameters of different primary suspension were different, which proved that primary parameter matching was not symmetric. But trend of the stiffness was the same, which displayed as "$\surd$". And the formulas of vertical dynamic stiffness were obtained as follows.

$$\begin{aligned}
\tilde{K}_{zp1} &= -0.0029f^4 + 0.051f^3 - 0.151f^2 - 0.184f + 1.092, \\
\tilde{K}_{zp2} &= -0.0021f^4 + 0.043f^3 - 0.168f^2 + 0.0375f + 0.855, \\
\tilde{K}_{zp3} &= -0.0024f^4 + 0.0452f^3 - 0.16f^2 - 0.03f + 0.94, \\
\tilde{K}_{zp4} &= -0.01f^4 + 0.111f^3 - 0.34f^2 + 0.356f + 0.15.
\end{aligned}$$

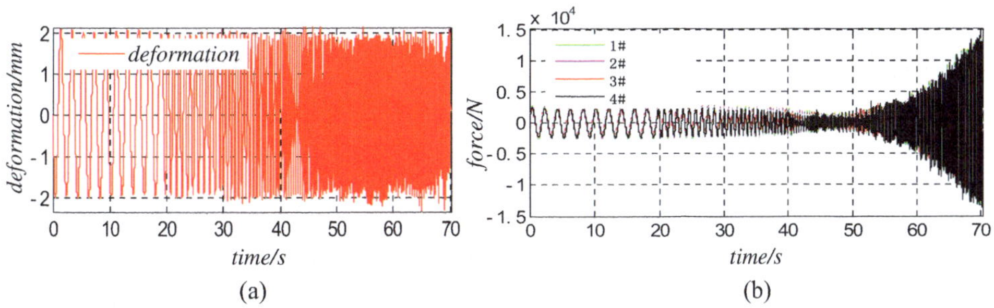

Fig. 7.6 Time domain signals of deformation and force. (**a**) Curve of deformation. (**b**) Curve of force

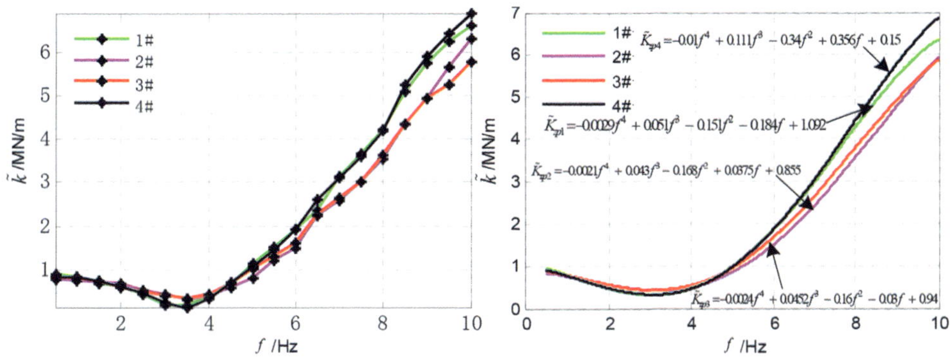

Fig. 7.7 Curve of vertical frequency-stiffness characteristic

7.6 Conclusion

This paper made a deep research on the axle-box suspension dynamic stiffness characteristics. And testing models including vertical and solving method based on stepping swept sine excitation were put forward. Then, experiment was taken on a certain bogie, and curves of frequency-stiffness were obtained based on the evaluation method for the dynamic parameters. And frequency characteristics curves of the suspension stiffness were solved. The results of this paper play a positive role to parameter matching and dynamic simulation.

References

1. Zhang, W.-h.: Research on measurement method for vehicle parameters[J]. Rolling Stock. **38**(12), 1–4 (2000)
2. Soize, C.A.: Model and numerical method in the medium frequency range for vibroacoustic prediction using the theory of structural fuzzy. Acoust. Soc Am. **94**, 849–865 (1993)
3. Yagi, T., Stensson, A., Hardell, C.: Simulation and visualization of the dynamic behavior of an overhead power system with contact breaking. Veh. Syst. Dyn. **25**(1), 31–49 (1996)
4. Wang, X., Su, J., Liu, Y., Zhou, D.: Research on dynamic measuring system and model of novel high speed railway vehicle bogie. In: ICCTP (2009)
5. Dukkipati, R.V., Narayana Swamy, S., Osman, M.: Independently rotating wheel systems for railway vehicles—a state of the art review. Veh. Syst. Dyn. **21**(21), 297–330 (1992)
6. Toyofuku, K.: Study on dynamic characteristic analysis of air spring with auxiliary chamber. JSAE Rev. **20**(3), 350–355 (1999)
7. Ewins, D.J.: Modal testing: theory and practice. Research Studies Press Ltd, Letchworh (1986)

Chapter 8
Test-Based Modeling, Source Characterization and Dynamic Substructuring Techniques Applied on a Modular Industrial Demonstrator

A. M. Steenhoek, M. W. van der Kooij, M. L. J. Verhees, D. D. van den Bosch, and J. M. Harvie

Abstract Dynamic Substructuring methods play a significant role in the analysis of complex systems and allow to assess the components individually and combine them in a modular fashion. By using compatible interfaces (through the Virtual Point Transformation), FRF-models from tests can be combined with numerically obtained FRF-models. In this paper a demonstrator structure consisting of a base-frame and several add-ons of varying complexity, both active vibration sources and passive components, is used for validation and comparison purposes. Test-based FRF models of the individual components are compared with numerically obtained results, a test-based FRF model of the full system obtained from test-based component models through substructuring is compared with numerically obtained results as well as a validation measurement on the full system. Finally, the system's response to an active vibration input is predicted using a blocked-force characterization of the source on a test-bench.

Keywords Frequency based substructuring · Dynamic substructuring · Test-based modeling · Hybrid modeling · Virtual point transformation · Source characterization · Blocked forces

8.1 Introduction

Modular modeling becomes increasingly important in many industries where an increasing number of end-products are based on reused subcomponents in different combinations. Dynamic Substructuring (DS) techniques have proven to give insight into the dynamics of a system or assembly by coupling the dynamics of the individual parts making up the system. One of the major challenges in combining component models is in finding the correct definition of the interfaces. The Virtual Point Transformation provides an effective method to transform measured Frequency Response Function (FRF) data to a 6-DoF *virtual node* which can be used for DS coupling [1, 2]. This allows one to use both test-based and numerically obtained FRFs successfully in a DS case, giving the engineer more freedom to pick the most suitable method of obtaining the component FRFs depending on the case. E.g., for complex parts one could use test-based models to get valid results at higher frequencies (>500 Hz) whereas simpler parts can be modeled with FEA techniques.

In this project a semi-industrial demonstrator is used to prove and evaluate the usefulness of a variety of (test-based) techniques to predict the response at a response of interest using modular FRF models and dynamic substructuring. The demonstrator consists of three parts: a base-frame, a bridge with an active vibration source and a passive L-shape with a sensor mounted on top, representing the response of interest. The base-frame is supported on the floor with three rubber mounts, the bridge is connected to the base-frame with a single bolt at both ends and the L-shape is connected with three bolts (in close proximity) to the base-frame. Figure 8.1a shows the complete assembly of the demonstrator, Fig. 8.1b shows the digital equivalent of the set-up which is used for measurement data processing.

The FRFs of all individual components are determined through various techniques involving test-based and hybrid modeling. The active vibration source is characterized using the blocked force technique [3], both from in-situ and free-free measurements. Using the models of the individual components, predictions at the response of interest are made through

A. M. Steenhoek (✉) · M. L. J. Verhees
ASML Netherlands B.V., Veldhoven, The Netherlands
e-mail: alexander.steenhoek@asml.com

M. W. van der Kooij · D. D. van den Bosch · J. M. Harvie
VIBES.technology, Delft, The Netherlands
e-mail: dvandenbosch@vibestechnology.com

A. Linderholt et al., *Dynamic Substructures, Volume 4*, Conference Proceedings of the Society for Experimental Mechanics Series,
https://doi.org/10.1007/978-3-030-12184-6_8

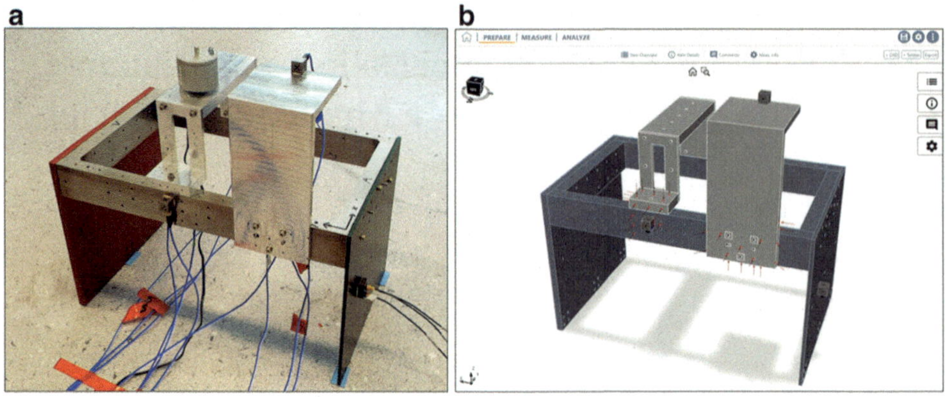

Fig. 8.1 (**a**) The assembled system consisting of the base-frame, the bridge including the active vibration source and the L-shape, and (**b**) the digital equivalent of the assembled system in the VIBES software for (live) data processing

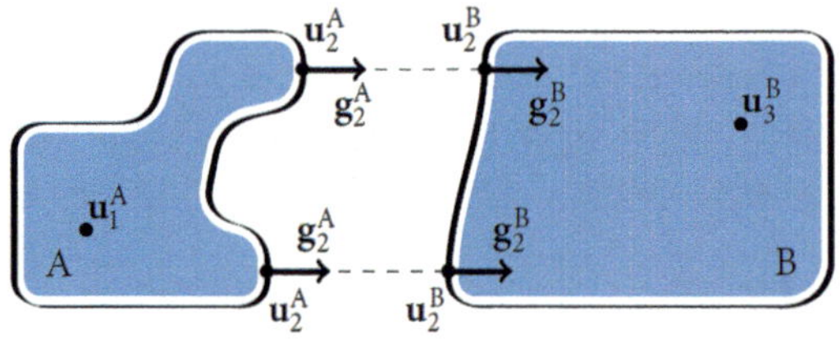

Fig. 8.2 Assembly of system A and system B

Dynamic Substructuring of the FRFs and application of the blocked forces to the system assembly. The results are then compared with validation measurements of the complete assembly.

8.2 Methodology

To allow assessment of the components individually and to combine them in a modular fashion, three methods are considered, i.e. the Virtual Point Transformation, Dynamic Substructuring and Source Characterization with Transfer Path Analysis using Blocked Forces. The combination of these methods ensures compatibility between component structures and allows for coupling and response synthesis, i.e. all required ingredients given the goal for modular modeling.

Consider an assembly consisting of two systems (A and B), as depicted in Fig. 8.2 above. When constituting this system in a "dual representation", all local component DoFs are explicitly considered. We can describe the equations of motion for both systems independently in the coupled situation, by means of Eq. (8.1).

$$\mathbf{u} = \mathbf{Y}\mathbf{f} \quad \text{EoM} \tag{8.1}$$

In this notation the force vector **f** contains all forces (external and interface) forces acting on the components and the system dynamics are governed by the system's FRFs **Y**. This system of equations can now be used to illustrate the methods mentioned above.

8.2.1 Virtual Points

For a component model obtained by measurement, all $n_u > n_q$ measured displacements **u** around an interface can be transformed to the virtual point by means of a kinematic relation between the measured DoFs **u** and the virtual DoFs, here denoted by **q**. This relation is governed by the so-called Interface Displacement Mode (IDM) matrix $\mathbf{R_u}$, such that: $\mathbf{u} = \mathbf{R_u}\mathbf{q}$. A similar relation $\mathbf{R_f}$ can be set up for the $n_f > n_m$ forces **f** and virtual point forces/moments, denoted by **m**. The inverted IDM matrices $(\mathbf{R_u})^+$ and $(\mathbf{R_f})^+$ are the actual transformation matrices that convert a $n_u \times n_f$ measured FRF matrix

$\mathbf{Y}(\omega)$ to a $n_q \times n_m$ virtual point FRF matrix as denoted by Eq. (8.2).

$$\mathbf{u} = \mathbf{Y}\mathbf{f} \Rightarrow \mathbf{q} = (\mathbf{R}_\mathbf{u})^+ \mathbf{Y}\left(\mathbf{R}_\mathbf{f}^T\right)^+ \mathbf{m} = \mathbf{Y}_{\mathbf{qm}}\mathbf{m} \quad \text{VP} \tag{8.2}$$

For more information on the virtual point transformation, the reader is referred to [1].

8.2.2 *Dynamic Substructuring*

Dynamic Substructuring (DS) is applied in this work with the Lagrange Multiplier Frequency Based Substructuring (LM-FBS) method, which works well in the receptance space and allows for direct implementation of (virtual point transformed) test-based results. This makes it the preferred DS method when working with models obtained by measurement.

Considering Eq. (8.1), a Boolean matrix $\mathbf{B}$ can be defined to enforce equilibrium between the systems, such that $\mathbf{Bu} = \mathbf{0}$. The interface forces can be now solved, resulting in the LM-FBS notation of the coupled system as denoted by Eq. (8.3).

$$\mathbf{u} = \mathbf{Y}\mathbf{f} - \mathbf{Y}\mathbf{B}^T\left(\mathbf{B}\mathbf{Y}\mathbf{B}^T\right)^{-1}\mathbf{B}\mathbf{Y}\mathbf{f} \quad \text{LM-FBS} \tag{8.3}$$

For more information on LM-FBS or the full derivation of equations, the reader is referred to [2].

8.2.3 *Source Characterization and TPA Using Blocked Forces*

We are considering an applied force on system A in the form of $\mathbf{f}_1^A$ and a response of interest on system B in the form of $\mathbf{u}_3^B$.

Blocked forces are a set of forces and moments that act on the interfaces of the active (A) and passive (B) subsystem, i.e. $\mathbf{f}_2$. They are able to represent the activity of the source system for what concerns responses on the passive side of the assembled system (AB). Blocked forces can be estimated for a source system mounted in its original assembly (or on a test-bench structure) using the in-situ method. Indicator responses $\mathbf{u}_4$ will be chosen such that the original excitation of the active system is fully observed and the matrix-inverse problem in Eq. (8.4) is overdetermined. By using the assembly receptance$\left(\mathbf{Y}_{42}^{\mathrm{AB}}\right)$ for forces on the interface DoFs ($\mathbf{f}_2$), to the indicators ($\mathbf{u}_4$), these blocked forces can be calculated. Next, using the assembly receptance to the response of interest, a response can be synthesized from these blocked forces according to Eq. (8.4).

$$\mathbf{f}_2^{\mathrm{bl}} = \left(\mathbf{Y}_{42}^{\mathrm{AB}}\right)^+ \mathbf{u}_4 \Rightarrow \mathbf{u}_3 = \mathbf{Y}_{32}^{\mathrm{AB}}\mathbf{f}_2^{\mathrm{bl}} \quad \text{BF-TPA} \tag{8.4}$$

For more information on source characterization and transfer path analysis, the reader is referred to [3].

8.3 Application

One objective of this project has been to obtain accurate component models in a broad frequency range (0–3 kHz). In order to do so a variety of approaches is used:

- Test-based modeling: An impact measurement in combination with the Virtual Point transformation can give an accurate component model compatible with other components.
- Test-based modeling incl. Transmission Simulators: In some cases, particular directions (DoFs) cannot be excited by impact measurement due to geometrical constraints. A transmission simulator can be used to regain access to these directions. Decoupling these Transmission Simulators from the component model is done by means of Dynamic Substructuring.
- Modal synthesis from measurement: In cases where it is desired to gain a better understanding of a structure (spatially) it can be beneficial to perform a modal analysis on the measurement and build the component FRF model through synthesis of the modes.

- Finite Element Modeling: Light-weight systems that consist of a homogeneous material (e.g. a single piece of milled aluminum) and thus show low damping properties can yield practical difficulties when measuring. When the geometry is straight-forward and the material properties are known, FEM can already produce an accurate model.
- System Equivalent Model Mixing [4]: In the case where a good spatial understanding is required, but a full modal measurement would be too extensive and the finite element model is not accurate enough, System Equivalent Model Mixing (SEMM) can provide an interesting alternative. SEMM leverages the spatial details of FEM and combines this with the frequency information from measurements on the most important nodes.

This high variety in component modeling methods allows to synthesize the full assembly model in many ways. These methods are compared per component, after which the most promising are coupled and validated against the measured full assembly model. Also, a variety of excitations is generated by the active vibration source mounted on the bridge. The characterization of this source is done by means of blocked forces (and free velocities), after which a response is synthesized on the assembly, which is again validated against the measured response on the full assembly model.

8.4 Discussion

In this presentation the above methods to obtain FRFs of components are compared and discussed. It is shown that each of the methods to obtain the component FRFs has clear (dis-)advantages depending on the characteristics of the component (e.g. mass, damping or the accessibility of coupling points). Secondly, the measured response signal on the point of interest due to the active vibration source is compared with predictions created from the individual, modular, building blocks which are combined through DS and TPA techniques. The quality of the resulting assembly model and synthesis is evaluated, with a discussion on the sources of discrepancies versus the validation. Finally, recommendations are given towards further industrial application of the used test-based methods.

References

1. van der Seijs, M.V. et al.: An improved methodology for the virtual point transformation of measured frequency response functions in dynamic substructuring. In: COMPDYN 2013 4th ECCOMAS Thematic Conference
2. de Klerk, D., et al.: General framework for dynamic substructuring: history, review, and classification of techniques. AIAA J. **46**(5), 1169–1181 (2008)
3. van der Seijs, M., et al.: General framework for transfer path analysis: history, theory and classification of techniques. Mech. Syst. Signal Process. **68–69**, 217–244 (2016)
4. Klaassen, S.W.B., et al.: System equivalent model mixing. Mech. Syst. Signal Process. **105**, 90–112 (2018)

Chapter 9
Development of a Low Cost Automatic Modal Hammer for Applications in Substructuring

Johannes Maierhofer, Ahmed El Mahmoudi, and Daniel J. Rixen

Abstract One key goal of Dynamic Substructuring (DS) is the coupling of measured components and simulated ones. This can be done using a frequency-based formulation of the system dynamics. For the experimental determination of the components dynamics, good and reliable measurements are extremely important to achieve correct results.

Usually the frequency response functions (FRF) are obtained using a modal hammer with a force sensor tip. Some of the problems that occur are that the excitation positions vary with every hit, that the angle is very hard to determine and furthermore that it is nearly impossible to bring the same energy into the system with every hit. This contribution gives a short motivation why the automatisation of modal analysis experiments could improve the method of experimental substructuring.

At the Chair of Applied Mechanics at TU-Munich, we developed a low-cost automatic modal hammer which is presented here. The whole device is positioned in front of the structure with a stand, so there is no need for readjustment in order to perform multiple impacts on the structure. The energy of the impacts can be adjusted by tuning the parameter settings of the automatic hammer.

The motion in the hammer is induced by an electromagnetic reluctance actuator. The principle is shown and a simple multi-body model set up in this paper. This model is used to tune parameters in a way to avoid double impacts and to predict the impact forces. The actuator is driven by some electronics with a microcontroller, whereby the acceleration time, voltage and the impulse series can be adjusted via PC. Consistence between the model and the real device are shown using a fully instrumented test rig. Furthermore, test series were carried out to prove repeatability.

Finally, a demonstration application on an academic structure is shown. Here the differences between classical modal analysis using a hand hammer and using the new automatic hammer are evident. Generally, the FRFs using the automatic hammer are less noisy and the coherence function is better.

Keywords Dynamic substructuring · Frequency based substructuring · Experimental substructuring · High quality FRF · Automatic modal hammer · AMimpact

9.1 Introduction

With the ever increasing effort to assure the quality and comfort of products, the simulation of and experimenting on full products has become more important than ever. One field of research is the combination of experimental evaluated subparts of a whole product with other simulated parts. A common technique is the coupling of the two parts in the frequency domain by a method called *Frequency Based Substructuring* (FBS). The following section gives a short insight into the FBS method and into the critical points in acquiring the necessary experimental data.

9.1.1 Frequency Based Substructuring

The substructuring technique considered in this paper is the Lagrange Multiplier Frequency Based Substructuring (LM-FBS) method. This method is suitable for both simulations and experiments. The coupling between the subsystems is based on calculated or measured frequency-dependent transfer functions, also known as frequency response function (FRF). This FRFs can be acquired by measuring the accelerations on a structure with a known force impact. The FRF generally refers

J. Maierhofer (✉) · A. El Mahmoudi · D. J. Rixen
Faculty of Mechanical Engineering, Technical University of Munich, Garching, Germany
e-mail: j.maierhofer@tum.de; a.mahmoudi@tum.de; rixen@tum.de

A. Linderholt et al., *Dynamic Substructures, Volume 4*, Conference Proceedings of the Society for Experimental Mechanics Series,
https://doi.org/10.1007/978-3-030-12184-6_9

to the ratio between output and input. In the context of experimental substructuring, for example, a FRF can be acquired by measuring an acceleration on a structure (output) with a known force (input).

It is common to use the admittance matrix $\boldsymbol{Y}^{(s)}$ to characterize the substructure (n), which includes the FRFs and is defined as the inverse of the dynamic stiffness matrix $\boldsymbol{Z}^{(s)}$ as shown in Eq. (9.1).

$$\boldsymbol{Y}^{(s)} = \boldsymbol{Z}^{(s)^{-1}} = \left(\boldsymbol{K}^{(s)} + \boldsymbol{C}(\boldsymbol{i}\boldsymbol{\omega})^{(s)} - \omega^2 \boldsymbol{M}^{(s)}\right)^{-1} \tag{9.1}$$

The dynamic stiffness matrix $\boldsymbol{Z}$ is composed of the mass matrix $\boldsymbol{M}$, damping matrix $\boldsymbol{C}$ and stiffness matrix $\boldsymbol{K}$. If the FRFs of the individual substructures were determined experimentally or simulatively, an analytical coupling according to the LM-FBS can be performed to obtain the assembled admittance $\boldsymbol{Y}_{\text{assembled}}$ of the total system.

$$\boldsymbol{u}_{\text{assembled}} = \overbrace{\boldsymbol{Y}\boldsymbol{f}}^{\boldsymbol{u}_{\text{uncoupled}}} \overbrace{- \boldsymbol{Y}\boldsymbol{B}^T \underbrace{(\boldsymbol{B}\boldsymbol{Y}\boldsymbol{B}^T)^{-1}}_{\text{int.flex.}} \underbrace{\boldsymbol{B}\boldsymbol{Y}\boldsymbol{f}}_{\Delta \boldsymbol{u}}}^{\boldsymbol{u}_{\text{coupling}}} = \boldsymbol{Y}_{\text{assembled}}\boldsymbol{f} \tag{9.2}$$

Equation (9.2) can be interpreted as the assembled response of a structure, which is a combination of the uncoupled system response $\boldsymbol{u}_{\text{uncoupled}}$ and coupling response $\boldsymbol{u}_{\text{coupling}}$, produced by the internal interface forces to keep the substructures together while be excited by the external forces. For a detailed derivation of the assembled admittance in the course of the LM-FBS procedure, see the corresponding literature [7].

9.1.2 Motivation for an Automated FRF Measurement

There are two motivating key points for automating the impact measurements, especially the impacting of the structure. The first one comes right out of the FBS method, the second one lies in possible non-linearities of the structure.

The measured FRFs are in a form that is equivalent to an admittance as one gives in a force in frequency domain and get an acceleration (also in frequency domain). Looking at Eq. (9.2), it's observed that the admittance has to be inverted for the LM-FBS-method. This inversion is somewhat problematic for the experimental side. The admittance is defined in form of an accelerance. Especially when the acceleration is very low, for example at anti resonances, very small errors in the force direction or the position of the impact cause very high errors in the FRFs. The problem here is that the force sensor is only unidirectional. Hitting the structure at a wrong angle means that the whole force is set to the assumed direction even if only one part of the force is in exactly that direction. The second effect is that a manual hammering changes the impulse position every time in a range of a few millimeters. However, for very light structures this displacement changes the measured FRF by quite an amount. When averaging multiple runs, this causes bad FRFs. This effect can be seen very well with the coherence of the FRFs which normally should be near 1. Looking at anti-resonances, the coherence often drops when using a classical manual impulse hammer.

If the structure has some nonlinearities which may come e.g. from mountings, material damping or anything else, there is a different answer to different levels of excitation force. Therefore, when measuring multiple times to make some averaging, it is necessary to excite the structure always with the same peak force. Manually it seems nearly impossible to impact lightweight structures in a free-free configuration multiple times with approximately the same force.

9.2 Existing Technology

The relevant literature seems to show some movement on the topic over recent years. This chapter gives an overview of the existing automated modal hammer systems and how they work. There are two fundamental categories of automated hammer systems. The linear acting and the rotating systems.

Linear Actuator Hammers The device in Fig. 9.1a built by the authors in [8] uses a solenoid to generate the impact. As this hammer stands on the structure, it is only suitable for very big structures, as the added mass influences the dynamic of the structure. In [9] a milling machine is measured using a solenoid on a special frame as seen in Fig. 9.1b. The AS-1220 Automated Impact Hammer depicted in Fig. 9.1c is sold by the company Alta Solutions. The device is built for automated testing in production lines, and is mainly intended for acoustic measurements [1].

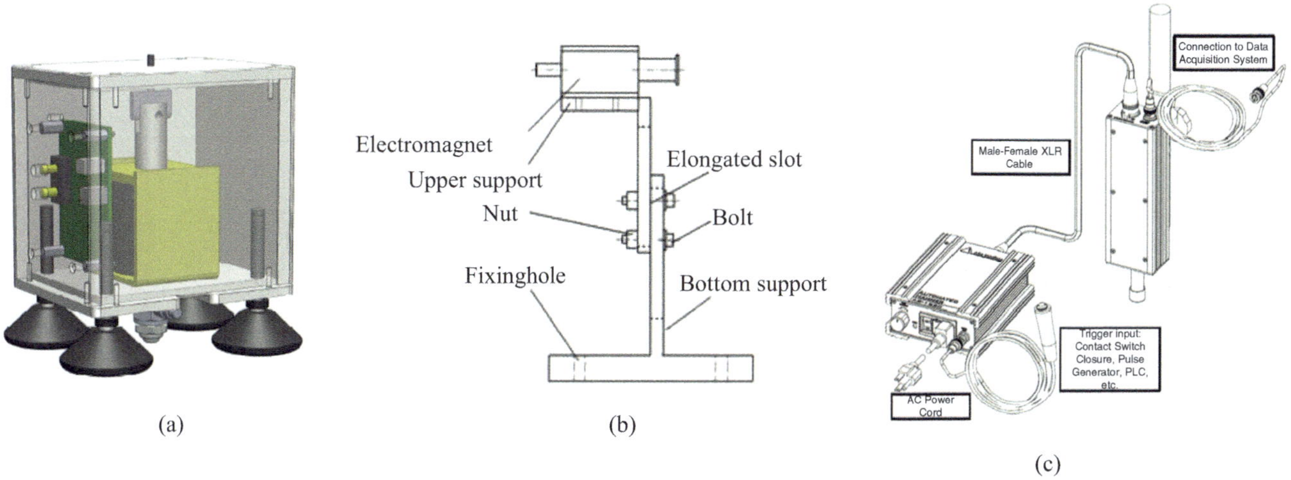

Fig. 9.1 Three different modal hammers with linear actuation. (**a**) Bridge hammer [8]. (**b**) Measuring a milling machine [9]. (**c**) AS1220 [1]

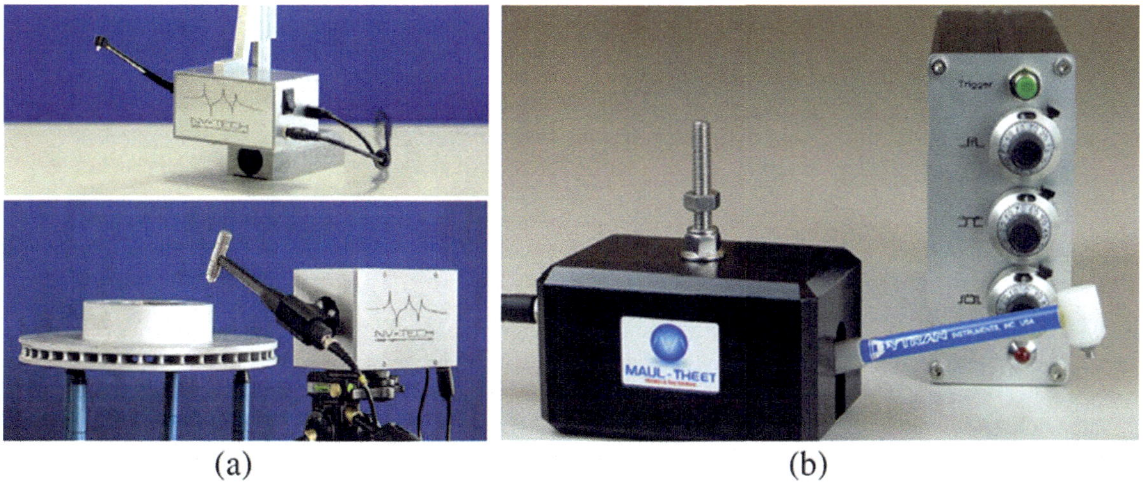

Fig. 9.2 Two different commercial modal hammers with rotational actuation. (**a**) NV-Tech Design—SAM [5]. (**b**) Maul-Theet—vImpact20 [2]

Beam-Bending Hammers The second type of hammer works in principle by bending a beam and using either the inertia force or the resetting spring force to generate an impact on the structure. The *Scalable Automatic Modal Hammer (SAM)* in Fig. 9.2a is produced by the company *NV Tech Design GmbH* and is specialized to characterize nonlinear structures. The device comes with software and claims very high reproducibility [5]. The working principle is to stop the elastic hammer shaft shortly before hitting the structure. Due to the inertia, the tip of the shaft bends through and touches the surfaces, i.e. the desired impact. The *vImpact* series from *Maul Theet* work with the same principal and are also available in different sizes. They are not controlled via a PC but with an external controller which allows the different settings [2]. The third hammer is described in [3] and works in a slightly different manner. The beam gets deflected using an electro magnet. The impact is then generated by the pullback through the resetting spring force and the following motion because of the mass inertia.

Rotating Hammer The last hammer presented here works passively. The tip is mounted on a pendulum which gets displaced and then released. Using gravity the pendulum accelerates and hits the structure. This is a very simple construction but is limited in the orientation of the working direction [6] (Fig. 9.3).

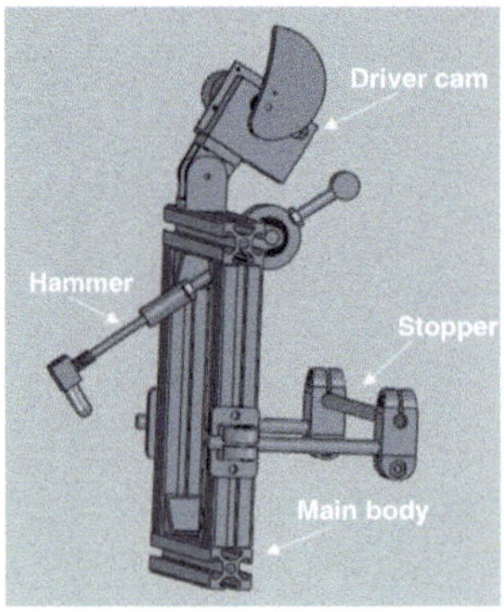

(a)

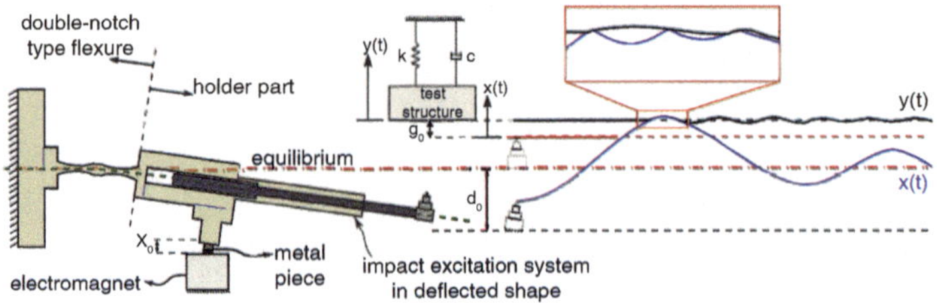

(b)

Fig. 9.3 Further principles of automated modal hammers. (**a**) Pendulum hammer [6]. (**b**) Beam deflection hammer [3]

9.3 AMimpact

All available automated impact hammers have one drawback: They are very high priced and very big. Therefore, they are quite cumbersome to mount in a way to excite the structure as desired. This becomes particularly obvious when working on structures on a whole car. The challenge was to set up a simple mechatronic system to fulfill the need for a handy and yet repeatable way of impacting light structures. After some calculations and tests the **AMimpact** was developed.

9.3.1 Description of the System

The whole system of the *AMimpact* is packed into a light 3D printed housing, which can be mounted on any available tripod. In order to use it, only a power source and a PC connected to the USB port are required. With a little software the user can set up all the possible parameters, which are explained later, and can even save them as presets. In Fig. 9.4a the finished system is shown, whereas Fig. 9.4b depicts the principle scheme of the AMimpact.

The core functionality comes from a magnetic linear actuator, which works on the principle of reluctance forces. As current runs through the coils, the bolt receives a forward driving force. On the front side of the bolt, a common piezo force sensor (Head of Model 086E80, PCB) is mounted. A spring is attached to the backside of the bolt and connected with the housing to reset the position of the bolt. The key idea is to switch off the current just the moment before the impact, so that the bolt flies to the structure ballistically. Due to the repulsion, double impacts are nearly impossible.

(a)

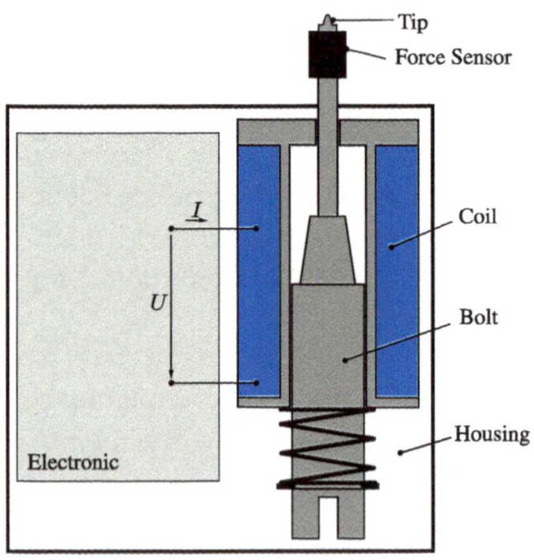

(b)

Fig. 9.4 The automated modal hammer of the Chair of Applied Mechanics, called AMimpact. (**a**) Picture of AMimpact. (**b**) Diagram of system

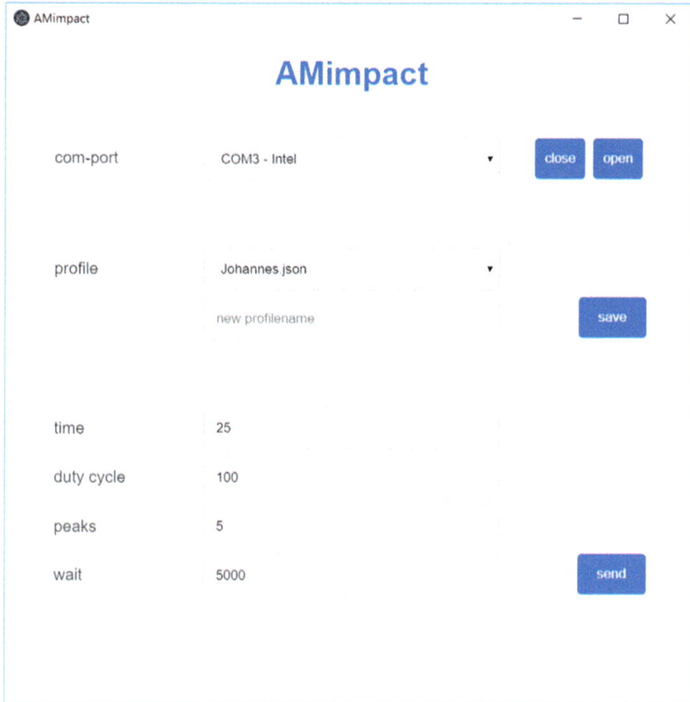

Fig. 9.5 Software for AMimpact

9.3.2 *Description of the Parameters*

The software Fig. 9.5 is platform-independent and so can run on macOS, Linux and Windows. The user has to select the port for communication. In the section *profile*, he can load or create his own profile of settings.

There are four parameters to set for the AMimpact. The parameter *time* describes how long the solenoid is under current. The *duty cycle* enables a chopping of the on-time with a pwm signal to lower the effective voltage, so as to lower the current in the coil. The number of impacts during one session can be set via *peaks*. In correlation with the number of peaks the *wait* time is the duration between each peak. All these parameters are used to set up the simulation in the following sections.

9.3.3 *Dynamics Model of AMimpact*

To precalculate the system properties and to estimate the impact forces, a dynamics model of the AMimpact is set up [4]. The minimal model of the mechatronic system is shown in Fig. 9.6. The system can be divided in two different states:

1. The tip of the hammer is between housing and structure
2. The tip touches the structure

As can be seen in Fig. 9.6, there is only one moving mass in the mechanism. This is the core of the solenoid magnet. The electric circuit is modeled with a voltage source, a resistor and an inductivity, which is dependent on the position of the shaft. Using this model the current can be calculated. The current results in a magnetic field that interacts with the iron core in the form of a reluctance force. The spring attached to the bolt generates the resetting backwards force.

Additionally a friction force is applied to the bolt. As no further information on the materials are available, a simple Coulomb friction is assumed with equal coefficients for static and sliding friction. To approximate the Coulomb friction law using a continuous function, the following equation is used.

$$F_R = \frac{2}{\pi} F_N \arctan(\frac{v}{\hat{v}}) \tag{9.3}$$

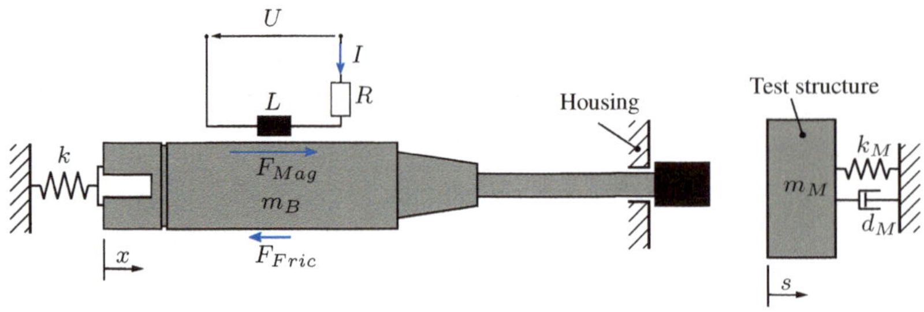

Fig. 9.6 Model of the AMimpact without contact [4]

Fig. 9.7 Model of the AMimpact with contact [4]

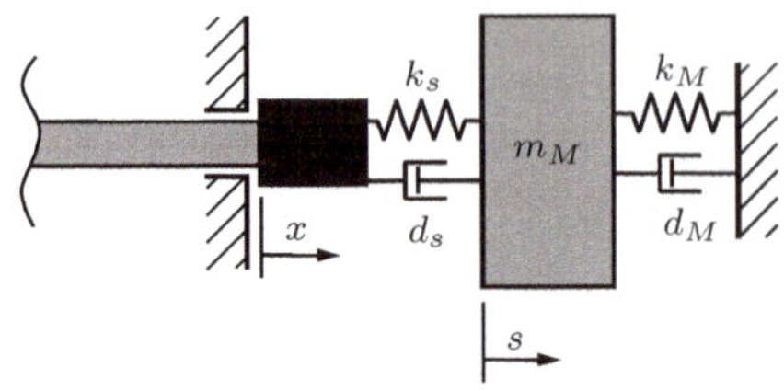

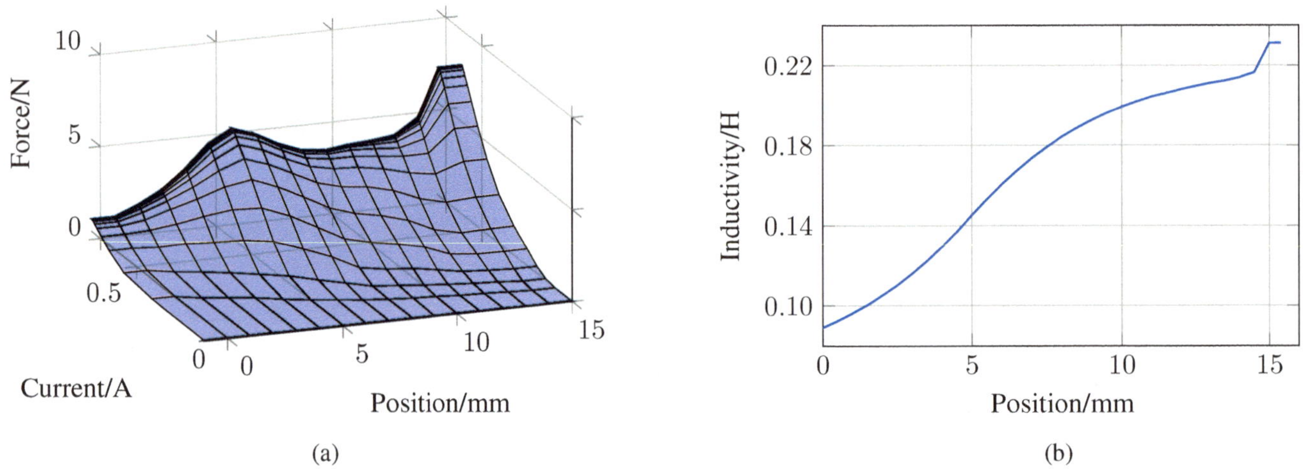

Fig. 9.8 Forces and inductivity for different positions. (**a**) Force map over position and current. (**b**) Inductivity over position

Here F_N describes the force normal to the surface, v is the relative velocity and $\hat{v}$ is a characteristic velocity that is much smaller than the relative velocity [10].

When the tip is in contact with the structure, Fig. 9.7 is valid. The test structure modeled here is a simple one degree of freedom (dof) spring-damper-mass system. The values for this are the modal values of the first eigenmode of the academic test structure presented later. The contact itself is approximated using the Hertzian theory and therefore also modeled with a spring-damper element. For the Hertzian theory, the following case is chosen: A sphere, representing the tip of the force sensor, is in contact with a plate, which represents the test structure. The pressure distribution is therefore a circle. To calculate the stiffness parameter the actual impact force is already needed to compute the intrusion of the tip into the structure. We insert an approximation for the force into the equation, calculate the stiffness and then the actual force. When the difference to the approximation is lower than the tolerance, the result is deemed to be valid.

9.4 Simulation of AMimpact

To perform the time integration of the modeled system, the parameters of the systems are needed.

As the force of the actuator is a function of the current in the coil, the material behavior, and the bolt position, the used solenoid was measured using a special test rig. In Fig. 9.8 the results are shown. It is obvious that the interrelations are very nonlinear. These measured forces are directly implemented to the simulation as characteristic maps. Also, the inductivity

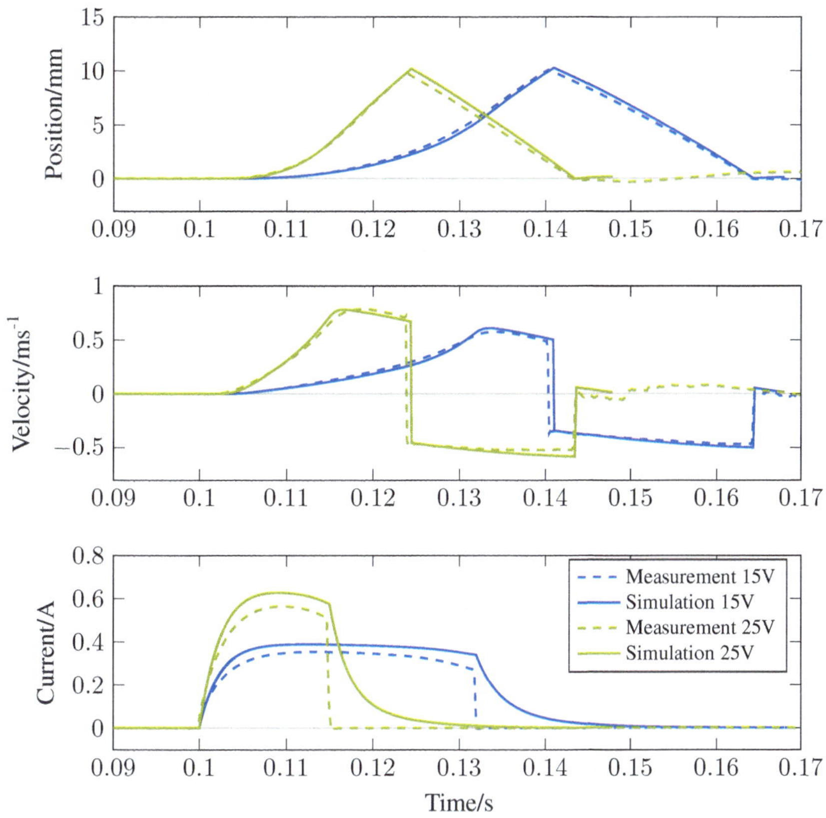

Fig. 9.9 Position, velocity and current for constant distance and varying voltage

depends on the position of the shaft in the actuator. The values are also taken experimentally. The result in Fig. 9.8b shows a quite linear relation between position and inductivity.

Putting all this information into the differential equations and performing time integration results in the following plots Fig. 9.9 for a constant distance between AMimpact and the test structure. Two driving voltages (15 V and 25 V) are considered and shown in green and blue. Also the whole AMimpact was instrumented with a laservibrometer, an accelerometer, the force sensor at the tip and a current sensor. The dashed lines represent the measured values which correspond quite well with the simulated ones. The current at the end of the active part seems to differ between simulation and measurement. The reason is that the measurement is taken before the flyback diode. This is why the measured current goes to zero immediately as the current runs through the diode. The flyback diode is connected in parallel with the coil, since an inductor cannot change it's current instantly. So, the flyback diode provides a path for the current when the coil is switched off. Otherwise, a voltage spike will occur causing damage to the switching transistors. All in all, it can be seen that the chosen model is quite accurate.

The plot in Fig. 9.10 shows the forces simulated and measured for hitting the test structure. The simulated force is slightly higher. The reason for this could lie in the material and friction parameters.

9.5 Results on Frequency Based Substructuring

The following results were obtained during a semester thesis [11] and are also discussed in further detail in [12]. Two different impact devices are taken into account for exciting the test structure depicted in Fig. 9.11: a manual impact hammer and the automatic hammer. The first one is a regular modal hammer (type: Model 086E80, PCB), while the second one is a prototype of the AMimpact hammer. The advantages and disadvantages of both devices are summarized in Table 9.1.

A simple academic beam as depicted in Fig. 9.11 is used as the test structure. The structure is made of aluminum. The L-shaped beam is monolithic to avoid friction effects. The beam is welded to the base plate, which can be screwed to ground.

Using both hammers—manual and automatic—an impact is applied to the interface, orthogonally to the structure's surface. Two different hammer tips are available for testing: a plastic and a steel one. Here, only the results for the steel

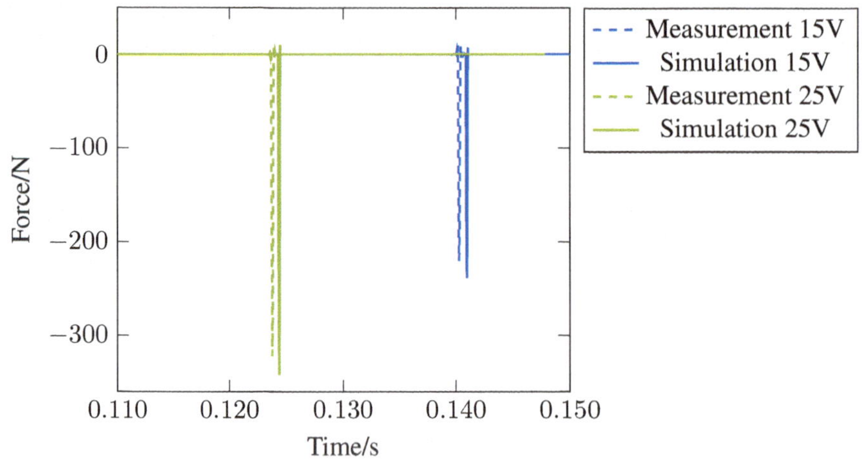

Fig. 9.10 Force over time for a constant distance and varying voltage

Fig. 9.11 Academic test structure

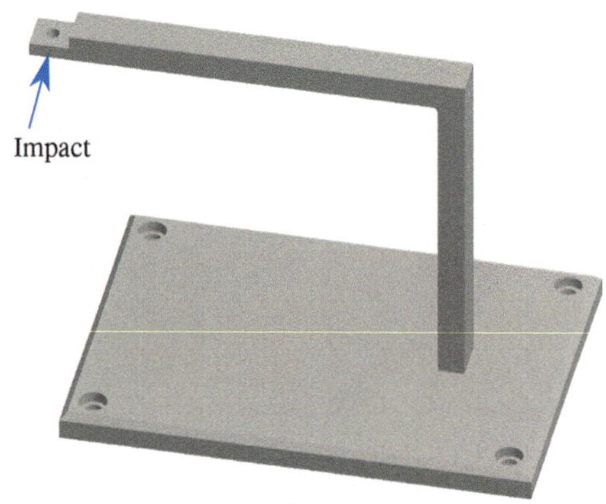

Table 9.1 Properties of impact devices

Type:	Manual hammer	AMimpact
Accuracy	−	++
Repeatability	−	++
User-friendliness	++	−
Size	+	−−
Time-consuming	+	−

tip are shown. In Fig. 9.13, the input spectrum and the FRFs results for two example points are shown. One point is the driving point, the other point is a non driving point. The Power Spectral Density (PSD) of the input for the automatic hammer appears clean and flat: a drop of 4 dB up to 800 Hz is observed. Conversely, the input spectrum of the manual hammer appears distorted due to some multiple hits issues. It is though possible to achieve a good input spectrum with the manual hammer. But most of the time the spectrum looks like shown in Fig. 9.12. For linear cases that input spectrum would be no big problem but for nonlinearities it is nearly impossible to generate reliable FRFs. The FRF in the automatic hammer case is generally less noisy over the entire frequency bandwidth. A more pronounced difference between the two cases is identified at very low frequencies, as highlighted by the coherence function. The distorted results of the non-driving point FRF and related coherence for the manual hammer are explained by the lower accuracy in the direction of the impact. After 800 Hz the signal is distorted by the anti-aliasing filter applied by the data acquisition system [11]. The AMimpact shows an optimal PSD of the impact (in terms of energy, flatness and cleanness of the spectrum) and appears to be much better than the manual hammer in the accuracy of the stroke (with regard position, direction and repeatability) (Fig. 9.13).

Observations that can be made with regard to difference between the plastic and the steel tip are mentioned briefly here. In both cases the input signal is flat and clean: a drop of 4 dB up to 800 Hz for plastic and 10 dB up to 3200 Hz for steel are seen. The plastic tip provides more energy at the lower frequencies, while the steel tip spreads the energy over a wider range.

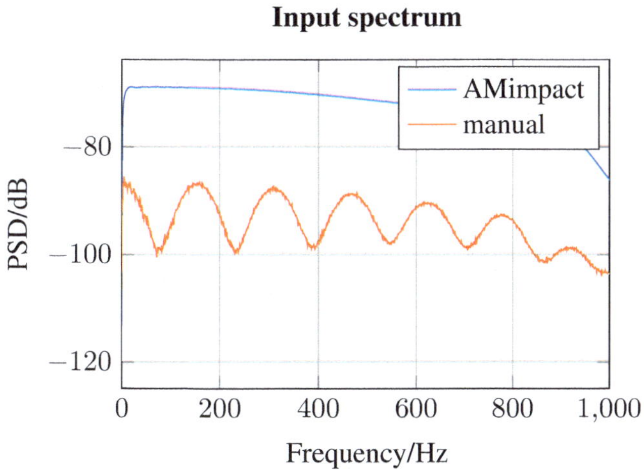

Fig. 9.12 Input spectrum of impact performed with manual hammer and AMimpact

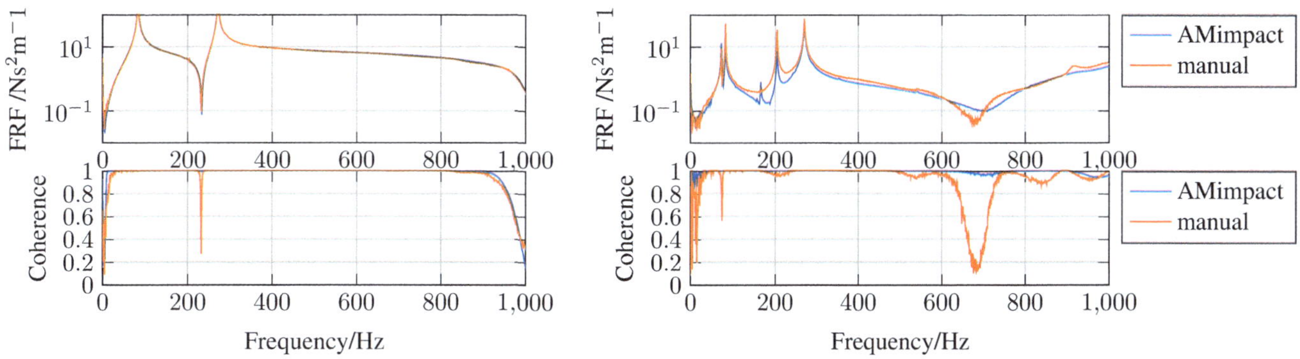

Fig. 9.13 Comparison between impacts performed with manual hammer and AMimpact. (**a**) Driving point FRF. (**b**) Non-driving point FRF

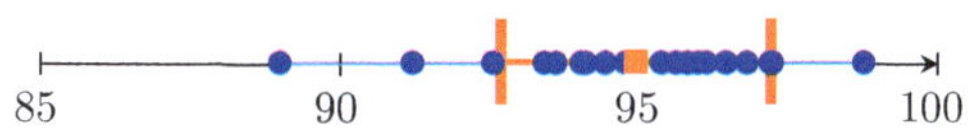

Fig. 9.14 Repeatability for one set of parameters using AMimpact

In the lower frequency range the quality of the FRF and coherence is higher for the plastic tip, especially in non-driving point directions. The results for the driving point are nearly identical. Therefore the use of a steel tip is necessary in order to increase the stiffness of the contact surfaces, thus exciting a broader range of frequencies. However, in the very low frequency range the use of a softer tip may be useful.

Tests regarding the repeatability show a good quality of the impulses. Therefore, a series of impacts are measured with different settings. The results for one series with 20 impulses on the test structure are shown in Fig. 9.14. The driving voltage is 15 V, the duration of the on-time 20 ms, the time between each impulse 3 s and the distance to the structure 2.5 mm. The mean excitation force on the structure results as 95 N. Most of the measurements are very close to that mean value. The standard deviation is 2.8 N, which is quite good. One explanation for the deviation could be the stick-slip effect between the bolt and its housing. The later the bolt gets into motion, the less is the kinetic energy which results in a lower force.

9.6 Conclusion

The presented paper shows why automated modal impulse hammers are a useful tool for improving measurements of FRFs in structural dynamics. One particular advantage comes into play when the hammer is combined with a laser scanner. As there is the need to scan many points with the same impact settings, the automation saves a significant amount of time. The laser can jump from point to point autonomous using software, and the AMimpact gives the same impact repeatedly without the need for any human interaction. Further work could include some optimization calculations to find the best impact settings regarding different quality functions. The simulation shows that multiple parameter sets may exist which lead qualitatively to the same impact force. Therefore one has to choose between these sets with regard to robustness of the impact and other outer constraints. Another future aspect will be the further development of the hardware, to increase usability and minimize the footprint of the AMimpact in reaching even more points in very complex geometries.

References

1. AS-1220 Automated Impact Hammer. Alta Solutions, Poway, CA (2013)
2. Automatischer Modalhammer — vImpact-20. Maul-Theet, Berlin
3. Bediz, B., Korkmaz, E., Ozdoganlar, O.B.: An impact excitation system for repeatable, high-bandwidth modal testing of miniature structures. J. Sound Vib. **333**(13), 2743–2761 (2014). https://doi.org/10.1016/j.jsv.2014.02.022
4. Bernhofer, T.: Mehrkörpersimulation eines automatischen Impulshammers. Bachelorthesis. Technische Universität München (2018)
5. Blaschke, P., Schneider, S., Kamenzky, R., Alarcón, D.J.: Non-linearity Identification of Composite Materials by Scalable Impact Modal Testing, pp. 7–14. Springer, New York (2017). https://doi.org/10.1007/978-3-319-54987-3_2
6. Brüggemann, T., Biermann, D., Zabel, A.: Development of an automatic modal pendulum for the measurement of frequency responses for the calculation of stability charts. Proc. CIRP **33**, 587–592 (2015). https://doi.org/10.1016/j.procir.2015.06.090
7. de Klerk, D., Rixen, D.J., de Jong, J.: The frequency based substructuring (FBS) method reformulated according to the dual domain decomposition method. In: 24th International Modal Analysis Conference, St.Louis, MO (2006)
8. Norman, P.E., Jung, G., Ratcliffe, C., Crane, R., Davis, C.: Development of an Automated Impact Hammer for Modal Analysis of Structures, September 2018. https://www.researchgate.net/publication/266278883_Development_of_an_Automated_Impact_Hammer_for_Modal_Analysis_of_Structures
9. Ning Liu, L., Guang Zhang, Y., Shi, Z., Zhanqiang, L.: Development of Electronic Impact Hammer and Its Application to Face Milling Cutter Modal Analysis, September 2013, vol. 797, pp. 585–591. https://doi.org/10.4028/www.scientific.net/AMR.797.585
10. Popov, V. Kontaktmechanik und Reibung: Von der Nanotribologie bis zur Erdbebendynamik. Springer, Berlin, Heidelberg (2016). ISBN: 9783662459751. https://doi.org/10.1007/978-3-662-45975-1
11. Trainotti, F.: Development of a proper FRF acquisition procedure for Experimental Dynamic Substructuring. Semester thesis. Technical University of Munich (2018)
12. Trainotti, F., Berninger, T.F.C., Rixen, D.J.: Use of laser vibrometry for precise FRF measurements in experimental substructuring. In: Proceedings of the 37th IMAC, A Conference and Exposition on Structural Dynamics (2019)

Chapter 10
Using SEMM to Identify the Joint Dynamics in Multiple Degrees of Freedom Without Measuring Interfaces

S. W. B. Klaassen and D. J. Rixen

Abstract As the number of models created in a modular fashion increase, the need for accurate identification of real joint dynamics rises. Since joint dynamics are a consequence of component-to-component interaction, they are only present in the assembled state. Yet, it is in the assembled state that measuring the interface degrees of freedom is practically infeasible. Nevertheless, the effects of the joint are present in measurements throughout the component, i.e. the joint dynamics are observable. In this work, system equivalent model mixing is used to expand an experimental measurement with interface degrees of freedom—either rotational or translational—extracted from a numerical model. Subsequently, joint dynamics can be obtained by applying classic frequency based decoupling methods. The strength of this method lies in the ability to test different interface configurations from a single measurement campaign, limited only by the actual number of sensor or impact locations. The paper shows that an updating scheme can be used to identify joint dynamics without directly measuring interfaces.

Keywords System equivalent model mixing · Interfaces · Joint identification · Optimization

10.1 Introduction

While the influence of joints are not always prevalent, they are one of a structure's main sources of uncertainty and unreproducibility. This is in part due to the lack of understanding of joint mechanisms which essentially result in *unidentified* dynamic influence. Although joints are often mistreated as an unfortunate consequences of today's assembly-based manufacturing, they are sometimes beneficial and in fact necessary: e.g. In turbine systems, the damping caused by friction contacts may help reduce the resonance amplitudes (which allows for smaller gaps between turbine blades and casing) and reduces the possibility of instability [1]. Identifying different joints will advance the understanding of joint mechanisms, which in turn allows us to steer towards these beneficial joint effects, rather than the unfavourable uncertainty.

In order to identify the joint, it must be observed. This is classically done by measuring physical effects or dynamic responses at the interfaces of the assembled system. Yet, in this assembled state it is highly impractical to perform measurements directly at, or even near the joint. To work around this problem a simplification is often introduced; common examples of such simplifications are reducing the DoF of the joint to a number and direction that can be measured, or by extrapolating the effects observed in the measurable areas (such as the sides of the assembly) over the total (unmeasurable) area of the interface.

Fortunately, the dynamic effect of the joint is observed in other responses (not on the interface) combined with the effects of the sub-components themselves. Therefore, if the effects of the components are removed, one can single out the joint dynamics. In [2, 3], joint identification is done by assuming a two-DoF joint[1] at the interface and using substructure decoupling to remove the effects of the known components from the assembled system. Even so, information is needed at the joint itself which requires impact locations and sensors to be placed there, referring back to the problem stated above.

If, however, the DoF at the interface are not measured but calculated by means of an expansion method, these limitations no longer hold. System Equivalent Model Mixing (SEMM) is a method based on frequency based substructuring that can be

[1]The deflection and rotation in a beam.

S. W. B. Klaassen (✉) · D. J. Rixen
Technische Universität München, Garching, Germany
e-mail: steven.klaassen@tum.de

A. Linderholt et al., *Dynamic Substructures, Volume 4*, Conference Proceedings of the Society for Experimental Mechanics Series,
https://doi.org/10.1007/978-3-030-12184-6_10

used to expand a measurement's DoF-set by coupling the measurement-based model to an equivalent—yet not identical—model with the required boundary DoF [4]. Note that this model only needs to have the required DoF, and not the correct joint dynamics (since these are provided by the measurement).

10.1.1 Outline of the Paper

In this work, SEMM is used to expand the DoF-set of assembled-system measurements to include the boundary DoF required to identify the joint. In Sect. 10.2 the theory will be explained: The theory starts with Sect. 10.2.1 which covers the Lagrange Multiplier Frequency Based Substructuring (LM-FBS) method, including weakly-formulated interface problems in order to lay-down the basis for SEMM as well as explain the additional steps required to perform joint identification. Once LM-FBS is covered, the SEMM method will be highlighted in Sect. 10.2.2. In Sect. 10.2.3, an optimisation scheme is introduced which deals with the inherent errors of SEMM expansion.

A proof of concept is given in Sect. 10.3 where a simple numerical case is presented to showcase the abilities and limitations of the method. Finally, in Sects. 10.4 and 10.5, a critical discussion and conclusion are provided alongside a prognosis for work induced by this concept.

10.2 The Joint Identification Method

A multitude of joint-types exist and thus the term is used broadly. For the context of this paper however, a joint is simply a cause of additional dynamics which exists only when two substructures are coupled. Therefore, where others would differentiate between bolts, welds, adhesives, contact-friction, etc. in this work a joint encompasses all. This is because the method discussed in this paper assumes that the *joint* is a black-box, and by removing the *known* component dynamics from the system dynamics this black-box is identified. It is important to note that the linearity assumption of the frequency based methods applies to the joint as well, and thus only linear (or linearised) properties of the joint can be identified.

Additionally, an important distinction is made between the joint and the interface. A joint is an actual source of dynamics whereas the interfaces are the sets of degrees of freedom on each side of the sub-component between which the joint acts. For example, in Fig. 10.1 a schematic of a standard (read: rigid) coupling problem is depicted. The interfaces are described by the red markers and the joint (in this case a rigid link) acts between the interfaces.

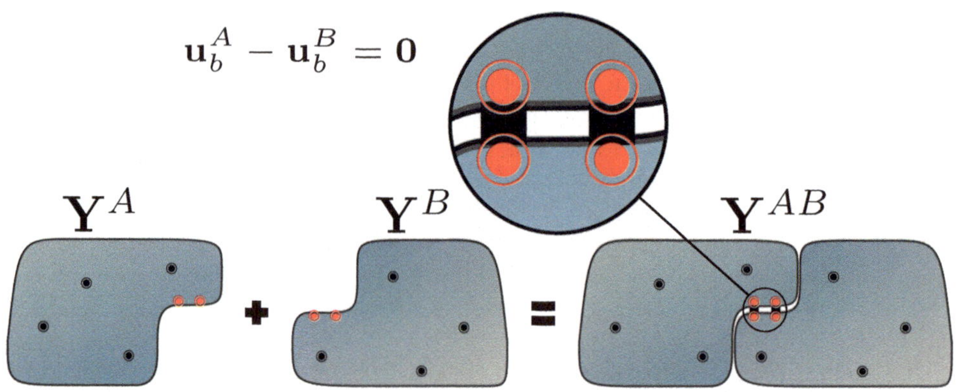

Fig. 10.1 Component A is connected to component B via a rigid connection. The black markers represent the internal DoF $\mathbf{u}_i$ and the red markers represent the boundary DoF $\mathbf{u}_b$. Due to the rigid connection the compatibility condition states that $\mathbf{u}_b^A - \mathbf{u}_b^B = \mathbf{0}$

10.2.1 *Frequency Based Substructuring: Coupling and Decoupling with a Weak-Formulated Joint*

Frequency based substructuring (FBS), or more specifically Lagrange Multiplier Frequency Based Substructuring (LM-FBS) is a dual-method that operates in the frequency domain and in admittance space; it therefore allows the user to directly implement measured Frequency Response Functions (FRF) [5]. Considering that performing measurements is the only way to identify the *unknown* joint dynamics, it is advantageous to use methods that agree with *measurement data*, i.e. FRF.

The equation of motion in the frequency domain for a general dynamic component s is given as follows:

$$\mathbf{u}^s(\omega) = \mathbf{Y}^s(\omega)\left(\mathbf{f}^s(\omega) + \mathbf{g}^s(\omega)\right) \tag{10.1}$$

where $\mathbf{u}^s(\omega)$ are the responses to the (known) external forces $\mathbf{f}^s(\omega)$ and (unknown) internal forces $\mathbf{g}^s(\omega)$ through the admittance matrix $\mathbf{Y}^s(\omega)$. In Fig. 10.1, components A and B are shown, their equations of motion can be derived in the form of Eq. (10.1). The implicit dependency on the frequency ω is omitted here for clarity, and will be done so for the remainder of the paper.

$$\mathbf{u}^{\mathrm{A}} = \mathbf{Y}^{\mathrm{A}}\left(\mathbf{f}^{\mathrm{A}} + \mathbf{g}^{\mathrm{A}}\right) \tag{10.2}$$

$$\mathbf{u}^{\mathrm{B}} = \mathbf{Y}^{\mathrm{B}}\left(\mathbf{f}^{\mathrm{B}} + \mathbf{g}^{\mathrm{B}}\right) \tag{10.3}$$

These can then be combined into block-form:

$$\mathbf{u} = \mathbf{Y}(\mathbf{f} + \mathbf{g}) \quad \text{where}: \quad \mathbf{Y} = \begin{bmatrix} \mathbf{Y}^{\mathrm{A}} & \\ & \mathbf{Y}^{\mathrm{B}} \end{bmatrix}, \quad \mathbf{u} = \begin{bmatrix} \mathbf{u}^{\mathrm{A}} \\ \mathbf{u}^{\mathrm{B}} \end{bmatrix}, \quad \mathbf{f} = \begin{bmatrix} \mathbf{f}^{\mathrm{A}} \\ \mathbf{f}^{\mathrm{B}} \end{bmatrix}, \quad \mathbf{g} = \begin{bmatrix} \mathbf{g}^{\mathrm{A}} \\ \mathbf{g}^{\mathrm{B}} \end{bmatrix} \tag{10.4}$$

The models are coupled via the internal forces $\mathbf{g}$ which act only on the boundary DoF (c.f. the red markers in Fig. 10.1), i.e. $\mathbf{g}_i = \mathbf{0}$. In order to derive these internal forces two conditions are set on the overall system. The first is the compatibility condition which states that the responses on either interface are equal:

$$\mathbf{u}_b^{\mathrm{A}} - \mathbf{u}_b^{\mathrm{B}} = \mathbf{0} \quad \rightarrow \quad \mathbf{B}\mathbf{u} = \mathbf{0} \tag{10.5}$$

where $\mathbf{B}_{M\times N}$ is a signed Boolean matrix which denotes for the N total global DoF, the M equality conditions that need to be satisfied to impose compatibility on the interface. Note that M is now also the number of compatibility constraints placed on the system.

Next, the equilibrium condition ensures that a force equilibrium exists on the boundary. In other words, it states that the forces on one side of the interface, are equal but opposite to the forces on the other side of the interface. In LM-FBS, these forces are represented by a set of unknown Lagrange multipliers $\boldsymbol{\lambda}_{M\times 1}$. Note that it can be cast in matrix form using the same signed Boolean matrix $\mathbf{B}$.

$$\mathbf{g}_b^{\mathrm{A}} = -\mathbf{g}_b^{\mathrm{B}} = \boldsymbol{\lambda} \quad \rightarrow \quad \mathbf{B}^T\boldsymbol{\lambda} = -\mathbf{g} \tag{10.6}$$

Equation (10.6) is substituted into Eq. (10.4) which is then pre-multiplied by $\mathbf{B}$ to enforce compatibility:

$$\mathbf{B}\mathbf{u} = \mathbf{B}\mathbf{Y}\left(\mathbf{f} - \mathbf{B}^T\boldsymbol{\lambda}\right) = \mathbf{0} \tag{10.7}$$

This is then solved for the Lagrange multipliers (read: boundary forces), i.e. the forces needed to enforce the compatibility:

$$\boldsymbol{\lambda} = \left(\mathbf{B}\mathbf{Y}\mathbf{B}^T\right)^{-1}\mathbf{B}\mathbf{Y}\mathbf{f} \tag{10.8}$$

which in turn, are resubstituted into the equation of motion to derive the coupled responses using relation (10.6):

$$\mathbf{u} = \mathbf{Y}^{\mathrm{AB}}\mathbf{f} = \left[\mathbf{Y} - \mathbf{Y}\mathbf{B}^T\left(\mathbf{B}\mathbf{Y}\mathbf{B}^T\right)^{-1}\mathbf{B}\mathbf{Y}\right]\mathbf{f} \tag{10.9}$$

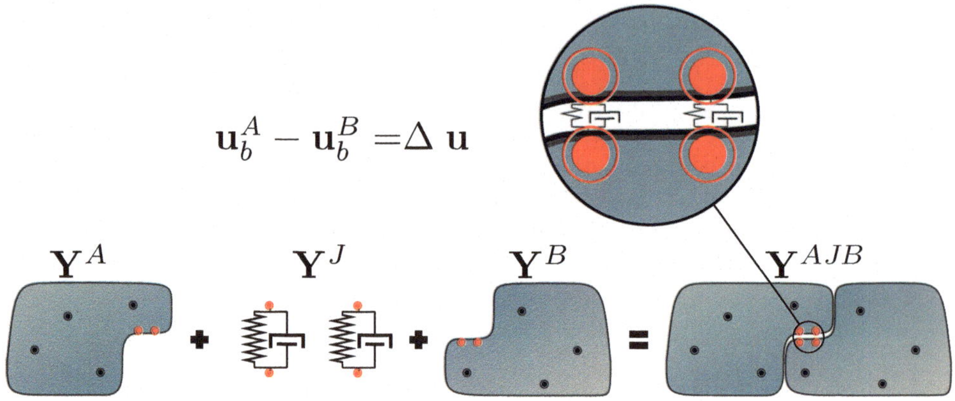

Fig. 10.2 Component A is connected to component B via a mass-less joint represented by $\mathbf{Y}^{\mathrm{J}}$. Due to the weakened connection the compatibility condition states that $\mathbf{u}_b^{\mathrm{A}} - \mathbf{u}_b^{\mathrm{B}} = \mathbf{\Delta u}$

The coupled admittance matrix $\mathbf{Y}^{\mathrm{AB}}$ is then:

$$\mathbf{Y}^{\mathrm{AB}} = \mathbf{Y} - \mathbf{Y}\mathbf{B}^T\left(\mathbf{B}\mathbf{Y}\mathbf{B}^T\right)^{-1}\mathbf{B}\mathbf{Y} \tag{10.10}$$

Equation (10.10) is a single-line equation of LM-FBS to couple models. Note that, although only two models were coupled in the presented example, the equation holds for multiple components.

10.2.1.1 Weakening in the Interface: Adding Joint Dynamics

Note that in the previous part the LM-FBS method is derived with strict compatibility and equilibrium between the components, and thus a rigid connection. In order to add a linear flexible joint one of two things can be done: Either the joint is added as a separate substructure into Eq. (10.10), which as explained before, is done easily. Or a joint is added as a relaxation of the compatibility condition between components A and B; this method is extensively described in [6] but will be shortly repeated here.

In Fig. 10.2 a flexible joint $\mathbf{Y}^{\mathrm{J}}$ is added between the interfaces of component A and B. Because of the joint, a gap can occur between the boundary DoF of the two components altering the compatibility condition from Eq. (10.5):

$$\mathbf{u}_b^{\mathrm{A}} - \mathbf{u}_b^{\mathrm{B}} = \mathbf{\Delta u} \quad \rightarrow \quad \mathbf{B}\mathbf{u} = \mathbf{\Delta u} \tag{10.11}$$

This gap is a response to the boundary forces $\boldsymbol{\lambda}$ which act on the joint:

$$\mathbf{\Delta u} = \mathbf{Y}^{\mathrm{J}}\boldsymbol{\lambda} \tag{10.12}$$

Substituting these relations into Eq. (10.4) results in:

$$\mathbf{B}\mathbf{u} = \mathbf{B}\mathbf{Y}\left(\mathbf{f} - \mathbf{B}^T\boldsymbol{\lambda}\right) = \mathbf{\Delta u} = \mathbf{Y}^{\mathrm{J}}\boldsymbol{\lambda} \tag{10.13}$$

which again, is solved for $\boldsymbol{\lambda}$:

$$\boldsymbol{\lambda} = \left(\mathbf{B}\mathbf{Y}\mathbf{B}^T + \mathbf{Y}^{\mathrm{J}}\right)^{-1}\mathbf{B}\mathbf{Y}\mathbf{f} \tag{10.14}$$

Similar to above, the weakly-coupled admittance matrix $\mathbf{Y}^{\mathrm{AJB}}$ is found to be:

$$\mathbf{Y}^{\mathrm{AJB}} = \mathbf{Y} - \mathbf{Y}\mathbf{B}^T\left(\mathbf{B}\mathbf{Y}\mathbf{B}^T + \mathbf{Y}^{\mathrm{J}}\right)^{-1}\mathbf{B}\mathbf{Y} \tag{10.15}$$

With Eq. (10.15) a coupled model can be created by weakening the compatibility condition. When including the joint in this manner, as compared to when including the joint as its own substructure, some extra assumptions are made:

- The joint is mass-less. This assumption is based on the fact that the equilibrium condition introduced in Eq. (10.6) remains unaltered. Therefore, the forces acting on the boundary DoF of component A and B are still equal but opposite *at all times*. Since this is no longer the case if a mass exists between the DoF (as this introduces counter-acting inertial forces), the joint is required to be mass-less. Alternatively, mass *could* be included beforehand by coupling (parts of) the joint-mass to either side of components A and B using standard LM-FBS.
- The joint model $\mathbf{Y}^{\mathrm{J}}_{M\times M}$ is constructed such that, when the boundary forces act on it, the response $\mathbf{\Delta u}$ is the difference in response between the M boundary DoF on component A and B. There is therefore no information pertaining to the relation between DoF on any one side, i.e. it is assumed that there exists no coupling between the DoF on *one* interface which might not always be the case.

These assumptions are recognized as similar to those found in the joint identification first introduced as inverse-substructuring, explained in detail in [7, 8]. These assumptions may, on first glance, limit the scope of joints that one might be interested in. However, both assumptions are valid for lightweight and small joint (such as friction contacts, glued contacts, welds, or even bolts when applied to a large structure). In fact, the linearity assumption accompanied by frequency based methods (small displacements and rotations) may already exclude the effects of these cross-coupling terms since they are bound to be non-linear.

10.2.1.2 Decoupling the Components to Obtain the Joint

LM-FBS can be used to decouple components from full-systems as easily as it can be used to couple components to systems. It can be shown that decoupling is as easy as adding a negative model [5]. In this case specifically, the interest is in obtaining the joint dynamics from the system $\mathbf{Y}^{\mathrm{AJB}}$, therefore decoupling is done by simply reversing Eq. (10.15). Figure 10.3 shows the process, which is indeed the reverse of what is shown in Fig. 10.2. It can be shown that standard decoupling is analogous to this equation inversion.

This is done by first pre- and post-multiplying Eq. (10.15) by $\mathbf{B}$ and $\mathbf{B}^T$ respectively and then solving it for $\mathbf{Y}^{\mathrm{J}}$ to obtain (10.16).

$$\mathbf{Y}^{\mathrm{J}} = -\mathbf{BYB}^T - \mathbf{BYB}^T \left(\mathbf{B\Delta YB}^T\right)^{-1} \mathbf{BYB}^T \tag{10.16}$$

where $\mathbf{\Delta Y}$ is the difference in FRF between the uncoupled and coupled model:

$$\mathbf{\Delta Y} = \mathbf{Y}^{\mathrm{AJB}} - \mathbf{Y} \tag{10.17}$$

10.2.2 System Equivalent Model Mixing

It is shown in the previous sections that, with LM-FBS, components can be coupled with and without joints, and that a joint can be extracted from the system model. However, this still requires that the boundary DoF (c.f. the red markers in Fig. 10.3) are known. Measuring these DoF, especially in several directions and including rotational DoF, is often practically infeasible in the *assembled state*. So what if, alternatively, one can expand a measurable DoF-set to include boundary DoF? In this section an expansion method known as SEMM will be used to expand the measured internal DoF (c.f. the black markers in Fig. 10.3) to the required boundary DoF.

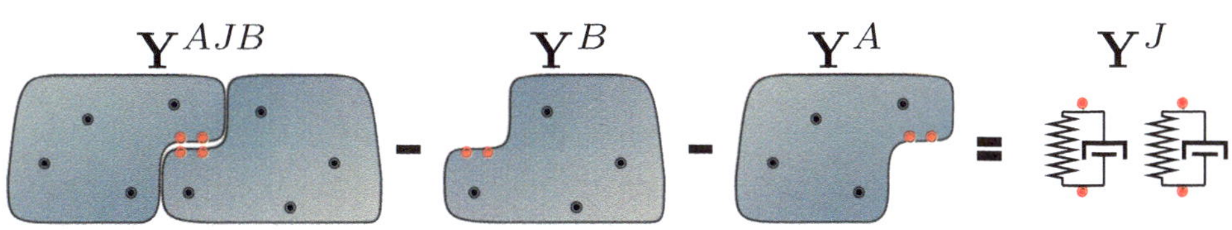

Fig. 10.3 The components $\mathbf{Y}^{\mathrm{A}}$ and $\mathbf{Y}^{\mathrm{B}}$ are decoupled from the full model $\mathbf{Y}^{\mathrm{AJB}}$. What is left is the dynamics of the joint represented by $\mathbf{Y}^{\mathrm{J}}$

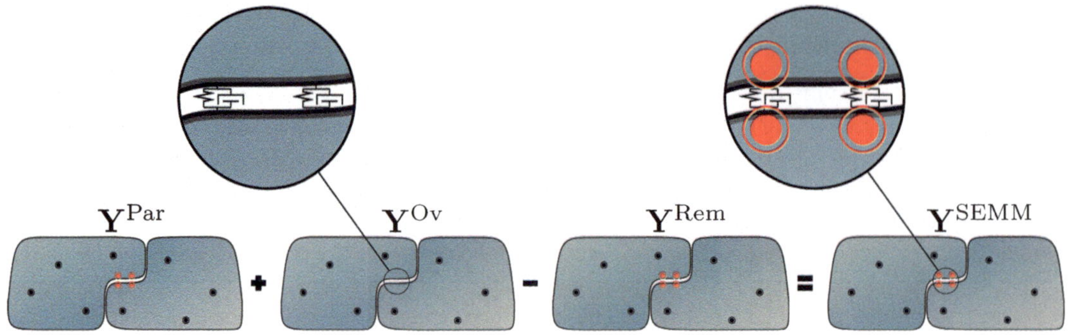

Fig. 10.4 The parent model and overlay model are coupled. The parent model $\mathbf{Y}^{\text{par}}$ contains all the DoF required but contains an erroneous joint. The overlay model $\mathbf{Y}^{\text{ov}}$ is a set of measurements of the system with the correct joint, but lacks the DoF required to identify this joint. From this, the removed model—in this case a copy of the parent model—is decoupled. The resulting SEMM model $\mathbf{Y}^{\text{SEMM}}$ has the *correct* joint (from the overlay model) and also has the required DoF to decouple (from the parent model)

System Equivalent Model Mixing (SEMM) is a method that uses substructuring to expand model dynamics contained in an *overlay* model $\mathbf{Y}^{\text{ov}}$ onto the DoF-space of a *parent* model $\mathbf{Y}^{\text{par}}$. In Fig. 10.4 the process is drawn schematically for component AB. Let us start by stating the equation of motion for the SEMM-system which directly follows from the schematic in Fig. 10.4.

$$\mathbf{u} = \mathbf{Y}(\mathbf{f} - \mathbf{g}), \quad \text{with} \quad \mathbf{Y} = \begin{bmatrix} \mathbf{Y}^{\text{par}} & & \\ & -\mathbf{Y}^{\text{rem}} & \\ & & \mathbf{Y}^{\text{ov}} \end{bmatrix}, \quad \mathbf{f} = \begin{bmatrix} \mathbf{f}^{\text{par}} \\ \mathbf{f}^{\text{rem}} \\ \mathbf{f}^{\text{ov}} \end{bmatrix}, \quad \mathbf{g} = \begin{bmatrix} \mathbf{g}^{\text{par}} \\ \mathbf{g}^{\text{rem}} \\ \mathbf{g}^{\text{ov}} \end{bmatrix} \tag{10.18}$$

Here, the models $\mathbf{Y}^{\text{par}}$, $\mathbf{Y}^{\text{rem}}$, $\mathbf{Y}^{\text{ov}}$ are the so-called parent, removed, and overlay models respectively. These are the building blocks for SEMM; they are derived below.

In order to apply the method in the joint-identification case we require:

1. FRF-based component models that contain the entire DoF-set including boundary DoF. These may be either numerical or experimental in nature (note: it may be possible to measure the boundary DoF in the *unassembled state*).
2. A set of measurements of an *assembled* full-system that observe the joint dynamics but do not have explicit boundary DoF.

Requirement one is the parent model: To compute the parent model the component models can either be left uncoupled (in block-diagonal form) or coupled with an *initial guess* joint model:

$$\mathbf{Y}^{\text{par}} = \left[\mathbf{Y}^{\text{AB}}_{gg}\right] = \begin{bmatrix} \mathbf{Y}^{\text{AB}}_{i_A i_A} & \mathbf{Y}^{\text{AB}}_{i_A i_B} & \mathbf{Y}^{\text{AB}}_{i_A b_A} & \mathbf{Y}^{\text{AB}}_{i_A b_B} \\ \mathbf{Y}^{\text{AB}}_{i_B i_A} & \mathbf{Y}^{\text{AB}}_{i_B i_B} & \mathbf{Y}^{\text{AB}}_{i_B b_A} & \mathbf{Y}^{\text{AB}}_{i_B b_B} \\ \mathbf{Y}^{\text{AB}}_{b_A i_A} & \mathbf{Y}^{\text{AB}}_{b_A i_B} & \mathbf{Y}^{\text{AB}}_{b_A b_A} & \mathbf{Y}^{\text{AB}}_{b_A b_B} \\ \mathbf{Y}^{\text{AB}}_{b_B i_A} & \mathbf{Y}^{\text{AB}}_{b_B i_B} & \mathbf{Y}^{\text{AB}}_{b_B b_A} & \mathbf{Y}^{\text{AB}}_{b_B b_B} \end{bmatrix} \tag{10.19}$$

where the (global) DoF-set g contains the boundary DoF-set b (c.f. the red markers in Fig. 10.4) and the internal DoF-set i (c.f. the black markers in Fig. 10.4) such that $g = \text{col}\left[i_A \; i_B \; b_A \; b_B\right]$ are all the DoF of component AB.

Note that if the joint is in fact rigid, then due to the compatibility equation (10.5) the third and fourth row, and third and fourth column are redundant since DoF $b_A = b_B$. It is important that the parent model, as postulated above, contains a flexible initial guess joint such that DoF $b_A \neq b_B$ for reasons which will be explained later. Note that an uncoupled model (infinitely flexible joint) is then also permitted. In this specific case the cross-terms between the components are all zero.

Next, the full-system measurements (which observe the joint dynamics) are required. These become the so-called overlay model:

$$\mathbf{Y}^{\text{ov}} = \begin{bmatrix} \mathbf{Y}^{\text{AJB}}_{i_A i_A} & \mathbf{Y}^{\text{AJB}}_{i_A i_B} \\ \mathbf{Y}^{\text{AJB}}_{i_B i_A} & \mathbf{Y}^{\text{AJB}}_{i_B i_B} \end{bmatrix} \tag{10.20}$$

As stated before, the measurements can only be done for the internal DoF i_A, i_B.

Finally, SEMM requires the *removed model* which is none other than (a reduced form of) the parent model (c.f. Fig. 10.4). One can choose to formulate the removed model as the parent model itself, or as a reduced form of the parent model which contain only the internal DoF i_A, i_B.[2] An explanation on the differences is omitted in this paper, but is given in [4]. In this application, it is chosen for the removed and parent model to be the same size and thus the same:

$$\mathbf{Y}^{\text{rem}} = \mathbf{Y}^{\text{par}} \tag{10.21}$$

To solve (10.18) the compatibility and equilibrium constraints are applied as in LM-FBS. Compatibility requires:

$$\mathbf{u}_g^{\text{par}} - \mathbf{u}_g^{\text{rem}} = \mathbf{0} \tag{10.22a}$$

$$\mathbf{u}_i^{\text{rem}} - \mathbf{u}_i^{\text{ov}} = \mathbf{0} \tag{10.22b}$$

The equilibrium condition reads:

$$\mathbf{g}_b^{\text{par}} + \mathbf{g}_b^{\text{rem}} = \mathbf{0} \tag{10.23a}$$

$$\mathbf{g}_i^{\text{par}} + \mathbf{g}_i^{\text{rem}} + \mathbf{g}_i^{\text{ov}} = \mathbf{0} \tag{10.23b}$$

Like before, the compatibility and equilibrium condition can be written in matrix-notation with the signed Boolean Matrix $\mathbf{B}$:

$$\text{Compatibility}: \mathbf{B}\mathbf{u} = \mathbf{0}, \quad \text{Equilibrium}: \mathbf{g} = \mathbf{B}^T\boldsymbol{\lambda}, \quad \text{with} \quad \mathbf{B} = \begin{array}{c} \begin{matrix} \mathbf{u}_i^{\text{par}} & \mathbf{u}_b^{\text{par}} & \mathbf{u}_i^{\text{rem}} & \mathbf{u}_b^{\text{rem}} & \mathbf{u}_i^{\text{ov}} \end{matrix} \\ \left[\begin{array}{cc|c|c|c} \mathbf{I} & \mathbf{0} & -\mathbf{I} & \mathbf{0} & \mathbf{0} \\ \mathbf{0} & \mathbf{I} & \mathbf{0} & -\mathbf{I} & \mathbf{0} \\ \mathbf{0} & \mathbf{0} & \mathbf{I} & \mathbf{0} & -\mathbf{I} \end{array}\right] \end{array}$$

The formulation exactly follows the LM-FBS notation described above, thus it is solved with Eq. (10.10). It can be shown that from this solution the single-line equation of SEMM can be obtained. The derivation is omitted in this paper, but can also be found in [4]:

$$\mathbf{Y}^{\text{SEMM}} = \mathbf{Y}_{gg}^{\text{par}} - \mathbf{Y}_{gg}^{\text{par}} \left(\mathbf{Y}_{ig}^{\text{par}}\right)^{+} \left(\mathbf{Y}_{ii}^{\text{par}} - \mathbf{Y}_{ii}^{\text{ov}}\right) \left(\mathbf{Y}_{gi}^{\text{par}}\right)^{+} \mathbf{Y}_{gg}^{\text{par}} \tag{10.24}$$

Here we made use of the relation (10.21) to substitute the terms of the removed model.

The SEMM model now contains the dynamics from the overlay model while including the boundary DoF's required to extract the joint model with Eq. (10.16).

When expanding a DoF-set, new information is extrapolated based on the information contained in the measurements of the overlay model. Unfortunately, such an extrapolation is generally erroneous. Consequently, any joint model extracted from the expanded SEMM model is also erroneous (although presumably less so than in the original parent model).

If one assumes that the overlay model contains the correct dynamics, then the only error made is this extrapolation error. It comes solely from the parent model's manifold (i.e. its modal content) differing from that of the *correct* overlay model. In other words, even though the SEMM method will alter the dynamics of the parent model to fit that of the overlay model, it can only do so based on the allowable modal directions of said parent model. This is because any deflection shape created by the parent model must still be a linear combination of its modeshapes.

Figure 10.5 illustrates this simple fact with a short example of a clamped-free beam which—for simplicity—has only the first two modes. These are depicted in (a). The first mode-shape is relatively the same for both parent (red) and overlay (blue) model, yet the second differs. The *normalized* deflection shapes of the parent (red), overlay (blue) and resulting SEMM (green) model for a given actuation are illustrated in (b). When looking at the SEMM model, even though the shared DoF match as per design, the internal DoF cannot since the required deflection shape is not a linear combination of the parent mode-shapes.

This is also the reason that no rigid connection should be used to couple the component models $\mathbf{Y}^{\text{A}}$ and $\mathbf{Y}^{\text{B}}$ to create the parent model $\mathbf{Y}^{\text{AB}}$. Since, if that were the case, the compatibility condition would ensure that $b_A = b_B$ in the entire modal

[2]Note that reduction in admittance space is done by simply removing the DoF from the matrix.

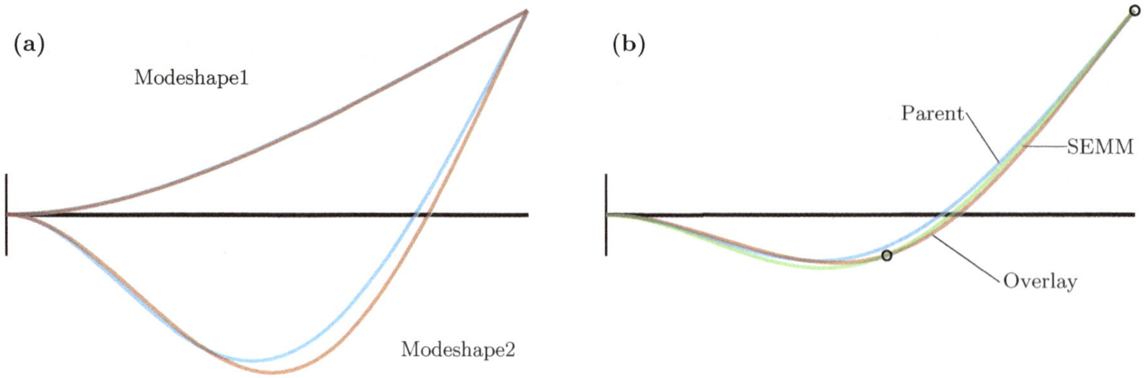

Fig. 10.5 (**a**) A clamped beam with two modeshapes. The first modeshape is equal for both the parent and overlay model. The second is different for the parent (blue) and overlay model (red). (**b**) The *normalized* deflection shapes of each model for a given actuation. The SEMM model (green) is equal to the overlay model on the control points (black circles) but differs at the other locations since its end result must be a linear combination of the parent modeshapes given in (**a**). This off-set is the expansion error ε

space; i.e. there exist no modal direction in which a gap—and thus joint dynamics—exists. Since the parent model can only adapt to the overlay model in the modal directions, forming a gap is a non-controllable direction. Consequently, the gap is never created and thus no joint would be able to be identified!

In the single-line Eq. (10.24) one can observe that any changes made to the parent model (including expansion error) is made in the second term of this equation. This second term is proportional to the difference between the parent and overlay model. The error ε of the SEMM expansion is therefore—in some form—proportional to the difference in overlay and parent model:

$$\varepsilon \propto \left\| \mathbf{Y}_{ii}^{\text{par}} - \mathbf{Y}_{ii}^{\text{ov}} \right\| \tag{10.25}$$

Also note that in the trivial solution when the overlay model and parent model are equal, there is no *observable* error in the SEMM model since $\mathbf{Y}_{ii}^{\text{SEMM}} = \mathbf{Y}_{ii}^{\text{ov}} = \mathbf{Y}_{ii}^{\text{par}}$. The term observable error is used here to accentuate that an error might still exist on the DoF which do not explicitly exist for the overlay model, and thus cannot be observed. Note that the reason enough internal DoF must be taken into account is to ensure the observability of this error and by extension the joint dynamics. If this is the case, one can derive from Eq. (10.25) that the parent and overlay model must converge to minimise the expansion error.

10.2.3 The Optimisation Scheme

Unfortunately the joint identified from the SEMM model is not perfect due to the expansion error discussed in the previous section. Now let us assume that the *only* difference between the parent and overlay model is due to the joint dynamics. If the joint model extracted from the SEMM model is closer (i.e. converged) to the joint implicit in the measurements, then surely this joint can be used to create a parent model which further resembles the overlay model. Due to this convergence, the expansion error of Eq. (10.25) is decreased which would result in a better identification of the joint. Repeating this, the process would converge to the identification of the true joint.

Following this conjecture, an iterative process can be described, its schematic is illustrated in Fig. 10.6:

1. Measurements are performed on the assembled system, these measurements have implicit information of the joint, but lack boundary DoF measurements: this becomes the overlay model $\mathbf{Y}^{\text{ov}}$.
2. An *initial guess* for the joint is constructed: $\mathbf{Y}_{n=0}^{J}$.
3. This joint is used to couple the component models A and B: this becomes the nth parent model $(\mathbf{Y}^{\text{par}})_n$.
4. SEMM is performed with the parent and overlay model in order to obtain the SEMM model $\mathbf{Y}_n^{\text{SEMM}}$.
5. Decoupling using Eq. (10.16) is performed to extract a new joint model from the SEMM model: $\mathbf{Y}_{n+1}^{J}$.
6. Steps 3–5 are repeated until the difference between parent and overlay model—and thus the expansion error of Eq. (10.25)—is minimised.

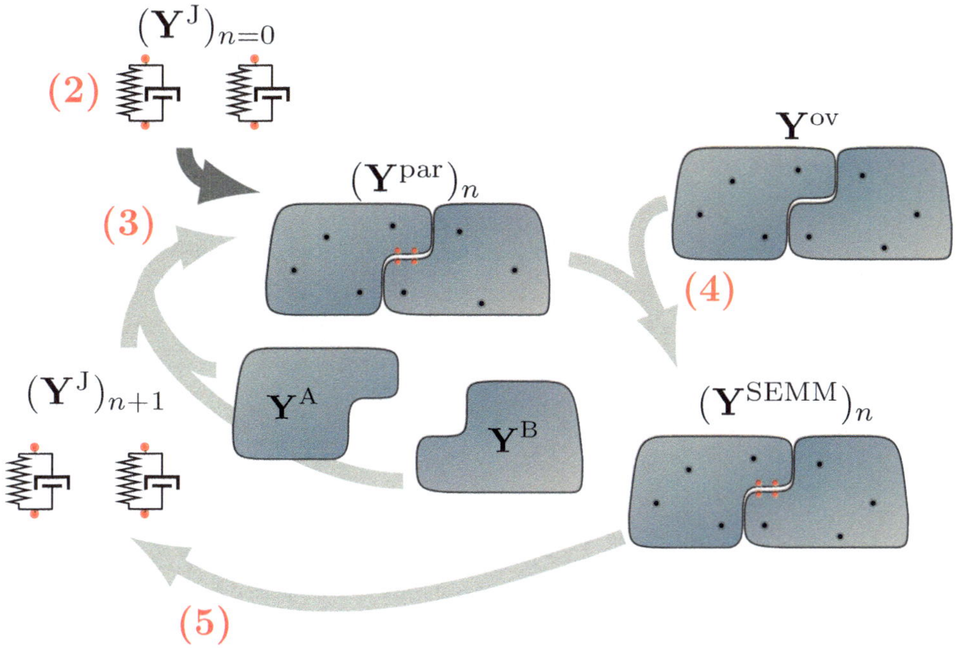

Fig. 10.6 The joint optimisation process. The initial joint (2) is used to create a parent model (3). The parent model is coupled to the overlay model in the SEMM process (4). From the new SEMM model a new joint is identified (5). This new joint is used to create a new parent model (3). The process is then repeated until parent and overlay model converge

There are some things that must be considered when implementing this scheme. In reality there are more error sources than just the SEMM expansion error (one can think of measurement errors, incorrect component models, incorrect DoF locations, etc.) and these errors will pollute the optimisation such that a true optimum may never be found in practice. Nevertheless, some techniques may be implemented to clean up the results such that a good estimation can still be achieved:

- If some notion of the joint geometry is known (such as there being no cross-couplings) a fitting can be done to acquire some (physical) properties of the joint. Knowing that the joint properties are—to some extent—frequency independent, one could implement this method at frequencies where the joint dynamics are more prevalent compared to the system dynamics. At these frequencies the observability of the joint is higher, resulting in a higher signal to noise ratio when decoupling. Such principles have been used successfully in the past [3].
- It may be that the method jumps between solutions, resulting in local optima. One can choose to average out past results such that the new joint $(\mathbf{Y}^{\mathrm{J}})_{n+1}$ is a mix between past results: $(\mathbf{Y}^{\mathrm{J}})_{n+1}$, $(\mathbf{Y}^{\mathrm{J}})_n$, $(\mathbf{Y}^{\mathrm{J}})_{n-1}$. This is in effect a relaxation scheme and can also be interpreted as analogous to implementing numerical damping in e.g. Newmark methods.
- Since only the joint is of importance, the quality of the component models is secondary. Yet, since this method does not discriminate between error in the joint model and error in the sub-components, any discrepancy between the sub-component models in the parent versus the overlay model is going to pollute the result. It might be important to ensure that, at the very least, the component models $\mathbf{Y}^{\mathrm{A}}$ and $\mathbf{Y}^{\mathrm{B}}$ are equal in both the parent and (implicit) overlay model. If the parent model is measured in the same manner (but disassembled) as the overlay model, then any measurement bias is also shared by both the parent and overlay model which will reduce this type of error.

10.3 A Numerical Study

A small numerical case is presented as a proof of concept. For this case the connection of two bar-truss structures shown in Fig. 10.7 is investigated. The truss structures are connected via a 6-DoF joint: 3 translational and 3 rotational DoF on a node in the centre (c.f. the blue node on the right side of Fig. 10.7) which acts as the interface. The centre-node is created by means of a RBE-3 spider element from the nodes at each structure's end. Note that the measurement locations (c.f. the red nodes in the right image of Fig. 10.7) are spread so that the joint dynamics are observable. Also note that none are on the interface area or inaccessible locations.

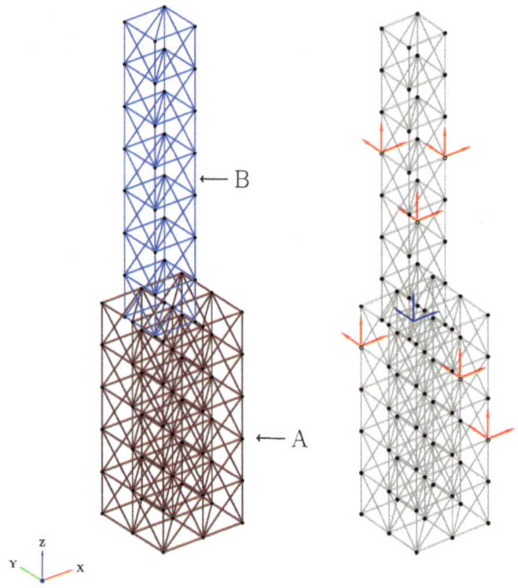

Bar Properties

E	7.80×10^{11}	Pa
ρ	5.00×10^{3}	$\mathrm{Kg/m^3}$
A	2.83×10^{-5}	$\mathrm{m^2}$
L	0.10	m

Fig. 10.7 The truss structure used in this test-case. On the left the two parts of the assembly are colour-coded. The bottom part (Structure A) is fixed at the bottom. Structure B is attached to structure A via a joint in 6-DoF. On the left the assembly AB is shown as a whole. The blue node is the location of the joint. The red nodes are the measurement locations, i.e. the internal DoF used in the SEMM-optimisation scheme. It is important to note that the structures are only connected via this 6-DoF connection point. The structure B is placed on A in such a way that symmetry is avoided

Table 10.1 The parameters (stiffness k and damping c coefficients) of the joints tested

	Strong joint		Weak joint		Initial guess	
DoF	k	c	k	c	k	c
x	1×10^{8}	100	1×10^{4}	100	1×10^{2}	0
y	1×10^{8}	100	1×10^{4}	100	1×10^{2}	0
z	1×10^{10}	1	1×10^{6}	1	1×10^{2}	0
θ_x	1×10^{10}	10	1×10^{6}	10	1×10^{2}	0
θ_y	1×10^{10}	10	1×10^{6}	10	1×10^{2}	0
θ_z	1×10^{8}	100	1×10^{3}	100	1×10^{2}	0

The parameters of the *initial-guess* joint used in the first iteration of the parent model is also provided. These parameters are used in both test-cases

Two joints are investigated: a *weak* joint and a *strong* joint. The *strong* joint's influence on the structure is noticeable but minimal. The *weak* joint is chosen such that it almost fully determines the assembly's dynamics. Only DoF-to-DoF damping and stiffness is investigated: no cross-coupling between DoF exist in the tested joints. Furthermore, the damping values are kept equal for both. The parameters of the joints are given in Table 10.1.

The optimisation scheme is run for several iterations; 20, 100 and 200 iterations are run which require approximately 2.5, 12.5, 25 s computation time respectively. In Fig. 10.8 the driving-point frequency response functions of the system $\mathbf{Y}^{\mathrm{AB}}$ for the boundary DoF θ_z are given; the graphs represent the results after the pre-determined number of iterations for both the case of the weak (left) and strong (right) joint. The reference FRF and the FRF of the rigidly connected $\mathbf{Y}^{\mathrm{AB}}$ are also provided as a comparison.

The FRF follow the reference throughout a large part of the frequency band. An interesting error occurs at or near the resonance frequencies of the *would-be* rigidly fixed system $\mathbf{Y}^{\mathrm{AB}}$. The results show that the optimising scheme cannot identify the joint near these frequencies.

Next, the linear joint parameters can be identified from the FRF of the joint $\mathbf{Y}^{\mathrm{J}}$. In order to do this, a frequency band is chosen such that it contains no resonance frequencies of the rigidly coupled $\mathbf{Y}^{\mathrm{AB}}$ in order to ensure that the previously observed error does not pollute these results. For this example the frequency band from 450 to 850 Hz is chosen. The joint parameters are extracted by means of a curve-fit of the driving-point FRF of $\mathbf{Z}^{\mathrm{J}}_n = (\mathbf{Y}^{\mathrm{J}}_n)^{-1}$

It is important to remember that this fitting does not influence the optimisation since the optimisation uses the raw model $\mathbf{Y}^{\mathrm{J}}_n$ for each subsequent step. In Table 10.2 the results of the fitting for $n = 20$, 100 and 200 are given. It is immediately apparent that these joint properties are non-physical! Most of the stiffness values are—with the exception of k_y—negative which indicates an active joint. However, it is important to note two things: first, even though the values are off, the FRF in Fig. 10.8 follow the reference FRF in amplitude and phase. Second, as the number of iterations increase, the joint parameters

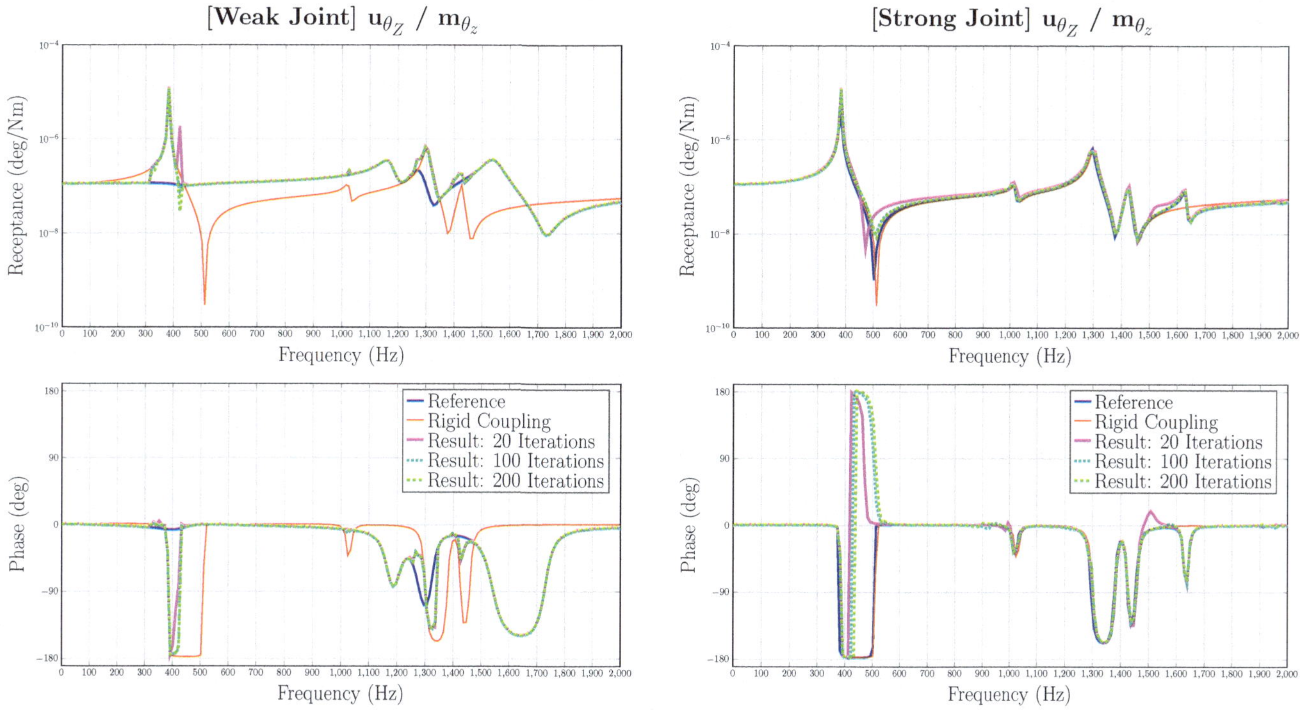

Fig. 10.8 The driving-point FRF on DoF θ_z. The FRF of the *weak* joint are given on the left and the FRF of the *strong* joint are given on the right. In both cases—and even after only a few iterations—the FRF seem to follow the reference; the exception being near the resonance frequencies of the rigid coupling. This results in spurious peaks near those frequencies

Table 10.2 The fitted parameters of the *strong* joint

Strong joint					
20 iterations		100 iterations		200 iterations	
k	Δ	k	Δ	k	Δ
-4.57×10^7	−0.340	-3.19×10^7	−0.496	-3.28×10^7	−0.485
7.25×10^6	1.140	3.13×10^7	0.505	4.56×10^7	0.341
-5.28×10^8	−1.277	-4.73×10^8	−1.325	-4.31×10^8	−1.366
-1.22×10^6	−3.915	-1.75×10^6	−3.756	-3.12×10^6	−3.505
-2.76×10^7	−2.558	-3.08×10^7	−2.511	-3.11×10^7	−2.508
-1.05×10^8	−0.020	-9.88×10^7	−0.005	-9.58×10^7	−0.019
c	Δ	c	Δ	c	Δ
1079.7	1.033	1571.60	1.196	2265.5	1.355
60.9	0.215	446.98	0.650	710.1	0.851
4053.8	3.608	4053.99	3.608	4100.8	3.613
11.7	0.067	30.68	0.487	−15.8	−0.198
163.4	1.213	106.00	1.025	11.6	0.064
685.8	0.836	491.74	0.692	377.1	0.576

Δ is the difference between the value and its reference given in orders of magnitude. The results for both the stiffness k and damping values c are off by up to a few orders

remain relatively the same. This indicates that the resulting joint parameters are approaching a (non-physical) local optimum, i.e. it indicates that multiple solutions exist that are indistinguishable from the chosen set of observation DoF.

Similarly the joint parameters of the weak joint can be identified. The same frequency band is used. Note that already after a few iterations the real joint parameters are found accurately for all DoF with the exception of the DoF θ_x. This may indicate that the observability of this DoF is poor. Nevertheless, as the iterations increase the optimisation converges to the true values.

Table 10.3 The fitted parameters of the *weak* joint

Weak joint					
20 iterations		100 iterations		200 iterations	
k	Δ	k	Δ	k	Δ
9.91×10^3	0.004	9.88×10^3	0.005	9.84×10^3	0.007
1.00×10^4	0.002	9.93×10^3	0.003	9.91×10^3	0.004
9.97×10^5	0.001	9.99×10^5	0.000	9.99×10^5	0.000
1.71×10^4	1.768	6.19×10^4	1.208	1.03×10^5	0.985
1.00×10^6	0.001	9.95×10^5	0.002	9.93×10^5	0.003
-7.98×10^2	−0.098	7.15×10^2	0.146	4.37×10^2	0.360
c	Δ	c	Δ	c	Δ
100.0	0.000	100.0	0.000	100.0	0.000
100.1	0.000	100.0	0.000	100.0	0.000
0.7	0.150	1.0	0.002	1.0	0.001
0.2	1.619	0.6	1.234	0.9	1.049
10.1	0.005	9.9	0.006	9.8	0.008
99.7	0.001	100.0	0.000	100.0	0.000

Δ is the difference between the value and its reference given in orders of magnitude. The results for both the stiffness k and damping values c are all very close to the true values. The exception is the DoF θ_x

10.4 Discussion

The FRF of Fig. 10.8 show that the joint identification results follow the reference FRF for frequencies other than at or near the resonance frequencies of the rigidly connected system $\mathbf{Y}^{\text{AB}}$. Furthermore, the extraction of joint parameters show that better results are obtained for the *weak* joint. This suggests that the joint identification can only be done when the joint dynamics are sufficiently observable.

Some additional thoughts are provided to contextualize the results:

- The optimisation scheme performs poorly at and near the resonance frequencies of the rigidly connected system $\mathbf{Y}^{\text{AB}}$. This results in spurious peaks; at these frequencies the optimisation is forced towards the rigid connection. As explained in Sect. 10.2.2: once the rigid connection is formed, the gap between the respective boundary DoF of structures A and B is closed. Remember that once closed, the gap cannot be opened since then the optimisation directions (the modal directions of the parent model) also contain no gap. A rigid connection can therefore be regarded as an—unwanted—local optimum.
- The structure used in this test-case is chosen for its simplicity. However, it might be that this case is too simple. The truss elements are created with bar elements[3] which hinders observability since it is hard to pinpoint the connectivity between the boundary and chosen internal DoF. For example, the results in Table 10.3 show that all the joint parameters are found accurately after only a few iterations, yet the stiffness and damping values of DoF θ_x were not. This indicates a lack of observability on this DoF which is curious considering the chosen locations of the internal DoF.

10.5 Conclusion

A SEMM-based method can be used to identify a joint without directly measuring the interface of an assembled system. The joint must be observed with the measurements performed and included in the overlay model. However, if this is the case any number of boundary DoF can be investigated within a single measurement campaign. That is to say: as long as the observability condition is met one could identify a joint with the same overlay model coupled to a parent model with any desired translational or rotational boundary DoF.

A numerical proof of concept is provided and for specific cases the joint is accurately identified but many envisioned error sources have been omitted. The additional error sources mentioned in Sect. 10.2.3 should first be investigated before any

[3]Many connections are therefore implicit, e.g. bending stiffness is created by the geometry of the truss, rather than in the element itself.

practical applications can realistically be performed. Nevertheless, the methodology is promising since it suddenly allows users to do a preliminary (read: linear) investigation of complex joints, unlimited by the inaccessibility of common interfaces.

Acknowledgements The project has received funding from the European Union's Horizon 2020 research and innovation programme under the Marie Sklodowska-Curie grant agreement No. 721865.

References

1. Ewins, D.J., Nowell, D., Petrov, E.: The importance of joints on the dynamics of gas turbine structures. In: Segalman, D.J., Bergman, L.A., Ewins, D.J. (eds.) Report on the SNL/NSF International Workshop on Joint Mechanics, Arlington, Virginia, pp. 23–32, Sandia National Laboratories, Albuquerque (2006)
2. Özşahin, O., Ertürk, A., Özgüven, H.N., Budak, E.: A closed-form approach for identification of dynamical contact parameters in spindle-holder-tool assemblies. Int. J. Mach. Tools Manuf. **49**(1), 25–35 (2009)
3. Tol, Ş, Özgüven, H.N.: Dynamic characterization of bolted joints using FRF decoupling and optimization. Mech. Syst. Signal Process. **54**, 124–138 (2015)
4. Klaassen, S.W.B., van der Seijs, M.V., de Klerk, D.: System equivalent model mixing. Mech. Syst. Signal Process. **105**, 90–112 (2018)
5. de Klerk, D., Rixen, D.J., de Jong, J.: The frequency based substructuring (FBS) method reformulated according to the dual domain decomposition method. In: IMAC-XXIV A Conference & Exposition on Structural Dynamics (2006)
6. Voormeeren, S.N., van der Valk, P.L.C., Rixen, D.J.: Practical aspects of dynamic substructuring in wind turbine engineering. In: Structural Dynamics and Renewable Energy, Vol. 1. IMAC-XXVIII A Conference & Exposition on Structural Dynamics, July 2011. Springer, New York/Jacksonville, FL (2010)
7. Meggitt, J.W.R., Moorhouse, A.T.: The in-situ decoupling of resiliently coupled sub-structures. In: *24th International Congress on Sound and Vibration*, July, London, pp. 1–8 (2017)
8. Haeussler, M., Klaassen, S.W.B., Rixen, D.J.: Comparison of Substructuring based techniques for dynamic property identification of rubber isolators. In: ISMA 2018 - International Conference on Noise and Vibration Engineering, Leuven, p. 1 (2018)

Chapter 11
Overview of Free Interface Substructuring Approaches for Systems with Arbitrary Viscous Damping in Dynamic Substructuring

Fabian M. Gruber, Dennis Berninger, and Daniel J. Rixen

Abstract Most classical substructuring methods yield great approximation accuracy if the underlying system is not damped. One approach is a fixed interface method, the Craig-Bampton method. In contrast, many other methods (e.g., MacNeal method, Rubin method, Craig-Chang method) employ free interface modes, (residual) attachment modes, and rigid body modes. None of the aforementioned methods takes any damping effects into account when performing the reduction. If damping significantly influences the dynamic behavior of the system, the approximation accuracy can be very poor. One procedure to handle arbitrarily viscously damped systems and to take damping effects into account is to transform the second-order differential equations into twice the number of first-order differential equations resulting in state-space representation of the system. Solving the corresponding eigenvalue problem allows the damped equations to be decoupled; however, complex eigenmodes and eigenvalues occur. The complex modes are used to build a reduction basis that includes damping properties.

The derivation of different Craig-Bampton substructuring methods (fixed interface) for viscously damped systems was presented in Gruber et al. (Comparison of Craig-Bampton approaches for systems with arbitrary viscous damping in dynamic substructuring). In contrast, we present here the derivation of different free interface substructuring methods for viscously damped systems in a comprehensible consistent manner. Craig and Ni suggested a method that employs complex free interface vibration modes (1989). De Kraker and van Campen give an extension of Rubin's method for general state-space models (1996). Liu and Zheng proposed an improved component modes synthesis method for nonclassically damped systems (2008), which is an extension of Craig and Ni's method. A detailed comparison between the different formulations will be given. Liu and Zheng's method can be considered as a second-order extension of Craig and Ni's method. We propose a third-order extension and a generalization to any given higher order. Moreover, a new method combining the reduction basis of Liu and Zheng's approach with the primal assembly procedure applied by de Kraker and van Campen is proposed. The presented theory and the comparison between the methods will be illustrated in different examples.

Keywords Dynamic substructuring · Component mode synthesis · Free interface methods · Damped systems · State-space formulation · Complex modes

11.1 Introduction

Dynamic substructuring techniques reduce the size of large models very efficiently. A large finite element model is divided into a number of substructures; each substructure is analyzed and reduced separately and then assembled into a low-order reduced model. This low-order reduced model approximates the original large model's behavior. During this process, each substructure's degrees of freedom (DOFs) are divided into internal DOFs (those not shared with any adjacent substructure) and boundary or interface DOFs (those shared with adjacent substructures and therefore forming the model's interface DOFs) [1]. Many substructuring methods that work with second-order equations of motion have been proposed in the past [2–9]. Only the substructures' mass and stiffness properties are commonly taken into account for the reduction by all those methods. Each substructure's reduction basis is thereby constructed exclusively on the properties of the mass matrix and stiffness matrix, but the properties of the damping matrix are ignored. The undamped equations of motion are assumed to correctly describe the substructures' dynamics: it is either assumed that there is no damping or that damping effects are completely negligible when building the reduction basis.

F. M. Gruber (✉) · D. Berninger · D. J. Rixen
Faculty of Mechanical Engineering, Technical University of Munich, Garching, Germany
e-mail: fabian.gruber@tum.de; dennis.berninger@tum.de; rixen@tum.de

A. Linderholt et al., *Dynamic Substructures, Volume 4*, Conference Proceedings of the Society for Experimental Mechanics Series,
https://doi.org/10.1007/978-3-030-12184-6_11

Substructuring methods working with the undamped equations of motion afford great approximation accuracy if the underlying system is damped only slightly or not at all [10]. The most popular approach is a fixed interface method, the Craig-Bampton method [2], which is based on fixed interface vibration modes and interface constraint modes. In contrast, many other methods (e.g., MacNeal method [3], Rubin method [4], Craig-Chang method [6]) employ free interface modes, (residual) attachment modes, and rigid body modes.

None of the aforementioned methods take any damping effects into account when performing the reduction. There is generally no justification for neglecting damping effects. If damping significantly influences the dynamic behavior of the system under consideration, then the approximation accuracy of these methods can be very poor. The damped equations of motion have to be taken into account to incorporate damping effects. One procedure to handle arbitrarily damped systems and to take damping effects into account is to transform the second-order differential equations into twice the number of first-order differential equations, resulting in a state-space representation of the system. Solving the corresponding eigenvalue problem allows the damped equations to be decoupled, but complex eigenmodes and eigenvalues occur. The complex modes are used to build a reduction basis that includes damping properties.

The derivation of different Craig-Bampton substructuring methods (fixed interface) for viscously damped systems is given in [11]. Thereby, Hasselman and Kaplan's approach [12], which is an extension of the Craig-Bampton method through its employment of complex component modes, Beliveau and Soucy's method [13], which replaces the real fixed interface normal modes of the second-order system with the corresponding complex modes of the first-order system, and an approach of de Kraker [14], which uses complex normal modes and modified static modes, were investigated [11].

In contrast, in this contribution, we wish to present the derivation of different free interface substructuring methods for viscously damped systems in a comprehensible consistent manner: Craig and Ni [15] presented a method that employs complex free interface vibration modes. De Kraker and van Campen [16] gave an extension of Rubin's method for general state-space models. Liu and Zheng [17] proposed an improved component modes synthesis method for nonclassically damped systems, which is an extension of Craig and Ni's method. All these methods will be briefly derived and a detailed comparison between the different formulations will be given. The presented theory and the comparison between the methods will be illustrated in different examples. Based on those insights, the following improvements will be proposed: Liu and Zheng's method can be considered as a second-order extension of Craig and Ni's method. We propose a third-order extension and a generalization to arbitrarily higher orders. Moreover, a new method combining the reduction basis of Liu and Zheng's approach with the primal assembly procedure applied by de Kraker and van Campen is proposed. The approximation accuracy of these new approaches will be compared to the existing methods and illustrated in different examples.

Section 11.2 recalls the superordinate governing equations, and the terminology and notation used throughout this paper is established up. In Sect. 11.3, a literature review of existing free interface substructuring approaches for viscously damped systems is given. Starting in Sect. 11.3.1, the first-order approach of Craig and Ni is presented. Following this, Liu and Zheng's approach is derived in Sect. 11.3.2, which is the second-order extension of the method of Craig and Ni. Section 11.3.3 recalls de Kraker and van Campen's approach, which is an extension of Rubin's method for general state-space models. Improvements of the existing free interface substructuring approaches are suggested in Sect. 11.4: a generalization to arbitrarily higher orders of Liu and Zheng's method is derived in Sect. 11.4.1. Following this, a new approach combining Liu and Zheng's reduction basis with a primal assembly procedure is demonstrated in Sect. 11.4.2. The properties of the presented methods are subsequently illustrated in detail in Sect. 11.5 using three different examples. Finally, a brief summary of findings and conclusions is given in Sect. 11.6.

11.2 Governing Equations

11.2.1 Undamped and Classically Damped Systems

Consider the equations of motion of a viscously damped linear system with m degrees of freedom (DOFs)

$$\boldsymbol{M}\ddot{\boldsymbol{u}} + \boldsymbol{C}\dot{\boldsymbol{u}} + \boldsymbol{K}\boldsymbol{u} = \boldsymbol{f} \tag{11.1}$$

with mass matrix $\boldsymbol{M}$, damping matrix $\boldsymbol{C}$, stiffness matrix $\boldsymbol{K}$, displacement vector $\boldsymbol{u}$, and external force vector $\boldsymbol{f}$. Associated with the undamped system of Eq. (11.1), i.e., $\boldsymbol{C} = \boldsymbol{0}$, is the eigenvalue problem

$$\left(-\omega_j^2 \boldsymbol{M} + \boldsymbol{K}\right)\boldsymbol{\theta}_j = \boldsymbol{0} \tag{11.2}$$

with eigenvalue ω_j^2 and corresponding eigenvector $\boldsymbol{\theta}_j$ for $j = 1, \ldots, m$. The eigenvectors $\boldsymbol{\theta}_j$ of Eq. (11.2) are called normal modes or natural modes and form the columns of the modal matrix $\boldsymbol{\Theta}$:

$$\boldsymbol{\Theta} = \left[\boldsymbol{\theta}_1 \, \boldsymbol{\theta}_2 \ldots \boldsymbol{\theta}_m\right] \tag{11.3}$$

If the modal matrix $\boldsymbol{\Theta}$ is normalized with respect to the mass matrix $\boldsymbol{M}$, then the normal modes' orthogonality conditions give

$$\boldsymbol{\Theta}^T \boldsymbol{M} \boldsymbol{\Theta} = \boldsymbol{I}, \qquad \boldsymbol{\Theta}^T \boldsymbol{K} \boldsymbol{\Theta} = \boldsymbol{\Omega}^2 = \operatorname{diag}\left(\omega_1^2, \, \omega_2^2, \, \ldots, \, \omega_m^2\right). \tag{11.4}$$

Using the modal transformation

$$\boldsymbol{u} = \boldsymbol{\Theta} \boldsymbol{p} \tag{11.5}$$

with the vector of modal coordinates $\boldsymbol{p}$, Eq. (11.1) takes the canonical form of the damped system [18]

$$\ddot{\boldsymbol{p}} + \boldsymbol{C}_{\text{modal}} \, \dot{\boldsymbol{p}} + \boldsymbol{\Omega}^2 \boldsymbol{p} = \boldsymbol{\Theta}^T \boldsymbol{f} \qquad \text{with} \qquad \boldsymbol{C}_{\text{modal}} = \boldsymbol{\Theta}^T \boldsymbol{C} \boldsymbol{\Theta}. \tag{11.6}$$

Mass matrix $\boldsymbol{M}$ and stiffness matrix $\boldsymbol{K}$ have been diagonalized by the modal transformation of Eq. (11.5) and the diagonal eigenvalue matrix $\boldsymbol{\Omega}^2$ is given in Eq. (11.4). Matrix $\boldsymbol{C}_{\text{modal}}$ is referred to as generalized damping matrix or modal damping matrix [1]. Any modal coupling in a linear system occurs exclusively through damping [19]. A damped system is termed "classically damped" if it can be decoupled by the modal matrix $\boldsymbol{\Theta}$ [1], i.e., $\boldsymbol{C}_{\text{modal}}$ is also diagonalized by the modal transformation of Eq. (11.5). A necessary and sufficient condition that the damped system can be decoupled and hence that the damping matrix $\boldsymbol{C}$ can be diagonalized by the modal matrix $\boldsymbol{\Theta}$ is [20]:

$$\boldsymbol{C} \boldsymbol{M}^{-1} \boldsymbol{K} = \boldsymbol{K} \boldsymbol{M}^{-1} \boldsymbol{C} \tag{11.7}$$

Condition (11.7) is usually not satisfied. Only under special conditions are the equations of motion completely diagonalized by the classical modal transformation [18], and these conditions appear to have little physical justification [21]. For instance, the often used but simplifying assumptions of mass proportional damping, stiffness proportional damping, Rayleigh damping [22], or modal damping fulfill condition (11.7) and diagonalize the generalized damping matrix $\boldsymbol{C}_{\text{modal}}$. Nevertheless, decoupling of Eq. (11.1) is not generally possible using classical modal analysis.

11.2.2 Nonclassically Damped Systems

A procedure to handle and decouple systems that are not classically damped, i.e., systems with general viscous damping, is to transform the m second-order equations of motion (11.1) into $2m$ first-order equations [1, 23–25]. The state-space vector of dimension $n = 2m$ is

$$\boldsymbol{z}(t) = \begin{bmatrix} \boldsymbol{u}(t) \\ \boldsymbol{v}(t) \end{bmatrix} \tag{11.8}$$

with $\boldsymbol{v}(t) = \dot{\boldsymbol{u}}(t)$. Adding the m redundant equations $\boldsymbol{M}\boldsymbol{v}(t) = \boldsymbol{M}\dot{\boldsymbol{u}}(t)$ to the equations of motion (11.1), the generalized state-space symmetric form

$$\boldsymbol{A}\dot{\boldsymbol{z}} + \boldsymbol{B}\boldsymbol{z} = \boldsymbol{F} \tag{11.9}$$

is obtained with

$$\boldsymbol{A} = \begin{bmatrix} \boldsymbol{C} & \boldsymbol{M} \\ \boldsymbol{M} & \boldsymbol{0} \end{bmatrix}, \qquad \boldsymbol{B} = \begin{bmatrix} \boldsymbol{K} & \boldsymbol{0} \\ \boldsymbol{0} & -\boldsymbol{M} \end{bmatrix}, \qquad \boldsymbol{F} = \begin{bmatrix} \boldsymbol{f} \\ \boldsymbol{0} \end{bmatrix}. \tag{11.10}$$

Associated with Eq. (11.9), which includes the damping matrix $\boldsymbol{C}$, is the eigenvalue problem

$$\left(\lambda_j \boldsymbol{A} + \boldsymbol{B}\right) \boldsymbol{\phi}_j = \boldsymbol{0} \tag{11.11}$$

with eigenvalue λ_j and corresponding state-space eigenvector $\boldsymbol{\phi}_j$ for $j = 1, \ldots, n$. Since matrices $\boldsymbol{A}$ and $\boldsymbol{B}$ are real, the n eigenvalues λ_j must either be real or they must occur in complex conjugate pairs [24]. For underdamped systems, the eigenvalues λ_j and corresponding eigenmodes $\boldsymbol{\phi}_j$ occur in complex conjugate pairs [26] and are termed "underdamped eigenvalues" and "underdamped eigenmodes", respectively. Since real eigenvalues indicate very high damping leading to overdamped modes, most structures have m complex conjugate pairs of eigenvalues and corresponding eigenvectors [1]. The eigenvectors $\boldsymbol{\phi}_j$ are referred to as complex normal modes.

Similar to the undamped case in Eq. (11.3), the state-space eigenvectors $\boldsymbol{\phi}_j$ form the complex state-space modal matrix

$$\boldsymbol{\Phi} = \left[\boldsymbol{\phi}_1\ \boldsymbol{\phi}_2 \ldots \boldsymbol{\phi}_n\right]. \tag{11.12}$$

An extensive investigation of properties and relations between complex normal modes of Eq. (11.12) and real normal modes of Eq. (11.3) is given in [27]. If the complex modal matrix $\boldsymbol{\Phi}$ is normalized with respect to the state-space matrix $\boldsymbol{A}$, the orthogonality conditions of the complex normal modes give

$$\boldsymbol{\Phi}^T \boldsymbol{A} \boldsymbol{\Phi} = \boldsymbol{I}, \qquad \boldsymbol{\Phi}^T \boldsymbol{B} \boldsymbol{\Phi} = -\boldsymbol{\Lambda} = -\text{diag}\left(\lambda_1, \lambda_2, \ldots, \lambda_n\right). \tag{11.13}$$

Complex modal matrix $\boldsymbol{\Phi}$ decouples the damped system if it is written in state-space format as in Eq. (11.9). Eigenvalue matrix $\boldsymbol{\Lambda}$ contains the eigenvalues λ_j of Eq. (11.11) for $j = 1, \ldots, n$ as diagonal entries. For the orthogonality conditions in Eq. (11.13), it is assumed that the system Eq. (11.9) is non-defective. The special case of defective systems is illustrated in detail in [28].

11.3 Literature Review of Free Interface Substructuring Approaches for Viscously Damped Systems

The equations of motion in state-space form of one substructure s are

$$\boldsymbol{A}^{(s)} \dot{\boldsymbol{z}}^{(s)} + \boldsymbol{B}^{(s)} \boldsymbol{z}^{(s)} = \boldsymbol{F}^{(s)} = \begin{bmatrix} \boldsymbol{f}^{(s)} + \boldsymbol{g}^{(s)} \\ \boldsymbol{0} \end{bmatrix}. \tag{11.14}$$

Equation (11.14) has $n^{(s)}$ DOFs and the superscript (s) is the label of the particular substructure s. State-space matrices $\boldsymbol{A}^{(s)}$ and $\boldsymbol{B}^{(s)}$ are constructed according to Eq. (11.10) whereby the substructure's mass matrix $\boldsymbol{M}^{(s)}$, damping matrix $\boldsymbol{C}^{(s)}$ and stiffness matrix $\boldsymbol{K}^{(s)}$ are used. Vector $\boldsymbol{f}^{(s)}$ corresponds to the vector of the external forces and vector $\boldsymbol{g}^{(s)}$ corresponds to the vector of internal reaction forces occurring at the interface due to the connection to adjacent substructures. The substructure state-space vector $\boldsymbol{z}^{(s)}$ is constructed according to Eq. (11.8) and can be divided into internal DOFs (those not shared with any adjacent substructure) and boundary DOFs (those shared with adjacent substructures and therefore forming the model's interface DOFs):

$$\boldsymbol{z}^{(s)} = \begin{bmatrix} \boldsymbol{u}^{(s)} \\ \boldsymbol{v}^{(s)} \end{bmatrix} = \begin{bmatrix} \boldsymbol{u}_i^{(s)} \\ \boldsymbol{u}_b^{(s)} \\ \boldsymbol{v}_i^{(s)} \\ \boldsymbol{v}_b^{(s)} \end{bmatrix} \tag{11.15}$$

The reduction is performed by using a substructure reduction matrix $\boldsymbol{\Phi}^{(s)}$. Thus, the $n^{(s)}$ DOFs of the substructure's state-space vector $\boldsymbol{z}^{(s)}$ are approximated by a reduced number of generalized DOFs $\boldsymbol{p}^{(s)}$:

$$\boldsymbol{z}^{(s)} \approx \begin{bmatrix} \boldsymbol{\Phi}_{\text{kept}}^{(s)} & \boldsymbol{\Phi}_a^{(s)} \end{bmatrix} \begin{bmatrix} \boldsymbol{p}_{\text{kept}}^{(s)} \\ \boldsymbol{p}_a^{(s)} \end{bmatrix} = \boldsymbol{\Phi}^{(s)} \boldsymbol{p}^{(s)} \tag{11.16}$$

In the reduction according to Eq. (11.16), $\boldsymbol{\Phi}_{\text{kept}}^{(s)}$ contains $n_{\text{kept}}^{(s)}$ kept eigenmodes that are solutions of the substructure's eigenvalue problem

$$\left(\lambda_j^{(s)} \boldsymbol{A}^{(s)} + \boldsymbol{B}^{(s)}\right) \boldsymbol{\phi}_j^{(s)} = \boldsymbol{0}. \tag{11.17}$$

Depending on the boundary conditions of the substructure, rigid body modes $\boldsymbol{\Phi}_r^{(s)}$ can also occur. Using the attachment modes $\boldsymbol{\Phi}_a^{(s)}$, forces occurring at the interface of connected substructures are represented statically completely.

The basic idea of model order reduction is the approximation by a reduced number of modes. Therefore, the matrix $\boldsymbol{\Phi}_{\text{kept}}^{(s)}$ contains only selected kept modes, where $n_{\text{kept}}^{(s)} < n^{(s)}$ holds. The selection is made according to a selection criterion. Usually, the kept modes are the eigenmodes corresponding to eigenvalues with the smallest magnitude, up to a limiting frequency, which depends on the application. This strategy is used in Craig and Ni's method [15] and in de Kraker and van Campen's method [16]. Liu and Zheng [17] proposed selecting the kept modes according to a weighting factor, which evaluates the influence of the respective mode on the transfer function. In this paper, the first selection criterion, i.e., keeping the modes corresponding to the eigenvalues with the smallest absolute magnitude, will be applied.

The reduced equations of motion of a substructure are obtained by inserting the reduction according to Eq. (11.16) into the substructure's equations of motion (11.14) and left multiplication with $\boldsymbol{\Phi}^{(s)^T}$. The reduced system can be written as

$$\boldsymbol{A}_{\text{red}}^{(s)} \dot{\boldsymbol{p}}^{(s)} + \boldsymbol{B}_{\text{red}}^{(s)} \boldsymbol{p}^{(s)} = \boldsymbol{\Phi}^{(s)^T} \boldsymbol{F}^{(s)} \tag{11.18}$$

with the reduced state-space matrices

$$\boldsymbol{A}_{\text{red}}^{(s)} = \boldsymbol{\Phi}^{(s)^T} \boldsymbol{A}^{(s)} \boldsymbol{\Phi}^{(s)} \qquad \text{and} \qquad \boldsymbol{B}_{\text{red}}^{(s)} = \boldsymbol{\Phi}^{(s)^T} \boldsymbol{B}^{(s)} \boldsymbol{\Phi}^{(s)}. \tag{11.19}$$

For the reduced substructure matrices, the following dimension properties apply:

$$\boldsymbol{A}_{\text{red}}^{(s)} \in \mathbb{C}^{(n_{\text{kept}}^{(s)}+n_r^{(s)}+n_a^{(s)})\times(n_{\text{kept}}^{(s)}+n_r^{(s)}+n_a^{(s)})} \quad \text{and} \quad \boldsymbol{B}_{\text{red}}^{(s)} \in \mathbb{C}^{(n_{\text{kept}}^{(s)}+n_r^{(s)}+n_a^{(s)})\times(n_{\text{kept}}^{(s)}+n_r^{(s)}+n_a^{(s)})},$$

whereby, $n_r^{(s)}$ is the number of state-space rigid body modes and $n_a^{(s)}$ is the number of attachment modes of substructure s. In the following, different substructuring methods for viscously damped systems are reviewed.

11.3.1 Craig and Ni's Method [15]

To start with, the first-order reduction according to Craig and Ni (CN) [15] is considered. Therefore, the reduction basis as proposed by Craig and Ni will be derived in Sect. 11.3.1.1, and afterwards the assembly procedure of substructures is recalled in Sect. 11.3.1.2.

11.3.1.1 Reduction Basis

Firstly, the substructures are divided into substructures without rigid body modes, and substructures with rigid body modes.

Substructures Without Rigid Body Modes

Eq. (11.16) shows that the approximation of the state-space vector z of a substructure[1] is obtained by the superposition of a dynamic part and a static part. The dynamic part is described using the kept eigenmodes $\boldsymbol{\Phi}_{\text{kept}}$. These are solutions of the corresponding eigenvalue problem of Eq. (11.17). The eigenmodes are orthonormalized, so that they satisfy the orthogonality conditions for complex modes

$$\boldsymbol{\Phi}^T \boldsymbol{A} \boldsymbol{\Phi} = \boldsymbol{I} \qquad\qquad \boldsymbol{\Phi}^T \boldsymbol{B} \boldsymbol{\Phi} = -\boldsymbol{\Lambda}. \tag{11.20}$$

[1]The indication of the substructure (s) is omitted in the Sects. 11.3.1.1, 11.3.2.1 and 11.3.3.1 for clarity.

Here, $\boldsymbol{\Lambda} \in \mathbb{C}^{n \times n}$ corresponds to the diagonal matrix, which contains the eigenvalues (see Sect. 11.2.2). The kept eigenmodes $\boldsymbol{\Phi}_{\text{kept}}$ are identical for all reduction methods considered in this paper. The static part is used to fully represent the connection forces that occur at the interface of adjacent substructures [10]. By considering only the static part, i.e., $\dot{\boldsymbol{z}} = \boldsymbol{0}$ in Eq. (11.14), and assuming that no external forces $\boldsymbol{f} = \boldsymbol{0}$ occur, the static problem can be written as

$$\boldsymbol{B} \boldsymbol{z}_{\text{stat}} = \boldsymbol{F}_a \qquad \text{with} \qquad \boldsymbol{F}_a = \begin{bmatrix} \boldsymbol{0}_i \\ \boldsymbol{I}_b \\ \hline \boldsymbol{0}_i \\ \boldsymbol{0}_b \end{bmatrix}. \tag{11.21}$$

Matrix $\boldsymbol{F}_a$ is a force matrix, which consists of unit-force vectors arranged in columns for each interface displacement DOF. The attachment modes $\boldsymbol{\Phi}_a$ result by solving $\boldsymbol{B}\boldsymbol{\Phi}_a = \boldsymbol{F}_a$ and can be symbolically represented as

$$\boldsymbol{\Phi}_a = \boldsymbol{B}^{-1} \boldsymbol{F}_a = \boldsymbol{G}_e \boldsymbol{F}_a. \tag{11.22}$$

Here, $\boldsymbol{G}_e$ corresponds to the flexibility matrix, which for substructures without rigid body modes is equal to the inverse of the state-space stiffness matrix $\boldsymbol{B}$. Eq. (11.22) shows that attachment modes are columns of the corresponding flexibility matrix. Based on the definition of the force matrix $\boldsymbol{F}_a$ according to Craig and Ni, the attachment modes of substructures without rigid body modes contain only entries in the displacement DOFs.

The basic idea of model order reduction is the approximation by keeping only a reduced number of modes $\boldsymbol{\Phi}_{\text{kept}}$. The information of these modes is already contained in the reduction basis. Therefore, the goal in the flexibility matrix is to solely consider information from the truncated modes $\boldsymbol{\Phi}_{\text{trunc}}$. The part of the flexibility matrix resulting from the truncated modes is called the residual flexibility matrix [1]. Using Eq. (11.20) this reads [29]

$$\boldsymbol{G}_{\text{res}} = \boldsymbol{\Phi}_{\text{trunc}} (-\boldsymbol{\Lambda}_{\text{trunc}})^{-1} \boldsymbol{\Phi}_{\text{trunc}}^T = \boldsymbol{G}_e + \boldsymbol{\Phi}_{\text{kept}} \boldsymbol{\Lambda}_{\text{kept}}^{-1} \boldsymbol{\Phi}_{\text{kept}}^T. \tag{11.23}$$

The advantage of the second representation is that only the eigenmodes corresponding to the smallest eigenvalues have to be calculated. From Eq. (11.23), the attachment modes $\boldsymbol{\Phi}_a$,CN according to Craig and Ni's method arise thus:

$$\boldsymbol{\Phi}_{a,\text{res}} = \boldsymbol{G}_{\text{res}} \boldsymbol{F}_a = \boldsymbol{\Phi}_{a,\text{CN}} \tag{11.24}$$

For Craig and Ni's attachment modes, $\boldsymbol{\Phi}_{a,\text{CN}} \in \mathbb{C}^{n \times n_{b,u}}$ applies. Here, $n_{b,u}$ corresponds to the number of interface displacement DOFs. The reduction of substructures without rigid body modes according to Craig and Ni's method is finally done by

$$z_{\text{CN}} \approx \underbrace{\begin{bmatrix} \boldsymbol{\Phi}_{\text{kept}} & \boldsymbol{\Phi}_{a,\text{CN}} \end{bmatrix}}_{\boldsymbol{\Phi}_{\text{CN}}} \begin{bmatrix} \boldsymbol{p}_{\text{kept}} \\ \boldsymbol{p}_a \end{bmatrix}. \tag{11.25}$$

Substructures with Rigid Body Modes

In the case of substructures with rigid body modes, the substructure stiffness matrix $\boldsymbol{K}$ is singular. The substructure has m_r physical rigid body modes $\boldsymbol{\theta}_r$ with zero eigenvalues, for which the following applies:

$$\boldsymbol{K} \boldsymbol{\theta}_r = \boldsymbol{0} \tag{11.26}$$

In state-space, the occurrence of a physical rigid body mode $\boldsymbol{\theta}_r$ leads to either one or two rigid body modes. For a detailed explanation, please refer to [28]. Thus, for the number n_r of rigid body modes in the state-space $m_r \leqslant n_r \leqslant 2m_r$ holds. Using the rigid body modes $\boldsymbol{\theta}_r$, two cases are distinguished [16, 17]:

Case 1: $\boldsymbol{C}\boldsymbol{\theta}_r = \boldsymbol{0}$

In this case, the rigid body motion does not cause any damping forces. The substructure is termed "defective". In the mathematical sense, a matrix of size $n \times n$ is deemed defective if it does not have a linearly independent set of n eigenvectors [1]. The corresponding rigid body movements in the state-space result in

$$\boldsymbol{\phi}_{\text{reg}} = \begin{bmatrix} \boldsymbol{\theta}_r \\ \boldsymbol{0} \end{bmatrix} \qquad \boldsymbol{\phi}_{\text{gen}} = \begin{bmatrix} \boldsymbol{0} \\ \boldsymbol{\theta}_r \end{bmatrix}. \tag{11.27}$$

Here, $\boldsymbol{\phi}_{\text{reg}}$ is referred to as regular rigid body mode and $\boldsymbol{\phi}_{\text{gen}}$ is referred to as a generalized rigid body mode.

Case 2: $\boldsymbol{C}\boldsymbol{\theta}_r \neq \boldsymbol{0}$
In this case, the rigid body motion causes damping forces. The substructure is non-defective. In contrast to defective matrices, matrices are called non-defective if the n-dimensional space is spanned by a complete set of independent eigenvectors [30]. Thus, there is only one regular rigid body mode $\boldsymbol{\phi}_{\text{reg}}$ and no generalized rigid body mode occurs in the state-space:

$$\boldsymbol{\phi}_{\text{reg}} = \begin{bmatrix} \boldsymbol{\theta}_r \\ \boldsymbol{0} \end{bmatrix}. \tag{11.28}$$

All state-space rigid body modes can be combined in the state-space rigid body matrix $\boldsymbol{\Phi}_r$. In the case of substructures with rigid body modes, an adaptation of the procedure is necessary to determine the attachment modes according to Eq. (11.24). As a first step, a projection matrix $\boldsymbol{P}$ is defined for substructures with rigid body modes [1, 15]

$$\boldsymbol{P} = \boldsymbol{I} - \boldsymbol{A}\boldsymbol{\Phi}_r \boldsymbol{A}_{rr}^{-1} \boldsymbol{\Phi}_r^T \qquad \text{with} \quad \boldsymbol{A}_{rr} = \boldsymbol{\Phi}_r^T \boldsymbol{A} \boldsymbol{\Phi}_r. \tag{11.29}$$

In Eq. (11.29), $\boldsymbol{A}_{rr}$ corresponds to the rigid body mass matrix in state-space. If an arbitrary force vector $\boldsymbol{F}$ is multiplied by the matrix $\boldsymbol{P}$, no rigid body modes are excited by the force vector [15]. Additionally, due to the singularity of the stiffness matrix $\boldsymbol{K}$, it is not possible to invert the state-space stiffness matrix $\boldsymbol{B}$. Therefore, as a second step, the n_i internal DOFs z_i of substructures with rigid body modes are divided into a part z_r corresponding to the number n_r of rigid body modes and a part z_e corresponding to the number n_e of elastic modes, where $n_i = n_r + n_e$ applies [16]

$$z = \begin{bmatrix} z_i \\ z_b \end{bmatrix} = \begin{bmatrix} z_r \\ z_e \\ z_b \end{bmatrix}. \tag{11.30}$$

The rigid body parts of the stiffness matrix $\boldsymbol{K}$ are fixed, from which the constrained stiffness matrix $\boldsymbol{K}_c$ follows:

$$\boldsymbol{K}_c = \begin{bmatrix} \boldsymbol{K}_{ee} & \boldsymbol{K}_{eb} \\ \boldsymbol{K}_{be} & \boldsymbol{K}_{bb} \end{bmatrix} \tag{11.31}$$

Matrix $\boldsymbol{K}_c$ is regular and can therefore be inverted. The pseudo flexibility matrix $\boldsymbol{G}$ for substructures with rigid body modes results in [15]

$$\boldsymbol{G} = \begin{bmatrix} \overline{\boldsymbol{K}}^{-1} & \boldsymbol{0} \\ \boldsymbol{0} & -\boldsymbol{M}^{-1} \end{bmatrix} \qquad \text{with} \qquad \overline{\boldsymbol{K}}^{-1} = \begin{bmatrix} \boldsymbol{0} & \boldsymbol{0} & \boldsymbol{0} \\ \boldsymbol{0} & \multicolumn{2}{c}{\multirow{2}{*}{$\begin{bmatrix} \boldsymbol{K}_{ee} & \boldsymbol{K}_{eb} \\ \boldsymbol{K}_{be} & \boldsymbol{K}_{bb} \end{bmatrix}^{-1}$}} \\ \boldsymbol{0} & & \end{bmatrix}. \tag{11.32}$$

Finally, the flexibility matrix $\boldsymbol{G}_e$ for substructures with rigid body modes is defined through the combination of both steps

$$\boldsymbol{G}_e = \boldsymbol{P}^T \boldsymbol{G} \boldsymbol{P}. \tag{11.33}$$

Analogous to substructures without rigid body modes, the residual flexibility matrix is defined as $\boldsymbol{G}_{\text{res}} = \boldsymbol{G}_e + \boldsymbol{\Phi}_{\text{kept}} \boldsymbol{\Lambda}_{\text{kept}}^{-1} \boldsymbol{\Phi}_{\text{kept}}^T$ and the residual attachment modes are obtained by

$$\boldsymbol{\Phi}_{a,\text{CN}} = \boldsymbol{G}_{\text{res}} \boldsymbol{F}_a. \tag{11.34}$$

Here, $\boldsymbol{\Lambda}_{\text{kept}}$ contains the eigenvalues of the kept eigenmodes without rigid body modes. Thus, the reduction of substructures with rigid body modes according to Craig and Ni's method is

$$z_{\text{CN}} \approx \underbrace{\begin{bmatrix} \boldsymbol{\Phi}_r & \boldsymbol{\Phi}_{\text{kept}} & \boldsymbol{\Phi}_{a,\text{CN}} \end{bmatrix}}_{\boldsymbol{\Phi}_{\text{CN}}} \begin{bmatrix} p_r \\ p_{\text{kept}} \\ p_a \end{bmatrix}. \tag{11.35}$$

11.3.1.2 Assembly Procedure (Elimination of Interface DOFs)

The assembly of two substructures α and β according to Craig and Ni is based on the interface displacement compatibility and the interface force equilibrium

$$\boldsymbol{u}_b^{(\alpha)} = \boldsymbol{u}_b^{(\beta)}, \tag{11.36}$$

$$\boldsymbol{g}_b^{(\alpha)} + \boldsymbol{g}_b^{(\beta)} = \boldsymbol{0}, \tag{11.37}$$

where the index b denotes the boundary DOFs. The interface forces $\boldsymbol{g}_b$ are approximated by the attachment modes. Thus, they correspond physically to the generalized attachment modes' DOFs $\boldsymbol{p}_a$ [31]. The condition

$$\boldsymbol{p}_a^{(\alpha)} = -\boldsymbol{p}_a^{(\beta)} \tag{11.38}$$

follows from the interface force equilibrium of Eq. (11.37) The assembly is done on the basis of the displacements. For this purpose, only the displacement part $\boldsymbol{u}$ of the approximation of the state-space vector in Eq. (11.25) is considered:

$$\boldsymbol{u} \approx \boldsymbol{\Phi}_{\text{kept},u}\ \boldsymbol{p}_{\text{kept}} + \boldsymbol{\Phi}_{a,u}\ \boldsymbol{p}_a \tag{11.39}$$

In Eq. (11.39), $\boldsymbol{\Phi}_{\text{kept},u}$ corresponds to the displacement part of the kept eigenmodes $\boldsymbol{\Phi}_{\text{kept}}$ and $\boldsymbol{\Phi}_{a,u}$ corresponds to the displacement part of the attachment modes $\boldsymbol{\Phi}_a$. Inserting the approximation (11.39) into the displacement compatibility condition (11.36) yields

$$\boldsymbol{\Phi}_{\text{kept},u,b}^{(\alpha)}\boldsymbol{p}_{\text{kept}}^{(\alpha)} + \boldsymbol{\Phi}_{a,u,b}^{(\alpha)}\boldsymbol{p}_a^{(\alpha)} = \boldsymbol{\Phi}_{\text{kept},u,b}^{(\beta)}\boldsymbol{p}_{\text{kept}}^{(\beta)} + \boldsymbol{\Phi}_{a,u,b}^{(\beta)}\boldsymbol{p}_a^{(\beta)}. \tag{11.40}$$

From Eq. (11.40), the following arises, by using condition (11.38) and rearranging:

$$\boldsymbol{\Phi}_{a,u,b}^{(\alpha)}\boldsymbol{p}_a^{(\alpha)} + \boldsymbol{\Phi}_{a,u,b}^{(\beta)}\boldsymbol{p}_a^{(\alpha)} = -\boldsymbol{\Phi}_{\text{kept},u,b}^{(\alpha)}\boldsymbol{p}_{\text{kept}}^{(\alpha)} + \boldsymbol{\Phi}_{\text{kept},u,b}^{(\beta)}\boldsymbol{p}_{\text{kept}}^{(\beta)}, \tag{11.41}$$

which can be written in matrix notation as

$$\left[\boldsymbol{\Phi}_{a,u,b}^{(\alpha)} + \boldsymbol{\Phi}_{a,u,b}^{(\beta)}\right]\boldsymbol{p}_a^{(\alpha)} = \left[-\boldsymbol{\Phi}_{\text{kept},u,b}^{(\alpha)}\ \ \boldsymbol{\Phi}_{\text{kept},u,b}^{(\beta)}\right]\begin{bmatrix}\boldsymbol{p}_{\text{kept}}^{(\alpha)}\\ \boldsymbol{p}_{\text{kept}}^{(\beta)}\end{bmatrix}. \tag{11.42}$$

Thus, it is possible to express the generalized attachment modes' DOFs $\boldsymbol{p}_a$ depending on the remaining DOFs $\boldsymbol{p}_{\text{kept}}$ of both substructures:

$$\boldsymbol{p}_a^{(\alpha)} = \left[\boldsymbol{\Phi}_{a,u,b}^{(\alpha)} + \boldsymbol{\Phi}_{a,u,b}^{(\beta)}\right]^{-1}\left[-\boldsymbol{\Phi}_{\text{kept},u,b}^{(\alpha)}\ \ \boldsymbol{\Phi}_{\text{kept},u,b}^{(\beta)}\right]\begin{bmatrix}\boldsymbol{p}_{\text{kept}}^{(\alpha)}\\ \boldsymbol{p}_{\text{kept}}^{(\beta)}\end{bmatrix} = -\boldsymbol{p}_a^{(\beta)} \tag{11.43}$$

The interface parameters are eliminated by means of

$$\begin{bmatrix}\boldsymbol{p}_{\text{kept}}^{(\alpha)}\\ \boldsymbol{p}_a^{(\alpha)}\\ \boldsymbol{p}_{\text{kept}}^{(\beta)}\\ \boldsymbol{p}_a^{(\beta)}\end{bmatrix} = \underbrace{\begin{bmatrix}\boldsymbol{I}_{\text{kept}}^{(\alpha)} & \boldsymbol{0}\\ \boldsymbol{D}_{1,u} & \boldsymbol{D}_{2,u}\\ \boldsymbol{0} & \boldsymbol{I}_{\text{kept}}^{(\beta)}\\ -\boldsymbol{D}_{1,u} & -\boldsymbol{D}_{2,u}\end{bmatrix}}_{\boldsymbol{\Phi}_{2,u}}\begin{bmatrix}\boldsymbol{p}_{\text{kept}}^{(\alpha)}\\ \boldsymbol{p}_{\text{kept}}^{(\beta)}\end{bmatrix} \tag{11.44}$$

with

$$\boldsymbol{D}_{1,u} = -\left[\boldsymbol{\Phi}_{a,u,b}^{(\alpha)} + \boldsymbol{\Phi}_{a,u,b}^{(\beta)}\right]^{-1}\boldsymbol{\Phi}_{\text{kept},u,b}^{(\alpha)} \quad \text{and} \quad \boldsymbol{D}_{2,u} = \left[\boldsymbol{\Phi}_{a,u,b}^{(\alpha)} + \boldsymbol{\Phi}_{a,u,b}^{(\beta)}\right]^{-1}\boldsymbol{\Phi}_{\text{kept},u,b}^{(\beta)}. \tag{11.45}$$

This finally results in the reduced overall system

$$\boldsymbol{A}_{\text{red}}\dot{\boldsymbol{p}}_{\text{kept}} + \boldsymbol{B}_{\text{red}}\boldsymbol{p}_{\text{kept}} = \boldsymbol{0} \tag{11.46}$$

with

$$\boldsymbol{A}_{\text{red}} = \boldsymbol{\Phi}_{2,u}^T \begin{bmatrix} \boldsymbol{\Phi}_{\mathsf{CN}}^{(\alpha)\,T}\boldsymbol{A}^{(\alpha)}\boldsymbol{\Phi}_{\mathsf{CN}}^{(\alpha)} & \boldsymbol{0} \\ \boldsymbol{0} & \boldsymbol{\Phi}_{\mathsf{CN}}^{(\beta)\,T}\boldsymbol{A}^{(\beta)}\boldsymbol{\Phi}_{\mathsf{CN}}^{(\beta)} \end{bmatrix} \boldsymbol{\Phi}_{2,u} \quad \text{and} \quad \boldsymbol{B}_{\text{red}} = \boldsymbol{\Phi}_{2,u}^T \begin{bmatrix} \boldsymbol{\Phi}_{\mathsf{CN}}^{(\alpha)\,T}\boldsymbol{B}^{(\alpha)}\boldsymbol{\Phi}_{\mathsf{CN}}^{(\alpha)} & \boldsymbol{0} \\ \boldsymbol{0} & \boldsymbol{\Phi}_{\mathsf{CN}}^{(\beta)\,T}\boldsymbol{B}^{(\beta)}\boldsymbol{\Phi}_{\mathsf{CN}}^{(\beta)} \end{bmatrix} \boldsymbol{\Phi}_{2,u}.$$

For the assembled reduced matrices, the following dimension properties apply:

$$\boldsymbol{A}_{\text{red}} \in \mathbb{C}^{(n_{\text{kept}}^{(\alpha)}+n_r^{(\alpha)}+n_{\text{kept}}^{(\beta)}+n_r^{(\beta)})\times(n_{\text{kept}}^{(\alpha)}+n_r^{(\alpha)}+n_{\text{kept}}^{(\beta)}+n_r^{(\beta)})} \quad \text{and} \quad \boldsymbol{B}_{\text{red}} \in \mathbb{C}^{(n_{\text{kept}}^{(\alpha)}+n_r^{(\alpha)}+n_{\text{kept}}^{(\beta)}+n_r^{(\beta)})\times(n_{\text{kept}}^{(\alpha)}+n_r^{(\alpha)}+n_{\text{kept}}^{(\beta)}+n_r^{(\beta)})}$$

This shows that the interface displacement DOFs and the interface forces are completely eliminated during assembly according to Craig and Ni. This is similar to the assembly procedure of MacNeal's method for the undamped case.

11.3.2 Liu and Zheng's Method [17]

11.3.2.1 Reduction Basis

The second-order reduction according to Liu and Zheng (LZ) [17] represents an extension of the first-order reduction according to Craig and Ni described in the previous section. For this reason, only the differences to the procedure in Sect. 11.3.1.1 are explained below. The exact representation (no reduction) of the state-space vector z considering all eigenmodes $\boldsymbol{\phi}_k$ is

$$z = \sum_{k=1}^{n} \boldsymbol{\phi}_k p_k = \sum_{k=1}^{n_{\text{kept}}} \boldsymbol{\phi}_{\text{kept},k} p_{\text{kept},k} + \sum_{k=n_{\text{kept}}+1}^{n} \boldsymbol{\phi}_{\text{trunc},k} p_{\text{trunc},k}. \tag{11.47}$$

The basic idea of model order reduction is the approximation by keeping only a reduced number n_{kept} of eigenmodes $\boldsymbol{\phi}_k$. Vector $\boldsymbol{\phi}_{\text{kept},k}$ in Eq. (11.47) denotes one kept eigenmode $\boldsymbol{\phi}_k$ and vector $\boldsymbol{\phi}_{\text{trunc},k}$ denotes one truncated eigenmode. According to Craig and Ni's reduction basis, the influence of the truncated modes on the dynamic behavior is statically approximated by means of the attachment modes. The extension in the second-order reduction according to Liu and Zheng refers to the additional determination of second-order attachment modes. In comparison to Craig and Ni's first-order reduction, a dynamic part is retained with the second-order attachment modes. Thus, the inertia and damping effects of the truncated modes are considered in the following [17]. The kept eigenmodes $\boldsymbol{\Phi}_{\text{kept}}$ used for the reduction are identical to the kept modes according to Craig and Ni's reduction basis.

Substructures Without Rigid Body Modes

In the case of substructures without rigid body modes, the attachment modes according to Liu and Zheng are defined as [17]

$$\boldsymbol{\Phi}_{a1,\mathsf{LZ}} = \boldsymbol{Q}_1\boldsymbol{F}_a \quad \text{and} \quad \boldsymbol{\Phi}_{a2,\mathsf{LZ}} = \boldsymbol{Q}_2\boldsymbol{F}_a \tag{11.48}$$

with

$$\boldsymbol{Q}_1 = \boldsymbol{B}^{-1} + \boldsymbol{\Phi}_{\text{kept}}\boldsymbol{\Lambda}_{\text{kept}}^{-1}\boldsymbol{\Phi}_{\text{kept}}^T \quad \text{and} \quad \boldsymbol{Q}_2 = -\boldsymbol{B}^{-1}\boldsymbol{A}\boldsymbol{B}^{-1} + \boldsymbol{\Phi}_{\text{kept}}\boldsymbol{\Lambda}_{\text{kept}}^{-2}\boldsymbol{\Phi}_{\text{kept}}^T. \tag{11.49}$$

Utilizing the relation $\boldsymbol{G}_e = \boldsymbol{B}^{-1}$, Eq. (11.49) can be rewritten as

$$\boldsymbol{Q}_1 = \boldsymbol{G}_e + \boldsymbol{\Phi}_{\text{kept}}\boldsymbol{\Lambda}_{\text{kept}}^{-1}\boldsymbol{\Phi}_{\text{kept}}^T = \boldsymbol{G}_{\text{res}} \quad \text{and} \quad \boldsymbol{Q}_2 = -\boldsymbol{G}_e\boldsymbol{A}\boldsymbol{G}_e + \boldsymbol{\Phi}_{\text{kept}}\boldsymbol{\Lambda}_{\text{kept}}^{-2}\boldsymbol{\Phi}_{\text{kept}}^T. \tag{11.50}$$

From Eq. (11.50), it can be seen that the first-order attachment modes $\boldsymbol{\Phi}_{a1,\mathsf{LZ}}$ according to Liu and Zheng are identical to the attachment modes $\boldsymbol{\Phi}_{a,\mathsf{CN}}$ as defined by Craig and Ni in Eq. (11.24):

$$\boldsymbol{\Phi}_{a1,\mathsf{LZ}} = \boldsymbol{\Phi}_{a,\mathsf{CN}} \tag{11.51}$$

The difference between this and the reduction according to Craig and Ni is the extension of the first-order attachment modes by the second-order attachment modes $\boldsymbol{\Phi}_{a2,\mathsf{LZ}}$ of Eq. (11.48). For the dimension of the attachment modes $\boldsymbol{\Phi}_{a,\mathsf{LZ}} = \left[\boldsymbol{\Phi}_{a1,\mathsf{LZ}}\ \boldsymbol{\Phi}_{a2,\mathsf{LZ}}\right]$ according to Liu and Zheng, $\boldsymbol{\Phi}_{a1,\mathsf{LZ}} \in \mathbb{C}^{n\times n_{b,u}}$, $\boldsymbol{\Phi}_{a2,\mathsf{LZ}} \in \mathbb{C}^{n\times n_{b,u}}$ and thus $\boldsymbol{\Phi}_{a,\mathsf{LZ}} \in \mathbb{C}^{n\times 2n_{b,u}}$ apply. The reduction of substructures without rigid body modes results in

$$z_{\mathsf{LZ}} \approx \underbrace{\left[\boldsymbol{\Phi}_{\text{kept}}\ \boldsymbol{\Phi}_{a1,\mathsf{LZ}}\ \boldsymbol{\Phi}_{a2,\mathsf{LZ}}\right]}_{\boldsymbol{\Phi}_{\mathsf{LZ}}} \begin{bmatrix} \boldsymbol{p}_{\text{kept}} \\ \boldsymbol{p}_a \\ \dot{\boldsymbol{p}}_a \end{bmatrix}. \tag{11.52}$$

Substructures with Rigid Body Modes

In the case of substructures with rigid body modes, the attachment modes are defined as [17]

$$\boldsymbol{\Phi}_{a1,\mathsf{LZ}} = \boldsymbol{G}_{\text{res}}\boldsymbol{F}_a = \boldsymbol{\Phi}_{a,\mathsf{CN}} \qquad \boldsymbol{\Phi}_{a2,\mathsf{LZ}} = -\boldsymbol{G}_e\boldsymbol{A}\boldsymbol{G}_e\boldsymbol{F}_a. \tag{11.53}$$

In Eq. (11.53), the matrices $\boldsymbol{G}_{\text{res}}$ and $\boldsymbol{G}_e$ are identical to the definition according to Craig and Ni in Eq. (11.33). In comparison to the attachment modes of substructures without rigid body modes, it can be seen that for substructures with rigid body modes, the dynamic term $\boldsymbol{\Phi}_{\text{kept}}\boldsymbol{\Lambda}_{\text{kept}}^{-2}\boldsymbol{\Phi}_{\text{kept}}^T$ is not considered for the computation of second-order attachment modes $\boldsymbol{\Phi}_{a2,\mathsf{LZ}}$ by Liu and Zheng [17]. This results in the reduction of substructures with rigid body modes according to Liu and Zheng as

$$z_{\mathsf{LZ}} \approx \underbrace{\left[\boldsymbol{\Phi}_r\ \boldsymbol{\Phi}_{\text{kept}}\ \boldsymbol{\Phi}_{a1,\mathsf{LZ}}\ \boldsymbol{\Phi}_{a2,\mathsf{LZ}}\right]}_{\boldsymbol{\Phi}_{\mathsf{LZ}}} \begin{bmatrix} \boldsymbol{p}_r \\ \boldsymbol{p}_{\text{kept}} \\ \boldsymbol{p}_a \\ \dot{\boldsymbol{p}}_a \end{bmatrix}. \tag{11.54}$$

11.3.2.2 Assembly Procedure (Elimination of Interface DOFs)

For assembly of the reduced substructures according to Liu and Zheng, the interface displacement compatibility (11.36) and the interface force equilibrium (11.37) are extended to the state-space form [17]:

$$\begin{bmatrix} \boldsymbol{u}_b^{(\alpha)} \\ \dot{\boldsymbol{u}}_b^{(\alpha)} \end{bmatrix} = \begin{bmatrix} \boldsymbol{u}_b^{(\beta)} \\ \dot{\boldsymbol{u}}_b^{(\beta)} \end{bmatrix}, \tag{11.55}$$

$$\begin{bmatrix} \boldsymbol{g}_b^{(\alpha)} \\ \dot{\boldsymbol{g}}_b^{(\alpha)} \end{bmatrix} + \begin{bmatrix} \boldsymbol{g}_b^{(\beta)} \\ \dot{\boldsymbol{g}}_b^{(\beta)} \end{bmatrix} = \boldsymbol{0} \tag{11.56}$$

Equations (11.55) and (11.56) show that for the assembly according to Liu and Zheng an additional velocity compatibility and an impulse equilibrium at the interface are used. Thus, it is possible to eliminate all interface DOFs. From the extended equilibrium of forces in Eq. (11.56), the state-space equilibrium condition

$$\begin{bmatrix} \boldsymbol{p}_a^{(\alpha)} \\ \dot{\boldsymbol{p}}_a^{(\alpha)} \end{bmatrix} + \begin{bmatrix} \boldsymbol{p}_a^{(\beta)} \\ \dot{\boldsymbol{p}}_a^{(\beta)} \end{bmatrix} = \boldsymbol{0} \tag{11.57}$$

follows. By inserting the approximation of Eq. (11.52) into the extended compatibility condition of Eq. (11.55), the following is obtained:

$$\boldsymbol{\Phi}_{\text{kept},b}^{(\alpha)}\boldsymbol{p}_{\text{kept}}^{(\alpha)} + \boldsymbol{\Phi}_{a1,b}^{(\alpha)}\boldsymbol{p}_a^{(\alpha)} + \boldsymbol{\Phi}_{a2,b}^{(\alpha)}\dot{\boldsymbol{p}}_a^{(\alpha)} = \boldsymbol{\Phi}_{\text{kept},b}^{(\beta)}\boldsymbol{p}_{\text{kept}}^{(\beta)} + \boldsymbol{\Phi}_{a1,b}^{(\beta)}\boldsymbol{p}_a^{(\beta)} + \boldsymbol{\Phi}_{a2,b}^{(\beta)}\dot{\boldsymbol{p}}_a^{(\beta)} \tag{11.58}$$

From Eq. (11.58) follows under utilization of Eq. (11.57) and rearrangement

$$\left[\boldsymbol{\Phi}_{a1,b}^{(\alpha)} + \boldsymbol{\Phi}_{a1,b}^{(\beta)}\ \ \boldsymbol{\Phi}_{a2,b}^{(\alpha)} + \boldsymbol{\Phi}_{a2,b}^{(\beta)}\right] \begin{bmatrix} \boldsymbol{p}_a^{(\alpha)} \\ \dot{\boldsymbol{p}}_a^{(\alpha)} \end{bmatrix} = \left[-\boldsymbol{\Phi}_{\text{kept},b}^{(\alpha)}\ \ \boldsymbol{\Phi}_{\text{kept},b}^{(\beta)}\right] \begin{bmatrix} \boldsymbol{p}_{\text{kept}}^{(\alpha)} \\ \boldsymbol{p}_{\text{kept}}^{(\beta)} \end{bmatrix}. \tag{11.59}$$

The concluding transformation is analogous to Craig and Ni's assembly

$$\begin{bmatrix} \boldsymbol{p}_{\text{kept}}^{(\alpha)} \\ \boldsymbol{p}_a^{(\alpha)} \\ \dot{\boldsymbol{p}}_a^{(\alpha)} \\ \boldsymbol{p}_{\text{kept}}^{(\beta)} \\ \boldsymbol{p}_a^{(\beta)} \\ \dot{\boldsymbol{p}}_a^{(\beta)} \end{bmatrix} = \underbrace{\begin{bmatrix} \boldsymbol{I}_{\text{kept}}^{(\alpha)} & \mathbf{0} \\ \boldsymbol{D}_1 & \boldsymbol{D}_2 \\ \mathbf{0} & \boldsymbol{I}_{\text{kept}}^{(\beta)} \\ -\boldsymbol{D}_1 & -\boldsymbol{D}_2 \end{bmatrix}}_{\boldsymbol{\Phi}_2} \begin{bmatrix} \boldsymbol{p}_{\text{kept}}^{(\alpha)} \\ \boldsymbol{p}_{\text{kept}}^{(\beta)} \end{bmatrix} \tag{11.60}$$

with

$$\boldsymbol{D}_1 = -\left[\boldsymbol{\Phi}_{a1,b}^{(\alpha)} + \boldsymbol{\Phi}_{a1,b}^{(\beta)}\ \boldsymbol{\Phi}_{a2,b}^{(\alpha)} + \boldsymbol{\Phi}_{a2,b}^{(\beta)}\right]^{-1} \boldsymbol{\Phi}_{\text{kept},b}^{(\alpha)} \quad \text{and} \quad \boldsymbol{D}_2 = \left[\boldsymbol{\Phi}_{a1,b}^{(\alpha)} + \boldsymbol{\Phi}_{a1,b}^{(\beta)}\ \boldsymbol{\Phi}_{a2,b}^{(\alpha)} + \boldsymbol{\Phi}_{a2,b}^{(\beta)}\right]^{-1} \boldsymbol{\Phi}_{\text{kept},b}^{(\beta)}. \tag{11.61}$$

For the assembled reduced matrices, the following dimension properties apply:

$$\boldsymbol{A}_{\text{red}} \in \mathbb{C}^{(n_{\text{kept}}^{(\alpha)}+n_r^{(\alpha)}+n_{\text{kept}}^{(\beta)}+n_r^{(\beta)})\times(n_{\text{kept}}^{(\alpha)}+n_r^{(\alpha)}+n_{\text{kept}}^{(\beta)}+n_r^{(\beta)})} \quad \text{and} \quad \boldsymbol{B}_{\text{red}} \in \mathbb{C}^{(n_{\text{kept}}^{(\alpha)}+n_r^{(\alpha)}+n_{\text{kept}}^{(\beta)}+n_r^{(\beta)})\times(n_{\text{kept}}^{(\alpha)}+n_r^{(\alpha)}+n_{\text{kept}}^{(\beta)}+n_r^{(\beta)})}.$$

Thus, the size of the reduced assembled system is identical to the size of the reduced assembled system according to Craig and Ni's method.

11.3.3 De Kraker and van Campen's Method [16]

De Kraker and van Campen's method (KC) [16] can be considered as an extension of Rubin's method for the case of arbitrarily viscously damped systems.

11.3.3.1 Reduction Basis

The reduction according to de Kraker and van Campen represents another approach for the determination of attachment modes. In their approach, the force matrix is extended to the velocity boundary DOFs [16]:

$$\boldsymbol{F}_{a,\mathsf{KC}} = \begin{bmatrix} \mathbf{0}_i & \mathbf{0}_i \\ \boldsymbol{I}_b & \mathbf{0}_b \\ \mathbf{0}_i & \mathbf{0}_i \\ \mathbf{0}_b & \boldsymbol{I}_b \end{bmatrix} \tag{11.62}$$

Substructures Without Rigid Body Modes

The attachment modes of substructures without rigid body modes according to de Kraker and van Campen are defined as follows

$$\boldsymbol{\Phi}_{a,\mathsf{KC}} = \boldsymbol{G}_e \boldsymbol{F}_{a,\mathsf{KC}}. \tag{11.63}$$

Since the force matrix $\boldsymbol{F}_{a,\mathsf{KC}}$ consists of a displacement and a velocity part, the resulting matrix $\boldsymbol{\Phi}_{a,\mathsf{KC}}$ of the attachment modes according to de Kraker and van Campen can symbolically be divided into a displacement and a velocity part:

$$\boldsymbol{\Phi}_{a,\mathsf{KC}} = \left[\boldsymbol{\Phi}_a\ \dot{\boldsymbol{\Phi}}_a\right] \tag{11.64}$$

In Eq. (11.64), the displacement attachment modes $\boldsymbol{\Phi}_a$ are identical to the attachment modes according to Craig and Ni as well as Liu and Zheng in Eq. (11.22). The velocity attachment modes $\dot{\boldsymbol{\Phi}}_a$ extend the reduction basis compared to the reduction approach according to Craig and Ni. For the attachment modes according to de Kraker and van Campen $\boldsymbol{\Phi}_a \in$

$\mathbb{C}^{n \times n_{b,u}}$, $\dot{\boldsymbol{\Phi}}_{a,\mathsf{KC}} \in \mathbb{C}^{n \times n_{b,u}}$ and thus $\boldsymbol{\Phi}_{a,\mathsf{KC}} \in \mathbb{C}^{n \times 2n_{b,u}}$ apply. Accordingly, the matrix $\boldsymbol{\Phi}_{a,\mathsf{KC}}$ has twice the number of columns compared to the matrix $\boldsymbol{\Phi}_{a,\mathsf{CN}}$ and has the same dimension as the attachment modes $\boldsymbol{\Phi}_{a,\mathsf{LZ}}$ according to Liu and Zheng. In addition, de Kraker only uses attachment modes and not residual attachment modes. De Kraker and van Campen's reduction is

$$z_{\mathsf{KC}} \approx \underbrace{\begin{bmatrix} \boldsymbol{\Phi}_{\text{kept}} & \boldsymbol{\Phi}_{a,\mathsf{KC}} \end{bmatrix}}_{\boldsymbol{\Phi}_{\mathsf{KC}}} \begin{bmatrix} \boldsymbol{p}_{\text{kept}} \\ \boldsymbol{p}_a \end{bmatrix}. \tag{11.65}$$

Substructures with Rigid Body Modes

For substructures with rigid body modes, de Kraker and van Campen define the pseudo flexibility matrix in a manner different from the reduction methods according to Craig and Ni as well as Liu and Zheng. From de Kraker and van Campen's definition results [16]

$$\boldsymbol{G}_{\mathsf{KC}} = \begin{bmatrix} \overline{\boldsymbol{K}}^{-1} & \mathbf{0} \\ \mathbf{0} & -\overline{\boldsymbol{M}}^{-1} \end{bmatrix} \quad \text{with} \tag{11.66}$$

$$\overline{\boldsymbol{K}}^{-1} = \begin{bmatrix} \mathbf{0} & \mathbf{0} & \mathbf{0} \\ \mathbf{0} & \multicolumn{2}{c}{\begin{bmatrix} \boldsymbol{K}_{ee} & \boldsymbol{K}_{eb} \\ \boldsymbol{K}_{be} & \boldsymbol{K}_{bb} \end{bmatrix}^{-1}} \\ \mathbf{0} & & \end{bmatrix} \quad \text{and} \quad \overline{\boldsymbol{M}}^{-1} = \begin{bmatrix} \mathbf{0} & \mathbf{0} & \mathbf{0} \\ \mathbf{0} & \multicolumn{2}{c}{\begin{bmatrix} \boldsymbol{M}_{ee} & \boldsymbol{M}_{eb} \\ \boldsymbol{M}_{be} & \boldsymbol{M}_{bb} \end{bmatrix}^{-1}} \\ \mathbf{0} & & \end{bmatrix}.$$

In Eq. (11.66), it can be seen that parts of the mass matrix $\boldsymbol{M}$ are also fixed. The attachment modes follow thus:

$$\boldsymbol{\Phi}_{a,\mathsf{KC}} = \boldsymbol{G}_{e,\mathsf{KC}} \boldsymbol{F}_a \qquad \text{with} \qquad \boldsymbol{G}_{e,\mathsf{KC}} = \boldsymbol{P}^T \boldsymbol{G}_{\mathsf{KC}} \boldsymbol{P} \tag{11.67}$$

The reduction of substructures with rigid body modes according to de Kraker and van Campen is

$$z_{\mathsf{KC}} \approx \underbrace{\begin{bmatrix} \boldsymbol{\Phi}_r & \boldsymbol{\Phi}_{\text{kept}} & \boldsymbol{\Phi}_{a,\mathsf{KC}} \end{bmatrix}}_{\boldsymbol{\Phi}_{\mathsf{KC}}} \begin{bmatrix} \boldsymbol{p}_r \\ \boldsymbol{p}_{\text{kept}} \\ \boldsymbol{p}_a \end{bmatrix}. \tag{11.68}$$

11.3.3.2 Assembly Procedure (Primal Assembly)

De Kraker and van Campen use a primal assembly approach for a reduction with free interface modes as proposed by Martinez et al. [32]. In their concept, the boundary displacement and velocity DOFs are to be retained in physical coordinates after the reduction. Hereby, superelements are generated out of the individual substructures, which are assembled in a primal way. The approach is based on the reduction of Eq. (11.68). By arranging the DOFs $z^{(s)}$ according to internal DOFs $z_i^{(s)}$ and boundary DOFs $z_b^{(s)}$, the approximation of the state-space vector can be written as

$$\begin{bmatrix} z_i^{(s)} \\ z_b^{(s)} \end{bmatrix} \approx \begin{bmatrix} \boldsymbol{\Phi}_{\text{kept},i}^{(s)} & \boldsymbol{\Phi}_{a,i}^{(s)} \\ \boldsymbol{\Phi}_{\text{kept},b}^{(s)} & \boldsymbol{\Phi}_{a,b}^{(s)} \end{bmatrix} \begin{bmatrix} \boldsymbol{p}_{\text{kept}}^{(s)} \\ \boldsymbol{p}_a^{(s)} \end{bmatrix}. \tag{11.69}$$

To consider the cases of substructures with and without rigid body modes at the same time, the rigid body modes $\boldsymbol{\Phi}_r^{(s)}$ are included in the matrix $\boldsymbol{\Phi}_{\text{kept}}^{(s)}$ in the following if the substructure possesses rigid body modes. The goal is now to have the boundary displacement and velocity DOFs as physical coordinates $z_b^{(s)}$ after the reduction. For this reason, the second line of Eq. (11.69) $z_b^{(s)} = \boldsymbol{\Phi}_{\text{kept},b}^{(s)} \boldsymbol{p}_{\text{kept}}^{(s)} + \boldsymbol{\Phi}_{a,b}^{(s)} \boldsymbol{p}_a^{(s)}$ is considered and resolved for $\boldsymbol{p}_a^{(s)}$:

$$\boldsymbol{p}_a^{(s)} = \boldsymbol{\Phi}_{a,b}^{(s)\,-1} z_b^{(s)} - \boldsymbol{\Phi}_{a,b}^{(s)\,-1} \boldsymbol{\Phi}_{\text{kept},b}^{(s)} \boldsymbol{p}_{\text{kept}}^{(s)} \tag{11.70}$$

From Eq. (11.70), the transformation matrix $\boldsymbol{\Phi}_2^{(s)}$ can be derived, which expresses the generalized DOFs $\boldsymbol{p}_a^{(s)}$ depending on the physical DOFs $z_b^{(s)}$ [32]:

$$\begin{bmatrix} \boldsymbol{p}_{\text{kept}}^{(s)} \\ \boldsymbol{p}_a^{(s)} \end{bmatrix} = \underbrace{\begin{bmatrix} \boldsymbol{I}_{\text{kept}}^{(s)} & \boldsymbol{0}_{\text{kept}}^{(s)} \\ -\boldsymbol{\Phi}_{a,b}^{(s)\,-1}\boldsymbol{\Phi}_{\text{kept},b}^{(s)} & \boldsymbol{\Phi}_{a,b}^{(s)\,-1} \end{bmatrix}}_{\boldsymbol{\Phi}_2^{(s)}} \begin{bmatrix} \boldsymbol{p}_{\text{kept}}^{(s)} \\ z_b^{(s)} \end{bmatrix}. \tag{11.71}$$

The final reduction follows from multiplying reduction matrix $\boldsymbol{\Phi}_{\text{KC}}^{(s)}$ and transformation matrix $\boldsymbol{\Phi}_2^{(s)}$:

$$z^{(s)} = \begin{bmatrix} z_i^{(s)} \\ z_b^{(s)} \end{bmatrix} \approx \boldsymbol{\Phi}_{\text{KC}}^{(s)}\boldsymbol{\Phi}_2^{(s)} \begin{bmatrix} \boldsymbol{p}_{\text{kept}}^{(s)} \\ z_b^{(s)} \end{bmatrix} = \begin{bmatrix} \boldsymbol{\Phi}_{\text{kept},i}^{(s)} - \boldsymbol{\Phi}_{a,i}^{(s)}\boldsymbol{\Phi}_{a,b}^{(s)\,-1}\boldsymbol{\Phi}_{\text{kept},b}^{(s)} & \boldsymbol{\Phi}_{a,i}^{(s)}\boldsymbol{\Phi}_{a,b}^{(s)\,-1} \\ \boldsymbol{0}_b^{(s)} & \boldsymbol{I}_b^{(s)} \end{bmatrix} \begin{bmatrix} \boldsymbol{p}_{\text{kept}}^{(s)} \\ z_b^{(s)} \end{bmatrix}. \tag{11.72}$$

It can be seen that only the internal DOFs are transformed into generalized coordinates and that the interface displacement and velocity DOFs are present in physical coordinates. Thus, it is possible to assemble the substructures in a primal way after the transformation and superelements are created. For the assembled reduced matrices, the following dimension properties apply:

$$\boldsymbol{A}_{\text{red}} \in \mathbb{C}^{(n_{\text{kept}}^{(\alpha)}+n_r^{(\alpha)}+n_{\text{kept}}^{(\beta)}+n_r^{(\beta)}+n_a^{(\alpha)})\times(n_{\text{kept}}^{(\alpha)}+n_r^{(\alpha)}+n_{\text{kept}}^{(\beta)}+n_r^{(\beta)}+n_a^{(\alpha)})} \quad \text{and}$$

$$\boldsymbol{B}_{\text{red}} \in \mathbb{C}^{(n_{\text{kept}}^{(\alpha)}+n_r^{(\alpha)}+n_{\text{kept}}^{(\beta)}+n_r^{(\beta)}+n_a^{(\alpha)})\times(n_{\text{kept}}^{(\alpha)}+n_r^{(\alpha)}+n_{\text{kept}}^{(\beta)}+n_r^{(\beta)}+n_a^{(\alpha)})}.$$

By keeping the boundary displacement and velocity DOFs, the size of the assembled system increases with the number of attachment modes $n_a^{(\alpha)}$ and $n_a^{(\beta)}$ compared to Craig and Ni's method and Liu and Zheng's method.

11.4 Improvements of Free Interface Substructuring Approaches

11.4.1 Third and Higher-Order Reduction Interface Flexibility Representation

The determination of higher-order attachment modes as in the reduction according to Liu and Zheng can be extended arbitrarily. Thus, the influence of the truncated modes is represented more accurately. However, the size of the reduced system also increases.

Substructures Without Rigid Body Modes
In the case of substructures without rigid body modes, the attachment modes of higher order are generated by

$$\boldsymbol{\Phi}_{a,k} = \boldsymbol{Q}_k \boldsymbol{F}_a \tag{11.73}$$

with

$$\boldsymbol{Q}_k = (-1)^{k-1}\boldsymbol{G}_e(\boldsymbol{A}\boldsymbol{G}_e)^{k-1} + \boldsymbol{\Phi}_{\text{kept}}\boldsymbol{\Lambda}_{\text{kept}}^{-k}\boldsymbol{\Phi}_{\text{kept}}^T \qquad \text{for} \qquad k = 1, \ldots, h.$$

Here, h corresponds to the highest order of the reduction.

Substructures with Rigid Body Modes
For substructures with rigid body modes, the attachment modes of higher order are generated by

$$\boldsymbol{\Phi}_{a,k} = \boldsymbol{Q}_k \boldsymbol{F}_a \tag{11.74}$$

with

$$Q_k = (-1)^{k-1} G_e (A G_e)^{k-1} \qquad \text{for} \qquad k = 1, \ldots, h.$$

Third-Order Reduction

Based on Eqs. (11.73) and (11.74), a third-order reduction (TO) will be explicitly given in the following. The third-order attachment modes for substructures without rigid body modes are defined as

$$\Phi_{a3} = \left((-1)^2 G_e (A G_e)^2 + \Phi_{\text{kept}} \Lambda_{\text{kept}}^{-3} \Phi_{\text{kept}}^T\right) F_a, \tag{11.75}$$

and the third-order attachment modes for substructures with rigid body modes are defined as

$$\Phi_{a3} = \left((-1)^2 G_e (A G_e)^2\right) F_a. \tag{11.76}$$

Thus, the reduction is

$$z_{\mathsf{TO}} \approx \underbrace{\left[\Phi_r \ \Phi_{\text{kept}} \ \Phi_{a1} \ \Phi_{a2} \ \Phi_{a3}\right]}_{\Phi_{\mathsf{TO}}} \begin{bmatrix} p_r \\ p_{\text{kept}} \\ p_a \\ \dot{p}_a \\ \ddot{p}_a \end{bmatrix}. \tag{11.77}$$

The first and second-order attachment modes correspond to the attachment modes according to Craig/Ni and Liu/Zheng. In Eq. (11.77), $\ddot{p}_a$ represents the third-order attachment modes' DOFs, which can be interpreted as the second derivative of the generalized displacements DOFs p_a and can conceptually be identified as accelerations. The assembly follows the assembly procedure according to Liu and Zheng in Sect. 11.3.2.2 on the basis of the displacements and velocities. The elimination of acceleration DOFs $\ddot{p}_a$ is not possible in state-space. Therefore, they are kept as independent DOFs and increase the size of the reduced system.

11.4.2 Combination of Liu/Zheng's Reduction Basis and Primal Assembly Procedure

Furthermore, a combination of Liu and Zheng's reduction basis as given in Sect. 11.3.2.1 and the primal assembly procedure as shown in Sect. 11.3.3.2 will be examined. In order to assemble the substructures in a primal fashion, the physical boundary displacement and velocity DOFs have to be retained in the reduced system. Thus, the assembled reduced system has a larger number of DOFs compared to Liu and Zheng's assembly procedure, which eliminates all boundary DOFs. Following the reduction approach of Sect. 11.3.2.1, the assembly has to be based on the Eq. (11.52). By arranging the displacement and velocity DOFs in the state-space vector $z^{(s)}$ according to internal DOFs $z_i^{(s)}$ and boundary DOFs $z_b^{(s)}$, Eq. (11.52) can be partitioned:

$$\begin{bmatrix} z_i^{(s)} \\ z_b^{(s)} \end{bmatrix} = \begin{bmatrix} \Phi_{\text{kept},i}^{(s)} & \Phi_{a1,i}^{(s)} & \Phi_{a2,i}^{(s)} \\ \Phi_{\text{kept},b}^{(s)} & \Phi_{a1,b}^{(s)} & \Phi_{a2,b}^{(s)} \end{bmatrix} \begin{bmatrix} p_{\text{kept}}^{(s)} \\ p_a^{(s)} \\ \dot{p}_a^{(s)} \end{bmatrix} \tag{11.78}$$

For clarity, $\Phi_{a1}^{(s)}$ and $\Phi_{a2}^{(s)}$ are combined to $\Phi_{a,\mathsf{LZ}}^{(s)} = \left[\Phi_{a1}^{(s)} \ \Phi_{a2}^{(s)}\right]$ and $p_a^{(s)}$ and $\dot{p}_a^{(s)}$ are combined to $p_{a,\mathsf{LZ}}^{(s)} = \left[p_a^{(s)} \ \dot{p}_a^{(s)}\right]$. In order to have the interface displacement and velocity DOFs in physical coordinates $z_b^{(s)}$ after the reduction, the second row of Eq. (11.78) is solved for $p_{a,\mathsf{LZ}}^{(s)}$:

$$p_{a,\mathsf{LZ}}^{(s)} = \Phi_{a,\mathsf{LZ},b}^{(s)}{}^{-1} z_b^{(s)} - \Phi_{a,\mathsf{LZ},b}^{(s)}{}^{-1} \Phi_{\text{kept},b}^{(s)} p_{\text{kept}}^{(s)}. \tag{11.79}$$

From Eq. (11.79), the transformation matrix $\boldsymbol{\Phi}_{2,\mathrm{LZ}}^{(s)}$ can be derived, which expresses the generalized DOFs $\boldsymbol{p}_{a,\mathrm{LZ}}^{(s)}$ depending on the physical DOFs $\boldsymbol{z}_b^{(s)}$ [32]

$$\begin{bmatrix} \boldsymbol{p}_{\mathrm{kept}}^{(s)} \\ \boldsymbol{p}_{a,\mathrm{LZ}}^{(s)} \end{bmatrix} = \underbrace{\begin{bmatrix} \boldsymbol{I}_{\mathrm{kept}}^{(s)} & \boldsymbol{0}_{\mathrm{kept}}^{(s)} \\ -\boldsymbol{\Phi}_{a,\mathrm{LZ},b}^{(s)}{}^{-1}\boldsymbol{\Phi}_{\mathrm{kept},b}^{(s)} & \boldsymbol{\Phi}_{a,\mathrm{LZ},b}^{(s)}{}^{-1} \end{bmatrix}}_{\boldsymbol{\Phi}_{2,\mathrm{LZ}}^{(s)}} \begin{bmatrix} \boldsymbol{p}_{\mathrm{kept}}^{(s)} \\ \boldsymbol{z}_b^{(s)} \end{bmatrix}. \tag{11.80}$$

The final reduction follows from multiplying reduction matrix $\boldsymbol{\Phi}_{\mathrm{LZ}}^{(s)}$ and transformation matrix $\boldsymbol{\Phi}_{2,\mathrm{LZ}}^{(s)}$:

$$\boldsymbol{z}^{(s)} = \begin{bmatrix} \boldsymbol{z}_i^{(s)} \\ \boldsymbol{z}_b^{(s)} \end{bmatrix} = \boldsymbol{\Phi}_{\mathrm{LZ}}^{(s)}\boldsymbol{\Phi}_{2,\mathrm{LZ}}^{(s)} \begin{bmatrix} \boldsymbol{p}_{\mathrm{kept}}^{(s)} \\ \boldsymbol{z}_b^{(s)} \end{bmatrix} = \begin{bmatrix} \boldsymbol{\Phi}_{\mathrm{kept},i}^{(s)} - \boldsymbol{\Phi}_{a,\mathrm{LZ},i}^{(s)}\boldsymbol{\Phi}_{a,\mathrm{LZ},b}^{(s)}{}^{-1}\boldsymbol{\Phi}_{\mathrm{kept},b}^{(s)} & \boldsymbol{\Phi}_{a,\mathrm{LZ},i}^{(s)}\boldsymbol{\Phi}_{a,\mathrm{LZ},b}^{(s)}{}^{-1} \\ \boldsymbol{0}_b^{(s)} & \boldsymbol{I}_b^{(s)} \end{bmatrix} \begin{bmatrix} \boldsymbol{p}_{\mathrm{kept}}^{(s)} \\ \boldsymbol{z}_b^{(s)} \end{bmatrix}. \tag{11.81}$$

It can be seen that only the internal DOFs are transformed into generalized coordinates and the boundary displacement and velocity DOFs are present in physical coordinates. Thus, the substructures can be assembled in a primal way after the transformation and superelements are created. For the assembled reduced matrices, the following dimension properties apply:

$$\boldsymbol{A}_{\mathrm{red}} \in \mathbb{C}^{(n_{\mathrm{kept}}^{(\alpha)}+n_r^{(\alpha)}+n_{\mathrm{kept}}^{(\beta)}+n_r^{(\beta)}+n_{a1}^{(\alpha)}+n_{a2}^{(\alpha)})\times(n_{\mathrm{kept}}^{(\alpha)}+n_r^{(\alpha)}+n_{\mathrm{kept}}^{(\beta)}+n_r^{(\beta)}+n_{a1}^{(\alpha)}+n_{a2}^{(\alpha)})} \quad \text{and}$$

$$\boldsymbol{B}_{\mathrm{red}} \in \mathbb{C}^{(n_{\mathrm{kept}}^{(\alpha)}+n_r^{(\alpha)}+n_{\mathrm{kept}}^{(\beta)}+n_r^{(\beta)}+n_{a1}^{(\alpha)}+n_{a2}^{(\alpha)})\times(n_{\mathrm{kept}}^{(\alpha)}+n_r^{(\alpha)}+n_{\mathrm{kept}}^{(\beta)}+n_r^{(\beta)}+n_{a1}^{(\alpha)}+n_{a2}^{(\alpha)})}$$

By keeping the boundary DOFs, the size of the assembled system increases with the number of attachment modes $n_{a1}^{(\alpha)} + n_{a2}^{(\alpha)}$ and $n_{a1}^{(\beta)} + n_{a2}^{(\beta)}$ compared to Liu and Zheng's method.

11.5 Numerical Applications

11.5.1 Beam Structure with Two Localized Dampers

To evaluate the proposed methods numerically, the clamped beam structure with two localized dampers shown in Fig. 11.1 is analyzed [16]. This example has already been used in [16]. The beam structure consists of 18 Euler-Bernoulli beam elements and only the bending vibration (no axial deformation) is considered. The total system has a length of 1.8 m and is divided into substructure α, consisting of ten beam elements and a length of 1.0 m, and substructure β, consisting of eight beam elements and a length of 0.8 m. A localized viscous damper with a damper constant $c = 1.0 \cdot 10^4\,\mathrm{N\,s\,m^{-1}}$ is attached to each substructure. The total system has $m = 36$ DOFs in the physical space and $n = 72$ DOFs in the state-space. Substructure α has $n^{(\alpha)} = 40$ DOFs in state-space and substructure β has $n^{(\beta)} = 36$ DOFs in the state-space. Furthermore, substructure β has two physical rigid body modes $m_r^{(\beta)} = 2$ (translation and rotation). In state-space representation, the physical rigid body modes lead to a single zero eigenvalue for the translational movement and to one zero eigenvalue with multiplicity two for the rotational movement. This can be traced back to the fact that the translation rigid body movement causes a damping force. On the other hand, the rotational rigid body movement is freely possible and doesn't cause a damping force. This results in a regular rigid body mode for the translation and a regular and a generalized rigid body mode for the rotation in the state-space. Accordingly, there are $n_r^{(\beta)} = 3$ state-space rigid body modes.

Figure 11.2 shows all 72 eigenvalues of the unreduced system in the complex plane. It can be seen that complex conjugate eigenvalues occur to a large extent . Altogether, there are 34 complex conjugate eigenvalue pairs. Thus, there are 68 complex eigenvalues with imaginary parts. In contrast, there are four real eigenvalues without imaginary parts (two of them with real part $\Re(\lambda) < -5000\,\mathrm{rad\,s^{-1}}$). This indicates very high damping of these eigenvalues [16].

Figure 11.3 shows the exact eigenvalues of both substructures. For substructure α, there are 19 eigenvalue pairs and 2 real eigenvalues. For substructure β, there are 16 eigenvalue pairs and one real eigenvalue as well as 3 rigid body modes with zero eigenvalue.

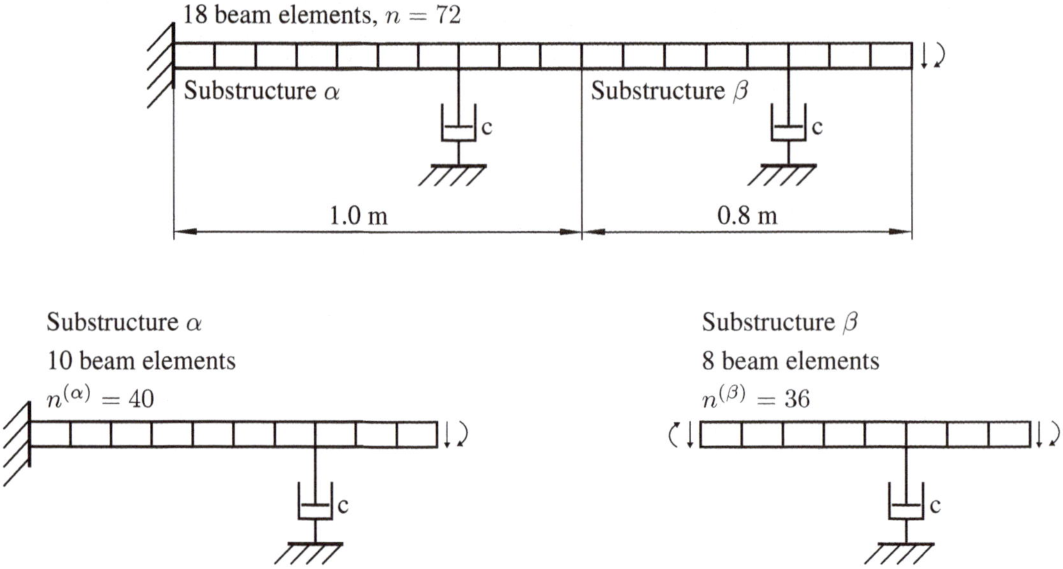

Fig. 11.1 Clamped beam structure with two localized dampers divided into two substructures [16]. The beam consists of 18 Euler-Bernoulli beam elements (Young's modulus $2.1 \cdot 10^{11}\,\mathrm{N\,m^{-2}}$, density $7.8 \cdot 10^{3}\,\mathrm{kg\,m^{-3}}$, cross-section $9.0 \cdot 10^{-4}\,\mathrm{m^{2}}$, moment of inertia $7.0 \cdot 10^{-8}\,\mathrm{m^{4}}$) and has $n = 72$ DOFs in state-space. The total length of the beam is 1.8 m. The length of substructure α is 1.0 m and the length of substructure β is 0.8 m. The damper constant $c =$ is $1.0 \cdot 10^{4}\,\mathrm{N\,s\,m^{-1}}$

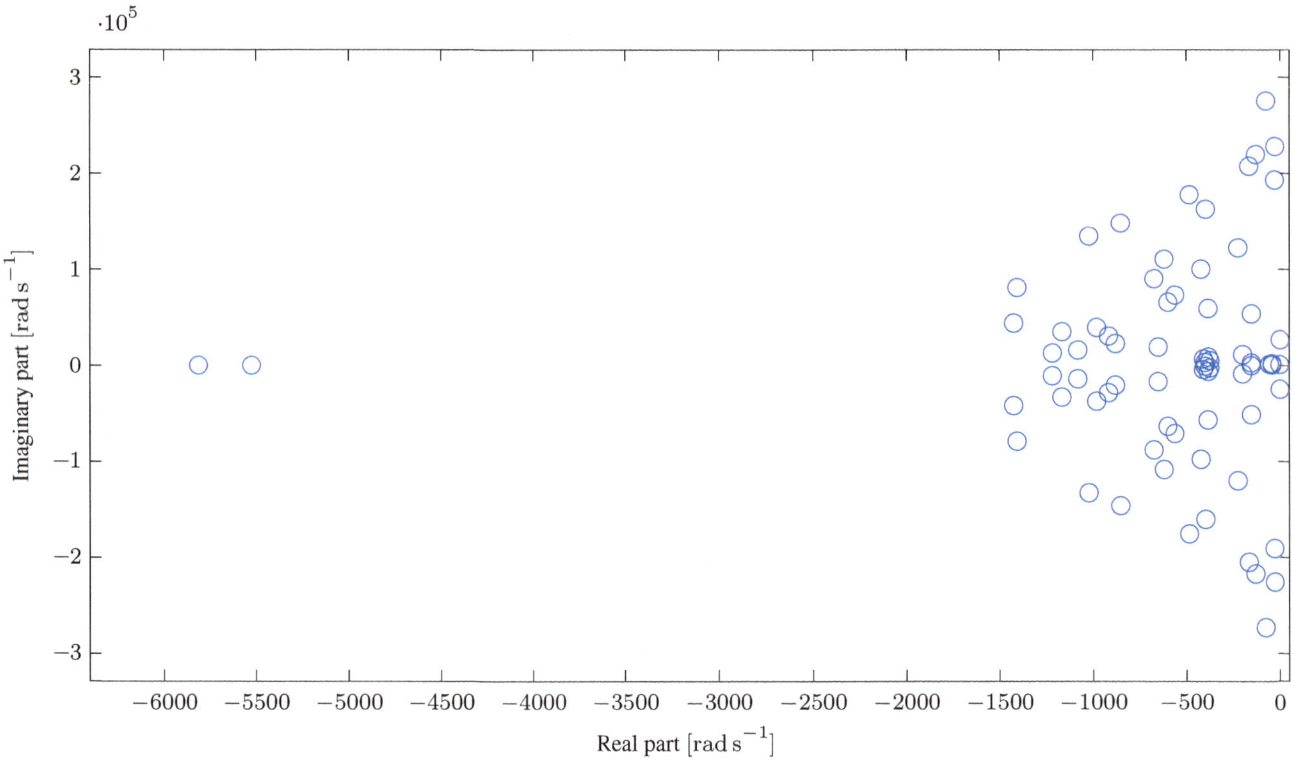

Fig. 11.2 Exact eigenvalues of the coupled unreduced system of Fig. 11.1

The clamped beam is reduced and assembled using the substructuring methods described in Sects. 11.3 and 11.4. For the reduction of substructure α, $n^{(\alpha)}_{\text{kept}} = 20$ eigenmodes belonging to the 20 eigenvalues with the lowest absolute value are kept. These are nine complex conjugate pairs and two real eigenvalues without imaginary parts. For substructure β $n^{(\beta)}_{\text{kept}} = 15$,

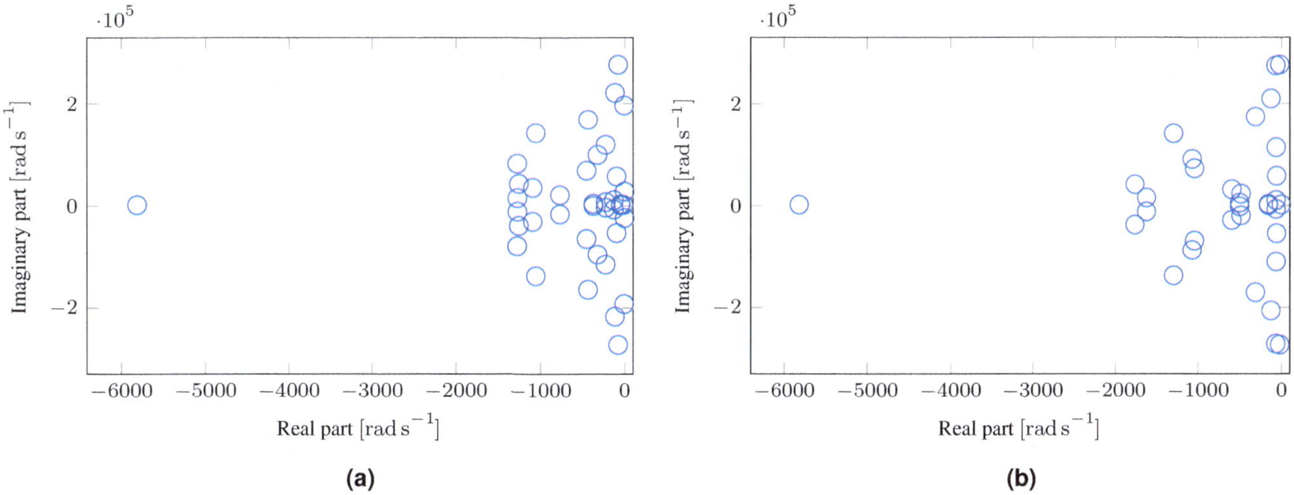

Fig. 11.3 Exact eigenvalues of the unreduced substructures of Fig. 11.1. (**a**) Substructure α. (**b**) Substructure β

Table 11.1 Modes used for reduction and resulting size of the reduced assembled system

	Craig/Ni's method		Liu/Zheng's method		de Kraker/van Campen's method		Liu/Zheng's basis with primal assembly		Third order reduction	
Substructure	α	β	α	β	α	β	α	β	α	β
Kept eigenmodes	20	15	20	15	20	15	20	15	20	15
Attachment modes	2	2	4	4	4	4	4	4	6	6
Rigid body modes	0	3	0	3	0	3	0	3	0	3
DOFs reduced system	38		38		42		42		42	

eigenmodes belonging to the 15 eigenvalues with the lowest absolute value are kept. These are seven complex conjugate pairs and one real eigenvalue without imaginary parts. Additionally, for the reduction of substructure β, the $n_r^{(\beta)} = 3$ rigid body modes are used.

For the reduction according to Craig and Ni, both attachment modes are determined for substructure α according to Eq. (11.24) and for substructure β according to Eq. (11.34). For the reduction according to Liu and Zheng and for the reduction according to de Kraker and van Campen, the four attachment modes are determined for substructure α according to Eqs. (11.48) and (11.63), respectively, and for substructure β according to Eqs. (11.53) and (11.67), respectively. For the third order approach, six attachment modes are determined according to Eq. (11.73) for substructure α and according to Eq. (11.74) for substructure β.

After assembly, the reduced system according to Craig and Ni's method (CN) and Liu and Zheng's method (LZ) has $n_{\text{red,CN}} = n_{\text{red,LZ}} = 38$ DOFs. According to de Kraker and van Campen's method (KC), Liu and Zheng's reduction basis with primal assembly (LZ,KC), and the third order approach (TO), the reduced system has $n_{\text{red,KC}} = n_{\text{red,LZ,KC}} = n_{\text{red,TO}} = 42$ DOFs. Table 11.1 summarizes the number of used modes and the size of the reduced assembled systems.

In order to quantify the differences of the individual methods, the relative error of the real parts and imaginary parts of the eigenvalue λ_k is used in the following. The eigenvalues of the full unreduced system $\lambda_{\text{full},k}$ are set in relation to the eigenvalues of the assembled reduced system $\lambda_{\text{red},k}$, which leads to the relative errors $\varepsilon_{\text{rel},\Re,k}$ of the real part and the relative errors $\varepsilon_{\text{rel},\Im,k}$ of the complex parts:

$$\varepsilon_{\text{rel},\Re,k} = \frac{\left|\Re\left(\lambda_{\text{red},k}\right) - \Re\left(\lambda_{\text{full},k}\right)\right|}{\Re\left(\lambda_{\text{full},k}\right)} \quad \text{and} \quad \varepsilon_{\text{rel},\Im,k} = \frac{\left|\Im\left(\lambda_{\text{red},k}\right) - \Im\left(\lambda_{\text{full},k}\right)\right|}{\Im\left(\lambda_{\text{full},k}\right)} \tag{11.82}$$

Figure 11.4 shows the relative errors of the real and imaginary parts corresponding to the 34 eigenvalues with the lowest absolute value for the various methods.For better distinguishability, only the relative errors $\varepsilon_{\text{rel}} > 3 \cdot 10^{-9}$ are depicted in the following figures. The relative errors of both eigenvalues belonging to a complex-conjugate pair are represented. The absence of a relative error of the imaginary part $\varepsilon_{\text{rel},\Im}$ implies that the associated eigenvalue is purely real and has no imaginary part. Relative errors increase with increasing eigenvalue. In addition, the imaginary parts show a slightly better agreement than the

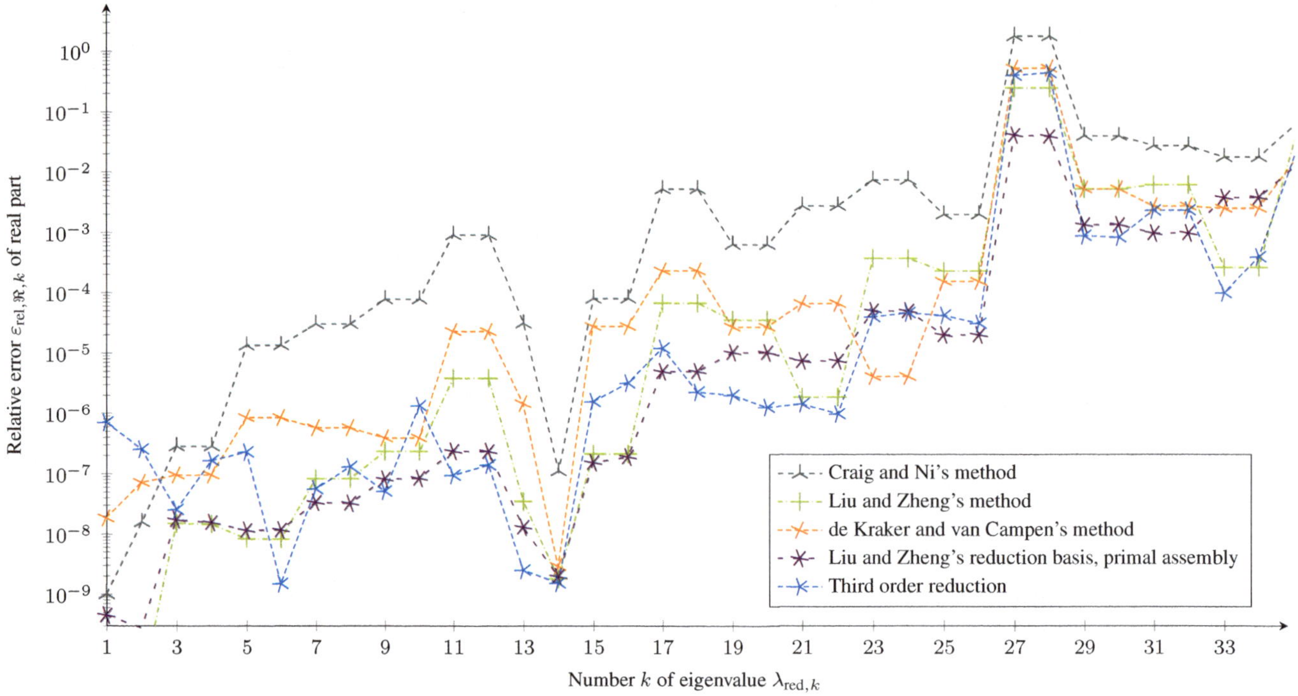

(a)

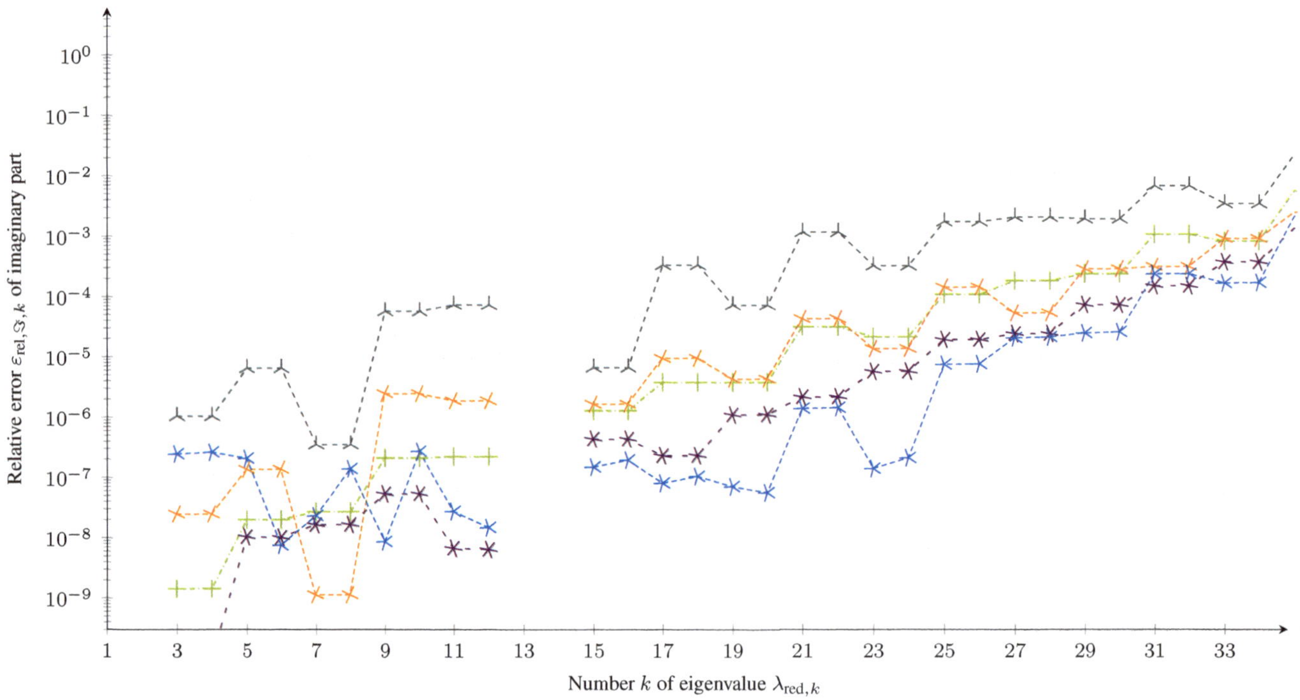

(b)

Fig. 11.4 Relative error of the real and imaginary parts of the approximated eigenvalues λ_{red} of the clamped beam of Fig. 11.1. The relative errors of the 34 eigenvalues with the lowest absolute value are shown using Craig and Ni's method (CN), Liu and Zheng's method (LZ) and de Kraker and van Campen's method (KC). Additionally, the combination of Liu and Zheng's reduction basis and primal assembly (LZ,KC) as well as the third order reduction (TO) are shown. The number of DOFs of the assembled system according to Craig and Ni's method as well as Liu and Zheng's method is $n_{\mathrm{red,CN}} = n_{\mathrm{red,LZ}} = 38$ and according to de Kraker and van Campen's method, the combination of Liu and Zheng's reduction basis and primal assembly, as well as the third order reduction is $n_{\mathrm{red,KC}} = n_{\mathrm{red,LZ,KC}} = n_{\mathrm{red,TO}} = 42$. (**a**) Relative error $\varepsilon_{\mathrm{rel},\Re,k}$ of real part to eigenvalue $\lambda_{\mathrm{red},k}$. (**b**) Relative error $\varepsilon_{\mathrm{rel},\Im,k}$ of imaginary part to eigenvalue $\lambda_{\mathrm{red},k}$

Table 11.2 Modes used for reduction and resulting size of the reduced assembled system relating to Fig. 11.5

	Craig/Ni's method		Liu/Zheng's method		de Kraker/van Campen's method		Liu/Zheng's basis with primal assembly		Third order reduction	
Substructure	α	β	α	β	α	β	α	β	α	β
Kept eigenmodes	22	17	22	17	20	15	20	15	20	15
Attachment modes	2	2	4	4	4	4	4	4	6	6
Rigid body modes	0	3	0	3	0	3	0	3	0	3
DOFs reduced system	42		42		42		42		42	

real parts. In the case of purely real eigenvalues, the real parts approximate the exact eigenvalues very well. Comparing the substructuring methods of Sect. 11.3, the eigenvalues are approximated one to two orders of magnitude more accurately by using Liu and Zheng's method and de Kraker and van Campen's method compared to Craig and Ni's method. Furthermore, no significant difference can be seen between Liu and Zheng's method and de Kraker and van Campen's method. Thus, Liu and Zheng's method approximates the exact eigenvalues at a smaller size of the reduced system as accurately as de Kraker and van Campen's method. Both approaches shown in Sect. 11.4 (the third order reduction in Sect. 11.4.1 and Liu and Zheng's reduction basis with primal assembly in Sect. 11.4.2) lead to an improvement of the approximation in comparison to the aforementioned methods. The overall best approximation is achieved by the third order reduction.

Comparison Based on the Identical Size of the Reduced System

In Fig. 11.4, the comparison is based on an identical number of kept eigenmodes for the reduction in all approaches. Thus, the reduced assembled systems have a different number of remaining DOFs, depending on the assembly procedure. In contrast, the relative errors for reduced systems of the same size is considered here. In order to achieve this, different numbers of eigenmodes are kept for the methods, in such a way that the size $n_{\text{red}} = 42$ of the reduced system is the same for all methods. For Craig and Ni's method as well as Liu and Zheng's method, 22 eigenmodes are kept. De Kraker and van Campen's method, Liu and Zheng's reduction and primal assembly, as well as the third order reduction, are based on the identical number of modes as in the preceding analysis. Table 11.2 shows the modes used for the reduction and the resulting size of the reduced assembled system. Figure 11.5 shows the relative errors of the clamped beam of Fig. 11.1 based on the identical size of the reduced system. It can be seen that this leads to improved results for Craig and Ni's method as well as Liu and Zheng's method. However, Liu and Zheng's reduction basis and primal assembly, as well as the third order reduction, still show better approximations of the exact eigenvalues.

11.5.2 Damped Free-Free Beam Model

In this section, the damped free-free beam structure shown in Fig. 11.6 is analyzed [17]. This example has already been used in [17]. The beam structure consists of 20 Euler-Bernoulli beam elements and the bending vibration (no axial deformation) is considered. The total system has a length of 1.8 m and is divided into substructure α, consisting of 12 beam elements and a length of 1.08 m, and substructure β, consisting of eight beam elements and a length of 0.72 m. The damping is proportional to the stiffness for each element. For elements 1–5 and 13–16 (left parts of both substructures), a damping matrix $\boldsymbol{C}_e = 0.0002\boldsymbol{K}_e$ is assumed. For elements 6–12 (right part of substructure α) the damping matrix is $\boldsymbol{C}_e = 0.0004\boldsymbol{K}_e$ and for elements 17–20 (right part of substructure β) the damping matrix is $\boldsymbol{C}_e = 0.0001\boldsymbol{K}_e$. Due to the different damping per element, the damping of both the substructures and the overall system is non-proportional overall.

The total system has $n = 42$ DOFs in the physical space and $m = 84$ DOFs in the state-space. Substructure α has $n^{(\alpha)} = 52$ DOFs in the state-space and substructure β has $n^{(\beta)} = 36$ DOFs in the state-space. Furthermore, both substructures possess two physical rigid body modes ($m_r^{(\alpha)} = m_r^{(\beta)} = 2$). In state-space representation, they lead to zero eigenvalues with multiplicity two for the translation as well as for the rotation for both substructures. This can be attributed to the fact that, due to the absence of local dampers, both the translational rigid body movements as well as the rotational rigid body movements do not cause any damping forces. Thus, for both substructures, one regular and one generalized rigid body mode apply with regard to the translation and also one regular and one generalized rigid body mode for the rotation result in the state-space, i.e., $n_r^{(\alpha)} = n_r^{(\beta)} = 4$.

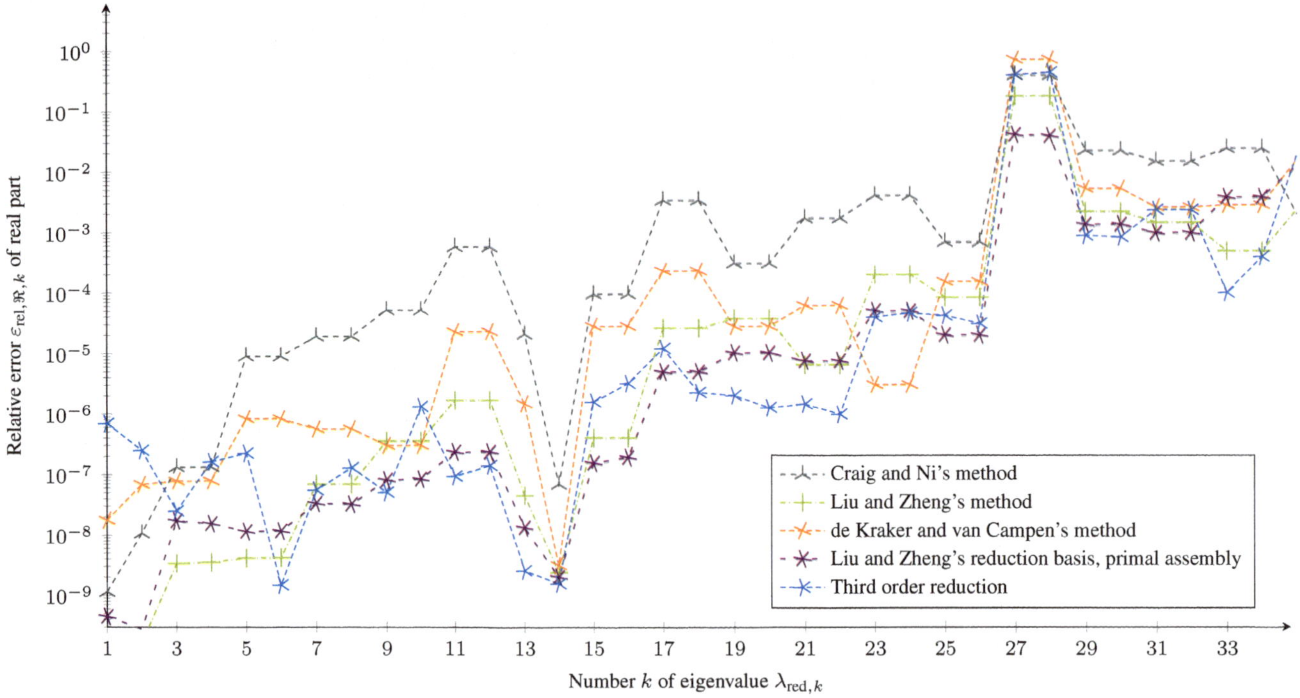

(a)

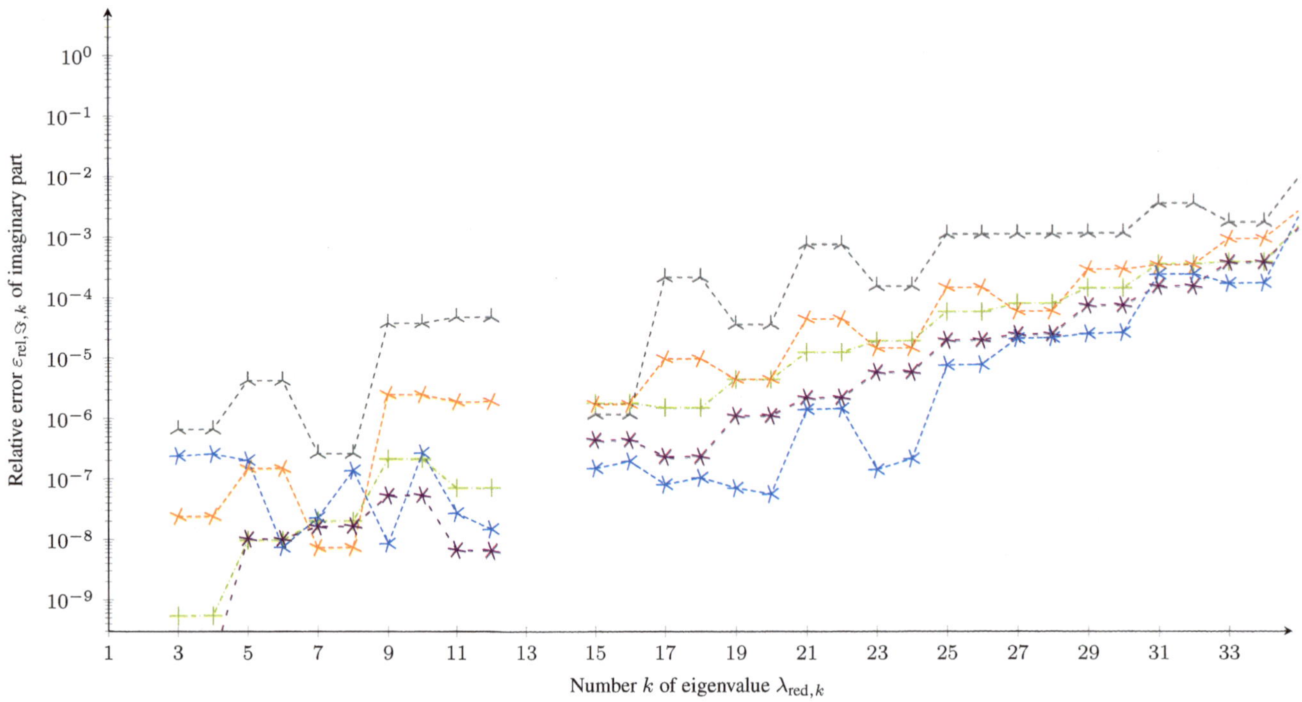

(b)

Fig. 11.5 Relative error of the real and imaginary parts of the approximated eigenvalues λ_{red} of the clamped beam of Fig. 11.1. The relative errors of the 34 eigenvalues with the lowest absolute value are shown using Craig and Ni's method (CN), Liu and Zheng's method (LZ) and de Kraker and van Campen's method (KC). Additionally, the combination of Liu and Zheng's reduction basis and primal assembly (LZ,KC), as well as the third order reduction (TO), are shown. The number of DOFs of the reduced assembled system is identical for all approaches $n_{\text{red,CN}} = n_{\text{red,LZ}} = n_{\text{red,KC}} = n_{\text{red,LZ,KC}} = n_{\text{red,TO}} = 42$. (**a**) Relative error $\varepsilon_{\text{rel},\Re,k}$ of real part to eigenvalue $\lambda_{\text{red},k}$. (**b**) Relative error $\varepsilon_{\text{rel},\Im,k}$ of imaginary part to eigenvalue $\lambda_{\text{red},k}$

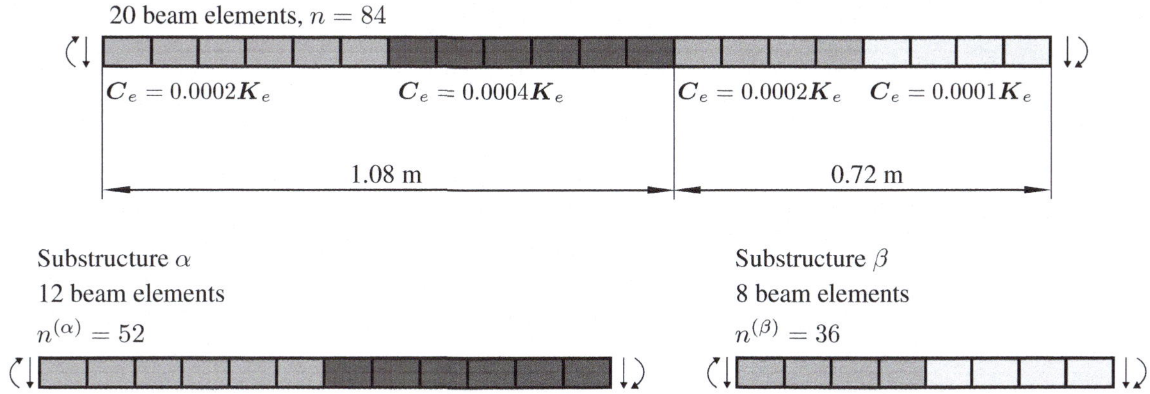

Fig. 11.6 Free-free beam structure divided into two substructures [17]. The beam consists of 20 Euler-Bernoulli beam elements (Young's modulus $7.0 \cdot 10^{10}\,\mathrm{N\,m^{-2}}$, density $2.7 \cdot 10^{3}\,\mathrm{kg\,m^{-3}}$, cross-section $5.0 \cdot 10^{-5}\,\mathrm{m^2}$, moment of inertia $1.04 \cdot 10^{-10}\,\mathrm{m^4}$) and has $n = 84$ DOFs in the state-space. The total length of the beam is 1.8 m. The length of substructure α is 1.08 m and the length of substructure β is 0.72 m. The damping for elements 1–5 and 13–16 is $\boldsymbol{C}_e = 0.0002\boldsymbol{K}_e$, for elements 6–12 $\boldsymbol{C}_e = 0.0004\boldsymbol{K}_e$ and for elements 17–20 $\boldsymbol{C}_e = 0.0001\boldsymbol{K}_e$

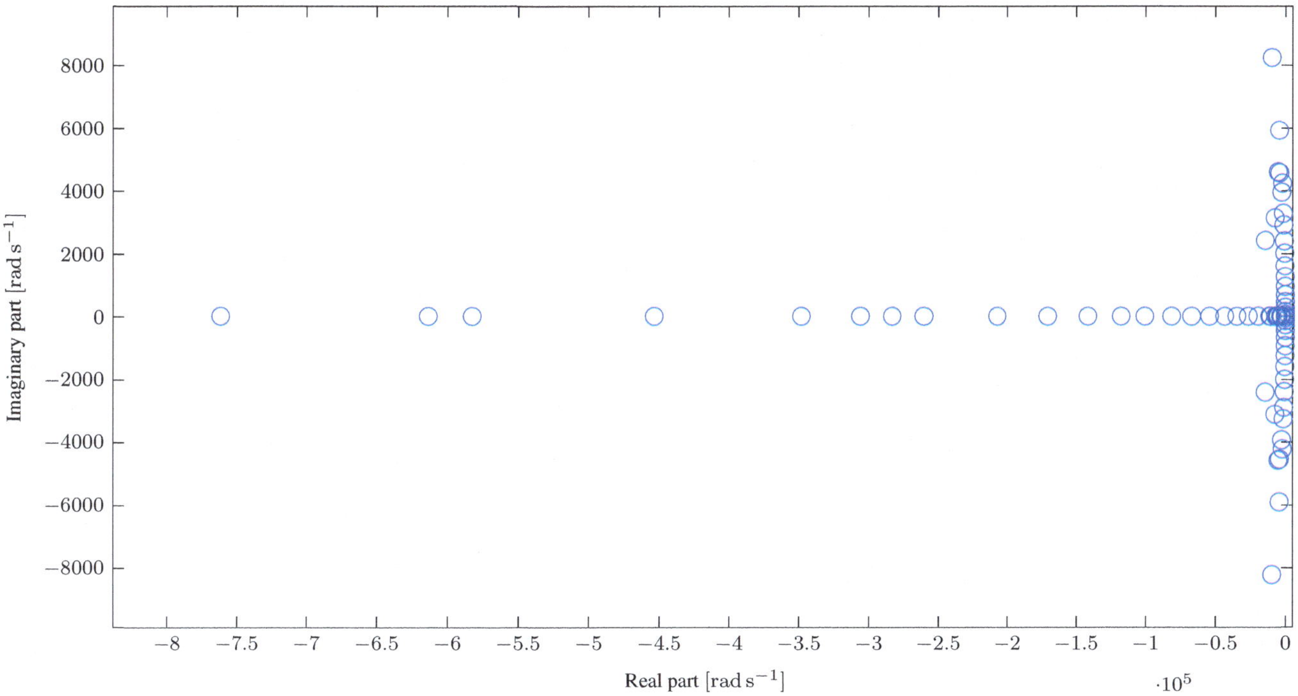

Fig. 11.7 Exact eigenvalues of the coupled unreduced system of Fig. 11.6

Figure 11.7 shows all 84 eigenvalues of the unreduced system in the complex plane. It can be seen that compared to the previous example in Fig. 11.2 significantly more real eigenvalues occur without imaginary parts. Altogether, there are 20 complex conjugate eigenvalue pairs. Thus, there are 40 complex eigenvalues with imaginary parts. In contrast, there are 40 real eigenvalues without imaginary parts. Additionally, four rigid body modes with zero eigenvalue occur. Furthermore, it can be seen that the maximum amount of the real parts is larger by two orders of magnitude compared to the previous example in Fig. 11.2, whereas the imaginary part is smaller by approximately two orders of magnitude.

Figure 11.8 shows the exact eigenvalues of the substructures. For substructure α there are 10 eigenvalue pairs and 28 real eigenvalues as well as 4 rigid body modes with zero eigenvalue. For substructure β there are 9 eigenvalue pairs and 14 real eigenvalues as well as 4 rigid body modes with zero eigenvalue.

The damped free-free beam is reduced and assembled using the substructuring methods described in Sects. 11.3 and 11.4. For the reduction of substructure α, $n_{\text{kept}}^{(\alpha)} = 11$ eigenmodes belonging to the 11 eigenvalues with the lowest absolute value

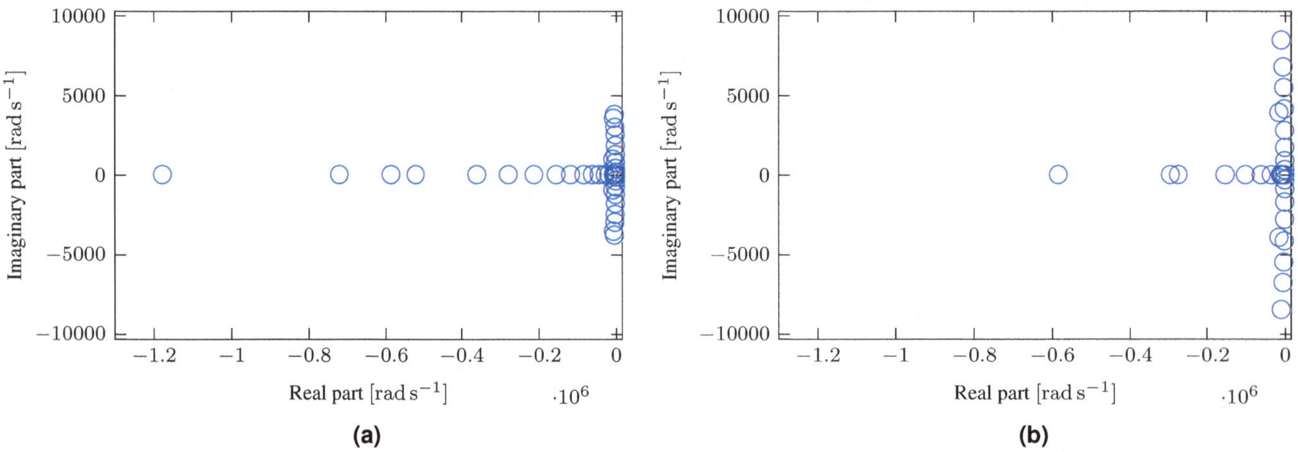

Fig. 11.8 Exact eigenvalues of the unreduced substructures of Fig. 11.6. (**a**) Substructure α. (**b**) Substructure β

Table 11.3 Modes used for reduction and resulting size of the reduced assembled system

	Craig/Ni's method		Liu/Zheng's method		de Kraker/van Campen's method		Liu/Zheng's basis with primal assembly		Third order reduction	
Substructure	α	β	α	β	α	β	α	β	α	β
Kept eigenmodes	11 (13)	10 (12)	11 (13)	10 (12)	11	10	11	10	11	10
Attachment modes	2	2	4	4	4	4	4	4	6	6
Rigid body modes	4	4	4	4	4	4	4	4	4	4
DOFs red. system	29 (33)		29 (33)		33		33		33	

The numbers in brackets are used for a comparison based on the identical size of the reduced system as shown in Fig. 11.10

are kept. These are 5 complex conjugate pairs and one real eigenvalue without imaginary parts. For substructure β, $n_{\mathrm{kept}}^{(\beta)} = 10$ eigenmodes belonging to the 10 eigenvalues with the lowest absolute value are kept. These are 5 complex conjugate pairs. Additionally, for the reduction of substructure α and substructure β the $n_r^{(\alpha)} = n_r^{(\beta)} = 4$ rigid body modes are used.

For the reduction according to Craig and Ni, both attachment modes are determined according to Eq. (11.34) for substructure α and substructure β. For the reduction according to Liu and Zheng and for the reduction according to de Kraker and van Campen, four attachment modes are determined per substructure according to Eqs. (11.53) and (11.67). For the third order approach, six attachment modes are determined per substructure according to Eq. (11.74). After assembly, the reduced system according to Craig and Ni's method (CN) and according to Liu and Zheng's method (LZ) has $n_{\mathrm{red,CN}} = n_{\mathrm{red,LZ}} = 29$ DOFs. According to de Kraker and van Campen's method (KC), Liu and Zheng's reduction basis with primal assembly (LZ,KC), as well as the third order reduction (TO), the reduced systems have $n_{\mathrm{red,KC}} = n_{\mathrm{red,LZ,KC}} = n_{\mathrm{red,TO}} = 33$ DOFs. Table 11.3 summarizes the number of used modes and the size of the reduced assembled system.

Additionally, the methods are compared based on the identical size of the reduced system. Therefore, an increased number of eigenmodes are kept for Craig and Ni's method as well as Liu and Zheng's method. The number of kept eigenmodes and the size of the reduced systems are given in brackets in Table 11.3.

Figures 11.9 and 11.10 show the relative errors of the real and imaginary parts belonging to the 21 eigenvalues with the lowest absolute value. In Fig. 11.9, Craig and Ni's method and Liu and Zheng's method have a smaller number of DOFs of the reduced system. In contrast, in Fig. 11.10, the number of kept eigenmodes for Craig and Ni's method and Liu and Zheng's method is increased compared to the other methods. Thus, the number of DOFs of the reduced system is identical for all approaches. The assembled system has four rigid body modes in the state-space. Therefore, no relative errors are determined for the eigenvalues 1–4.

Comparing the substructuring methods described in Sect. 11.3, Liu and Zheng's method shows the most accurate results. Despite the increased size of the reduced system, de Kraker and van Campen's method leads to larger relative errors. When considering the relative error $\varepsilon_{\mathrm{rel},\Im}$ of the imaginary parts of eigenvalue 5 and 6, the difference between Liu and Zheng's method and de Kraker and van Campen's method is four orders of magnitude. The worst approximation is obtained using Craig and Ni's method.

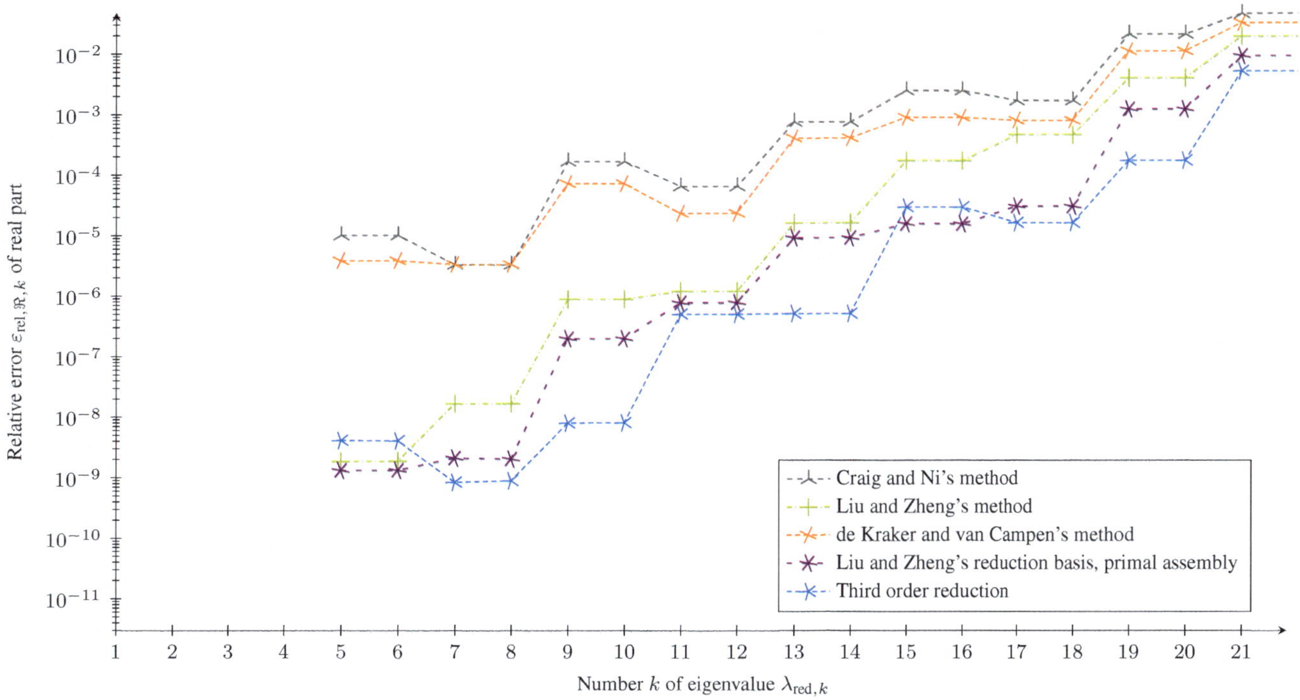

(a)

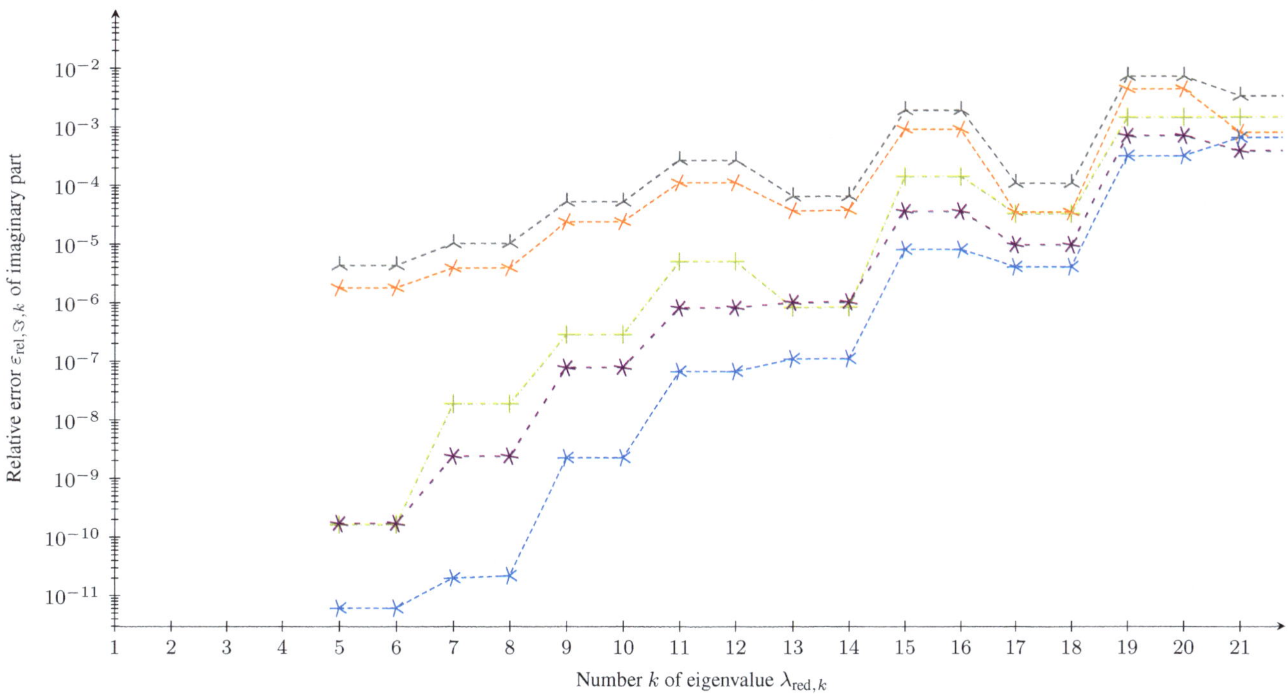

(b)

Fig. 11.9 Relative error of the real and imaginary parts of the approximated eigenvalues λ_{red} of the free-free beam of Figure 11.6. The relative errors of the 21 eigenvalues with the lowest absolute value are shown using Craig and Ni's method (CN), Liu and Zheng's method (LZ) and de Kraker and van Campen's method (KC). Additionally, the combination of Liu and Zheng's reduction basis and primal assembly (LZ,KC) as well as the third order reduction (TO) are shown. The number of DOFs of the assembled system according to Craig and Ni's method as well as Liu and Zheng's method is $n_{\mathrm{red,CN}} = n_{\mathrm{red,LZ}} = 29$ and according to de Kraker and van Campen's method, the combination of Liu and Zheng's reduction basis and primal assembly, as well as the third order reduction is $n_{\mathrm{red,KC}} = n_{\mathrm{red,LZ,KC}} = n_{\mathrm{red,TO}} = 33$. (**a**) Relative error $\varepsilon_{\mathrm{rel},\Re,k}$ of real part to eigenvalue $\lambda_{\mathrm{red},k}$. (**b**) Relative error $\varepsilon_{\mathrm{rel},\Im,k}$ of imaginary part to eigenvalue $\lambda_{\mathrm{red},k}$

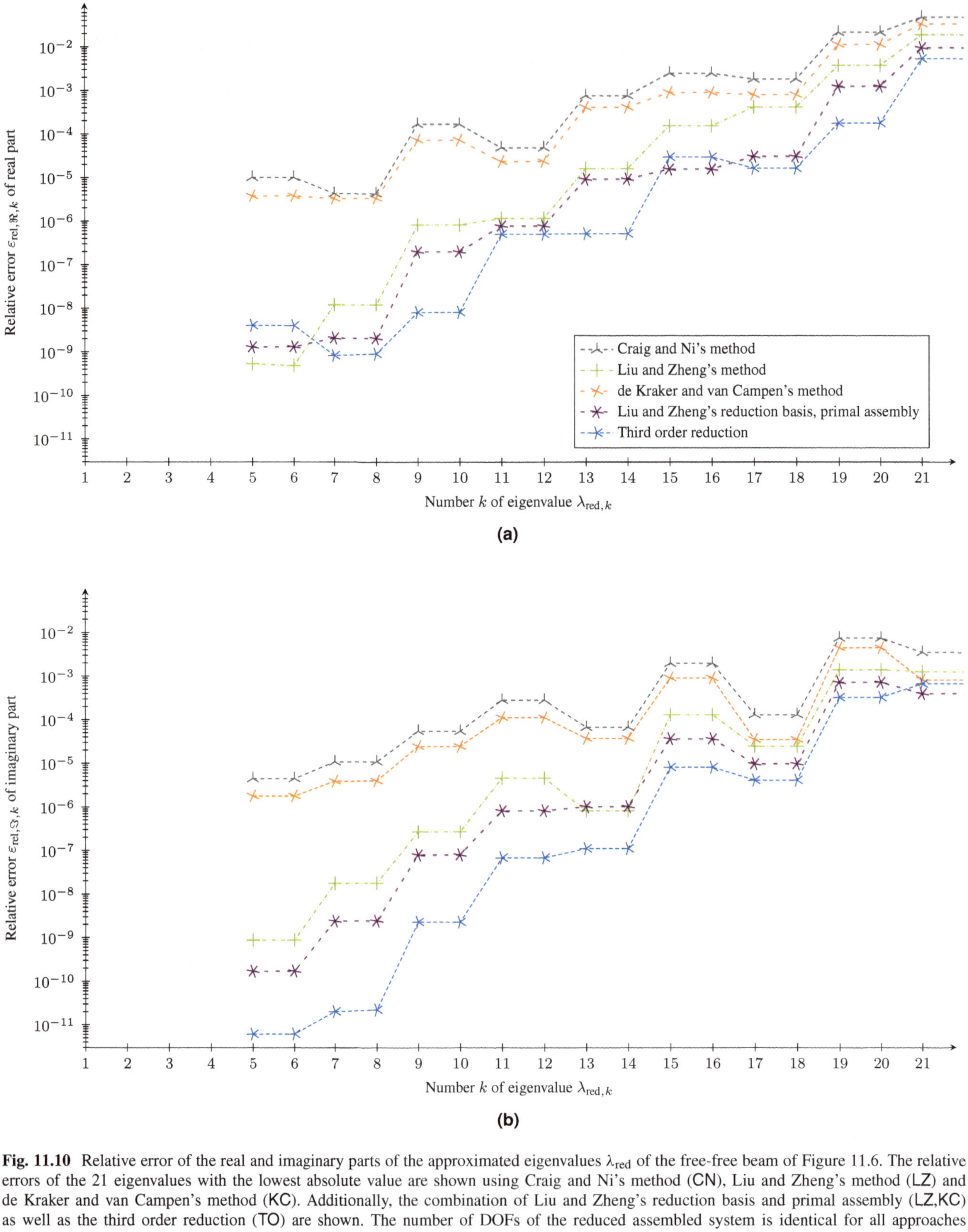

Fig. 11.10 Relative error of the real and imaginary parts of the approximated eigenvalues λ_{red} of the free-free beam of Figure 11.6. The relative errors of the 21 eigenvalues with the lowest absolute value are shown using Craig and Ni's method (CN), Liu and Zheng's method (LZ) and de Kraker and van Campen's method (KC). Additionally, the combination of Liu and Zheng's reduction basis and primal assembly (LZ,KC) as well as the third order reduction (TO) are shown. The number of DOFs of the reduced assembled system is identical for all approaches $n_{\mathrm{red,CN}} = n_{\mathrm{red,LZ}} = n_{\mathrm{red,KC}} = n_{\mathrm{red,LZ,KC}} = n_{\mathrm{red,TO}} = 33$. (**a**) Relative error $\varepsilon_{\mathrm{rel},\Re,k}$ of real part to eigenvalue $\lambda_{\mathrm{red},k}$. (**b**) Relative error $\varepsilon_{\mathrm{rel},\Im,k}$ of imaginary part to eigenvalue $\lambda_{\mathrm{red},k}$

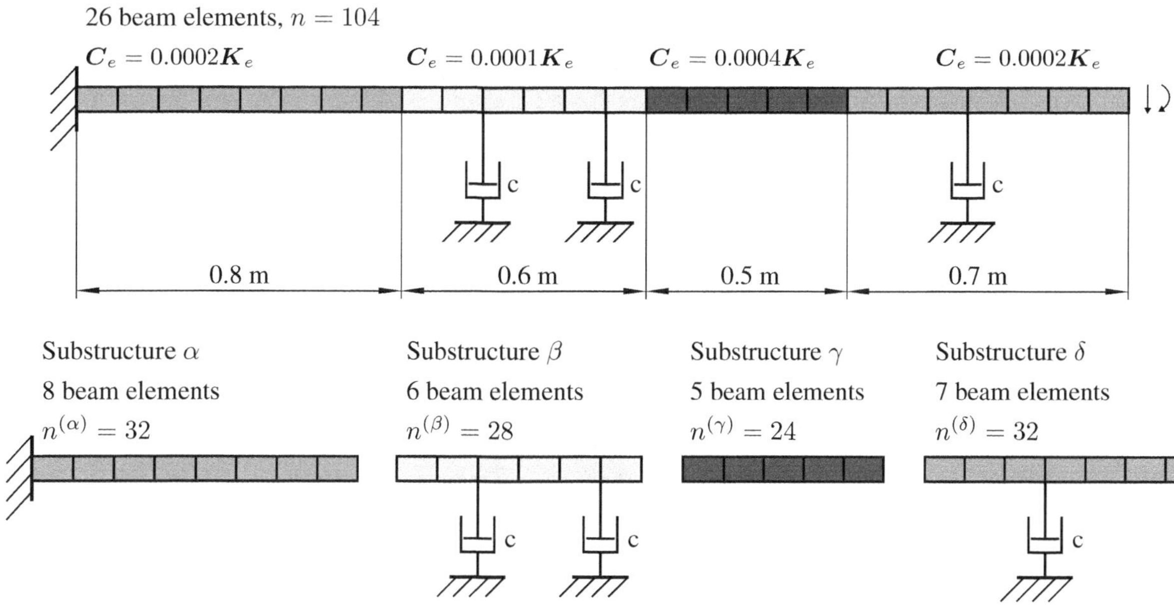

Fig. 11.11 Clamped beam structure with three local dampers divided into four substructures. It consists of 26 Euler-Bernoulli beam elements (Young's modulus $2.1 \cdot 10^{11}\,\mathrm{N\,m^{-2}}$, density $7.8 \cdot 10^3\,\mathrm{kg\,m^{-3}}$, cross-section $5.0 \cdot 10^{-5}\,\mathrm{m^2}$, moment of inertia $1.04 \cdot 10^{-10}\,\mathrm{m^4}$) and has $n = 104$ DOFs in the state-space. The total length of the beam is 2.6 m. The length of substructure α is 0.8 m, the length of substructure β is 0.6 m, the length of substructure γ is 0.5 m and the length of substructure δ is 0.7 m. The damper constant $c =$ is $1.0 \cdot 10^4\,\mathrm{N\,s\,m^{-1}}$. The damping for the elements 1–8 and 20–26 is $\boldsymbol{C}_e = 0.0002\boldsymbol{K}_e$, for the elements 9–14 $\boldsymbol{C}_e = 0.0001\boldsymbol{K}_e$ and for the elements 15–19 $\boldsymbol{C}_e = 0.0004\boldsymbol{K}_e$

Liu and Zheng's reduction basis with primal assembly leads to an improvement in the results compared to Liu and Zheng's method and de Kraker and van Campen's method. Compared to Liu and Zheng's method, the relative error according to Liu and Zheng's reduction with primal assembly is lower by a half to one order of magnitude.

The most exact approximation is achieved by the third order reduction. In comparison to Liu and Zheng's reduction and primal assembly, the exact eigenvalues are approximated more precisely by about one order of magnitude using the third order reduction.

11.5.3 Damped Beam Model with Four Substructures and Localized Dampers

As a concluding example, the damped beam structure with four substructures and localized dampers as shown in Fig. 11.11 is analyzed. The beam structure consists of 26 Euler-Bernoulli beam elements and the bending vibration (no axial deformation) is considered. The total system has a length of 2.6 m and is divided into substructure α, consisting of eight beam elements and a length of 0.8 m, substructure β, consisting of six beam elements and a length of 0.6 m, substructure γ consisting of five beam elements and a length of 0.5 m, and substructure δ consisting of seven beam elements and a length of 0.7 m. Two localized viscous dampers are attached to substructure β and one localized viscous damper to substructure δ, each with a damper constant of $c = 1.0 \cdot 10^4\,\mathrm{N\,s\,m^{-1}}$. In addition, the damping matrix per element is assumed to be proportional to the stiffness matrix. For elements 1–8 and 20–26, a damping matrix of $\boldsymbol{C}_e = 0.0002\boldsymbol{K}_e$ is assumed. $\boldsymbol{C}_e = 0.0001\boldsymbol{K}_e$ is prescribed for elements 9–14 and $\boldsymbol{C}_e = 0.0004\boldsymbol{K}_e$ for the elements 15–19. Thus, the damping of the overall system is not proportional, due to the localized dampers on the one hand and due to the different stiffness proportional damping properties per element on the other hand. The total system has $n = 52$ DOFs in physical space and $m = 104$ DOFs in state-space. Substructure α has $n^{(\alpha)} = 32$ DOFs in state-space, substructure β has $n^{(\beta)} = 28$ DOFs in state-space, substructure γ has $n^{(\gamma)} = 24$ DOFs in state-space and substructure δ has $n^{(\delta)} = 32$ DOFs in state-space. Furthermore, substructures β, γ and δ have two physical rigid body modes ($m_r^{(\beta)} = m_r^{(\gamma)} = m_r^{(\delta)} = 2$). For the state-space rigid body modes, different cases occur considering the three substructures β, γ and δ. For substructure β, each of the physical rigid body modes leads in state-space to a simple zero eigenvalue for the translation and for the rotation. This can be attributed to the fact that due to the two localized dampers, both the translational rigid body motion and the rotational rigid body motion cause damping forces. Considering substructure γ, there is one zero eigenvalue with multiplicity two for the translation as well as for the rotation in

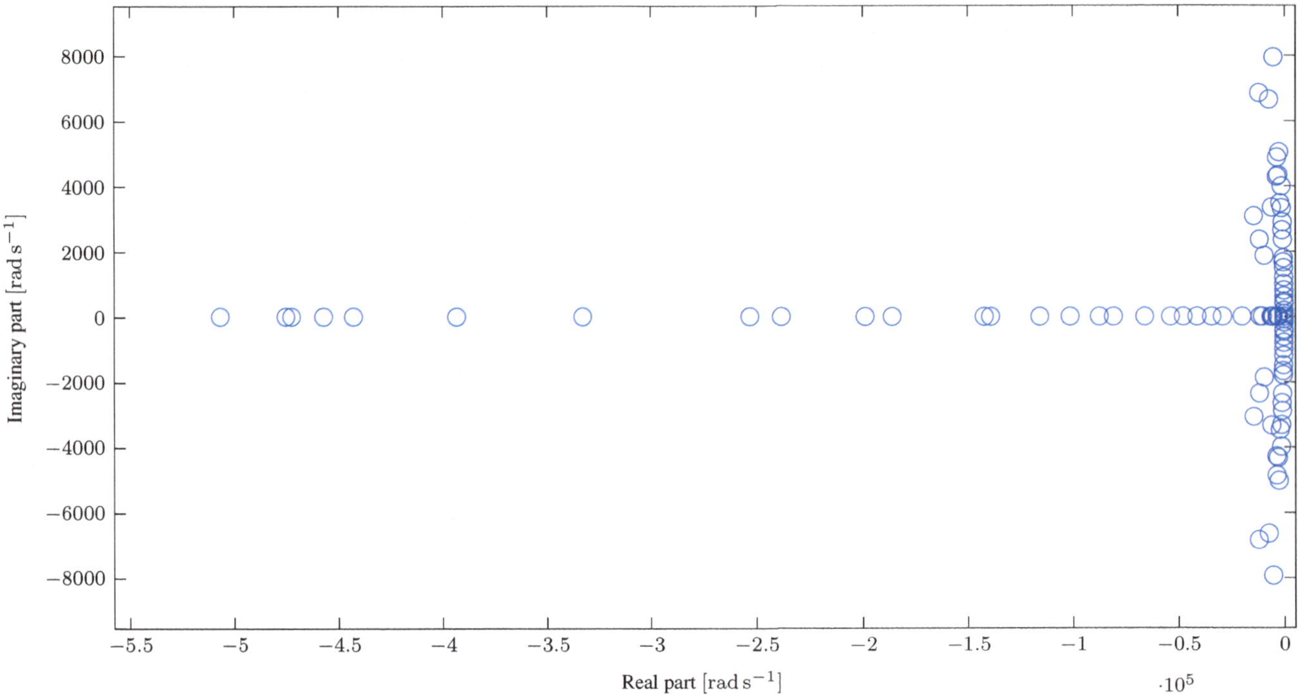

Fig. 11.12 Exact eigenvalues of the coupled unreduced system of Fig. 11.11

state-space. This follows from the fact that due to the absence of localized dampers, both the translational rigid body motion and the rotational rigid body motion do not cause any damping forces. For substructure δ, the translational rigid body motion causes a damping force, but the rotational rigid body motion is possible without causing any damping force. One single zero eigenvalue for the translation and one zero eigenvalue with multiplicity two for the rotation result for substructure δ in state-space. Thus, in state-space, the physical rigid body modes for substructure β lead to one regular state-space rigid body mode for the translation and one regular state-space rigid body mode for the rotation. For substructure γ, they lead to one regular and one generalized state-space rigid body mode for the translation and also one regular and one generalized state-space rigid body mode for the rotation. Considering substructure δ, one regular state-space rigid body mode for the translation and one regular and one generalized state-space rigid body mode for the rotation occur. Accordingly, $n_r^{(\beta)} = 2$, $n_r^{(\gamma)} = 4$ and $n_r^{(\delta)} = 3$ holds for the number of state-space rigid body modes. This example represents all possible combinations of regular and generalized state-space rigid body modes.

Figure 11.12 shows all 104 eigenvalues of the unreduced system in the complex plane. In total, there are 28 complex conjugate eigenvalue pairs. Thus, there are 56 complex eigenvalues with imaginary parts. In contrast, there are 48 real eigenvalues without imaginary parts. The real parts and imaginary parts are roughly in the order of magnitude of the preceding example in Fig. 11.7.

Figure 11.13 shows the exact eigenvalues of the substructures. For substructure α, there are 8 eigenvalue pairs and 16 real eigenvalues. For substructure β, there are 8 eigenvalue pairs and 10 real eigenvalues as well as 2 rigid body modes with zero eigenvalue. Substructure γ shows 3 eigenvalue pairs and 14 real eigenvalues as well as 4 rigid body modes with zero eigenvalue. For substructure δ, there are 7 eigenvalue pairs and 15 real eigenvalues as well as 3 rigid body modes with zero eigenvalue.

In comparison to the beam structure in Fig. 11.1, substructure α in Fig. 11.11 and substructure α Fig. 11.1 are basically identical to one another. Substructure δ in Fig. 11.11 and substructure β in Fig. 11.1 are also basically identical to each another. The main difference between both systems is that for the beam structure in Fig. 11.11, the damping matrix of each element is additionally proportional to the stiffness matrix while for the beam structure of Fig. 11.1 viscous damping occurs only due to the two localized dampers. When considering the eigenvalues of the substructures of the two different systems in Figs. 11.3 and 11.13, this leads to a increased number of purely real eigenvalues for the substructures with stiffness proportional damping. In addition, the eigenvalues show a significantly higher real part.

The clamped beam is reduced and assembled using the substructuring methods described in Sects. 11.3 and 11.4. For the reduction of substructure α, $n_{\text{kept}}^{(\alpha)} = 15$ eigenmodes belonging to the 15 eigenvalues with the lowest absolute value are

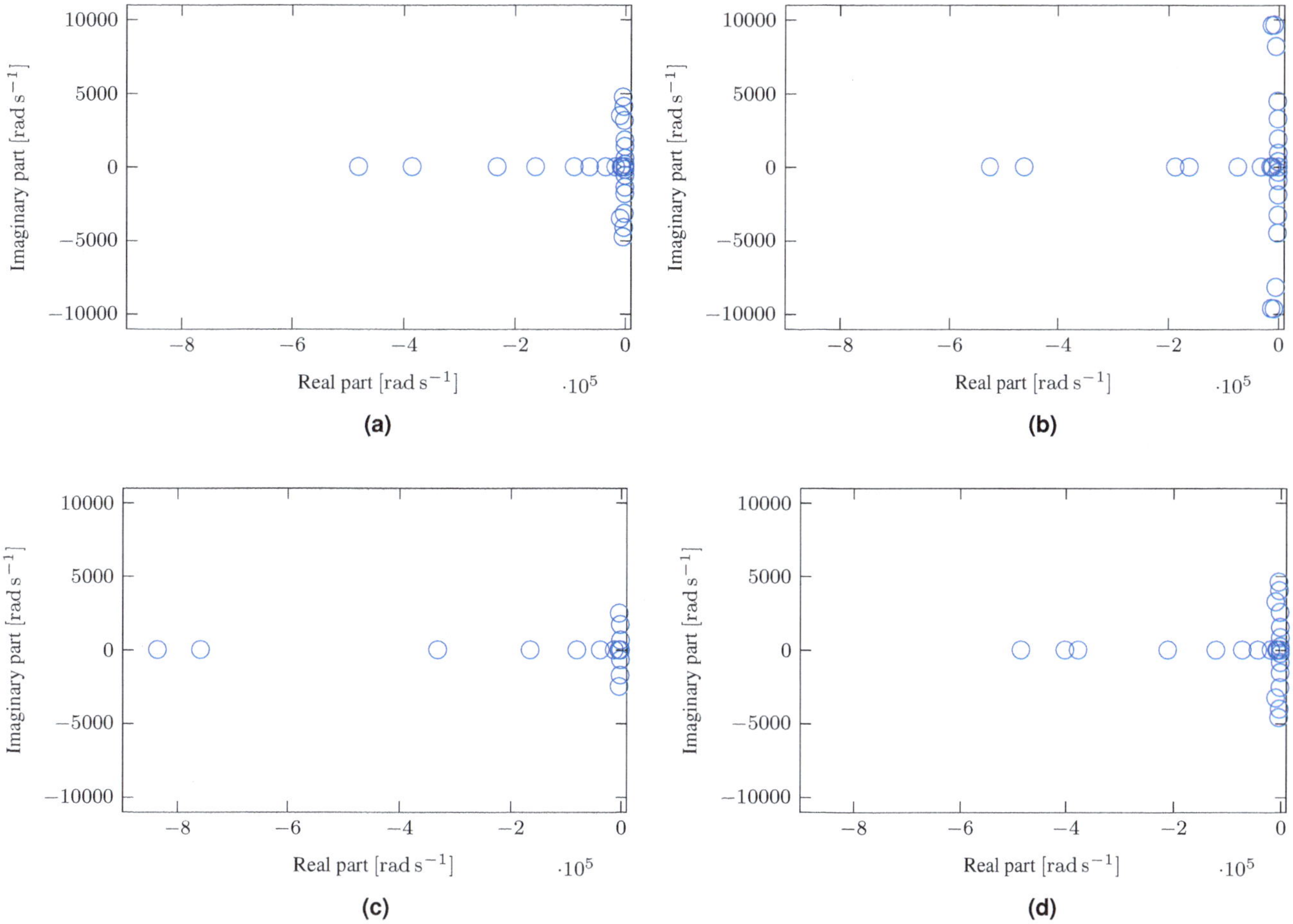

Fig. 11.13 Exact eigenvalues of the unreduced substructures of Figure 11.11. (**a**) Substructure β. (**b**) Substructure γ. (**c**) Substructure δ

kept. These are 6 complex conjugate pairs and 3 real eigenvalues without imaginary parts. For substructure β, $n_{\text{kept}}^{(\beta)} = 10$ eigenmodes belonging to the 10 eigenvalues with the lowest absolute value are kept. These are 5 complex conjugate pairs. For substructure γ, $n_{\text{kept}}^{(\gamma)} = 13$ eigenmodes belonging to the 13 eigenvalues with the lowest absolute value are kept. These are 3 complex conjugate pairs and 7 real eigenvalues without imaginary parts. For substructure δ, $n_{\text{kept}}^{(\delta)} = 12$ eigenmodes belonging to the 12 eigenvalues with the lowest absolute value are kept. These are 5 complex conjugate pairs and 2 real eigenvalues without imaginary parts. Additionally, for the reduction of substructure β the $n_r^{(\beta)} = 2$ rigid body modes, for substructure γ the $n_r^{(\gamma)} = 4$ and for substructure δ the $n_r^{(\delta)} = 3$ state-space rigid body modes are used.

For Craig and Ni's method, two attachment modes are determined for each boundary of each substructure. For Liu and Zheng's and for de Kraker and van Campen's method, four attachment modes are used for each boundary. For the third order approach, six attachment modes per interface are determined according to Eqs. (11.73) and (11.74). After assembly, the reduced system has $n_{\text{red,CN}} = n_{\text{red,LZ}} = 59$ DOFs for Craig and Ni's method (CN) and Liu and Zheng's method (LZ). The reduced system has $n_{\text{red,KC}} = n_{\text{red,LZ,KC}} = n_{\text{red,TO}} = 71$ DOF for de Kraker and van Campen's method (KC), Liu and Zheng's reduction with primal assembly and the third order reduction (TO). Due to the increased number of interface DOFs compared to the previous examples, there is a clear difference between the approaches. Table 11.4 summarizes the number of modes used and the size of the reduced assembled systems of all methods.

Additionally, the methods will be compared for the identical size of the reduced system in the following. Therefore, an increased number of eigenmodes are kept for Craig and Ni's method as well as for Liu and Zheng's method. The number of kept eigenmodes and the size of the reduced systems are given in brackets in Table 11.4.

Figures 11.14 and 11.15 show the relative errors of the real and imaginary parts corresponding to the 39 eigenvalues with the lowest absolute value. When comparing the approaches with the identical number of remaining DOFs of the reduced system in Fig. 11.14, Liu and Zheng's method consistently achieves more precise approximations than Craig and Ni's methods. The difference is between five orders of magnitude for low eigenvalues and two orders of magnitude for high

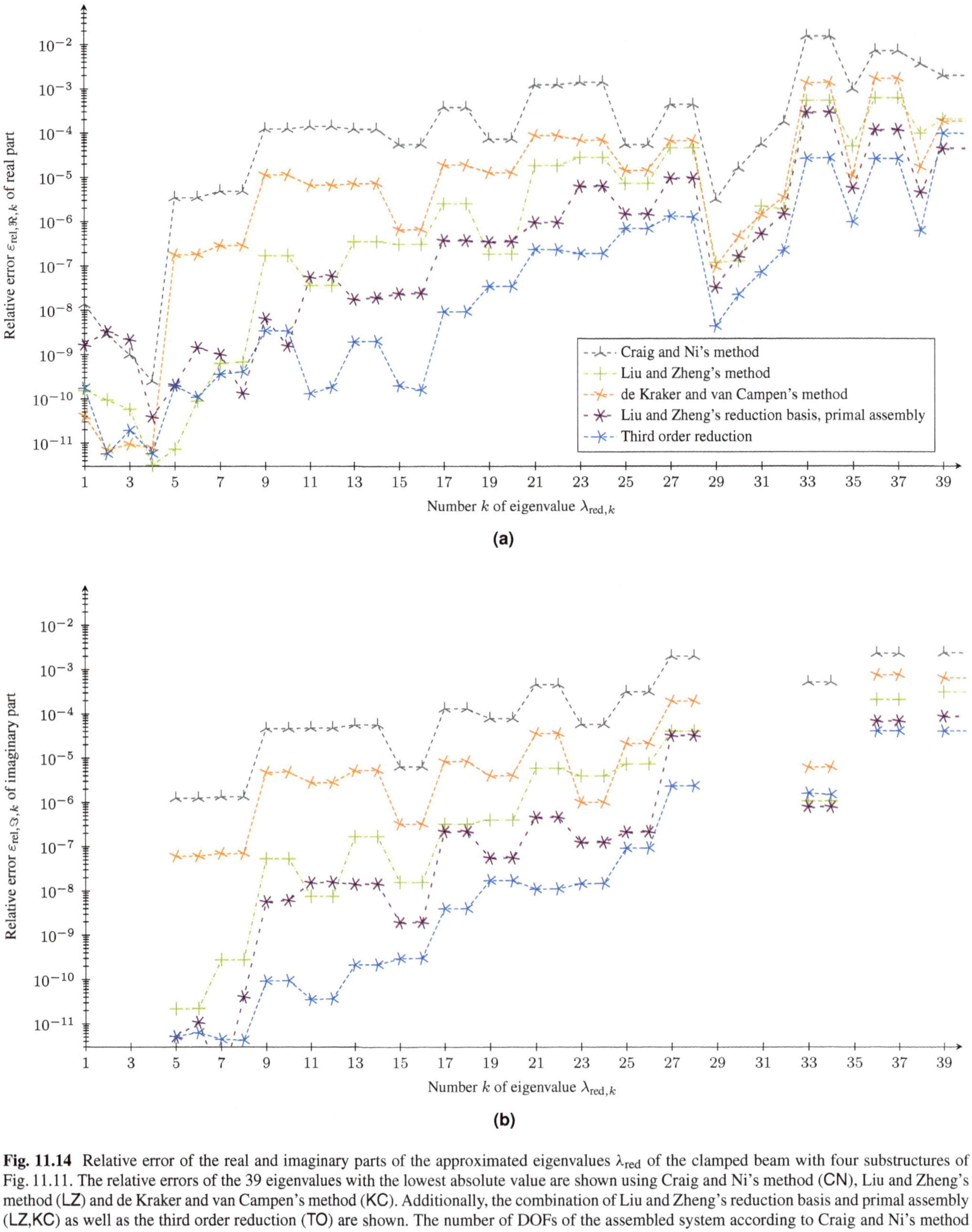

Fig. 11.14 Relative error of the real and imaginary parts of the approximated eigenvalues λ_{red} of the clamped beam with four substructures of Fig. 11.11. The relative errors of the 39 eigenvalues with the lowest absolute value are shown using Craig and Ni's method (CN), Liu and Zheng's method (LZ) and de Kraker and van Campen's method (KC). Additionally, the combination of Liu and Zheng's reduction basis and primal assembly (LZ,KC) as well as the third order reduction (TO) are shown. The number of DOFs of the assembled system according to Craig and Ni's method as well as Liu and Zheng's method is $n_{\mathrm{red,CN}} = n_{\mathrm{red,LZ}} = 59$ and according to de Kraker and van Campen's method, the combination of Liu and Zheng's reduction basis and primal assembly, as well as the third order reduction is $n_{\mathrm{red,KC}} = n_{\mathrm{red,LZ,KC}} = n_{\mathrm{red,TO}} = 71$. (**a**) Relative error $\varepsilon_{\mathrm{rel},\Re,k}$ of real part to eigenvalue $\lambda_{\mathrm{red},k}$. (**b**) Relative error $\varepsilon_{\mathrm{rel},\Im,k}$ of imaginary part to eigenvalue $\lambda_{\mathrm{red},k}$

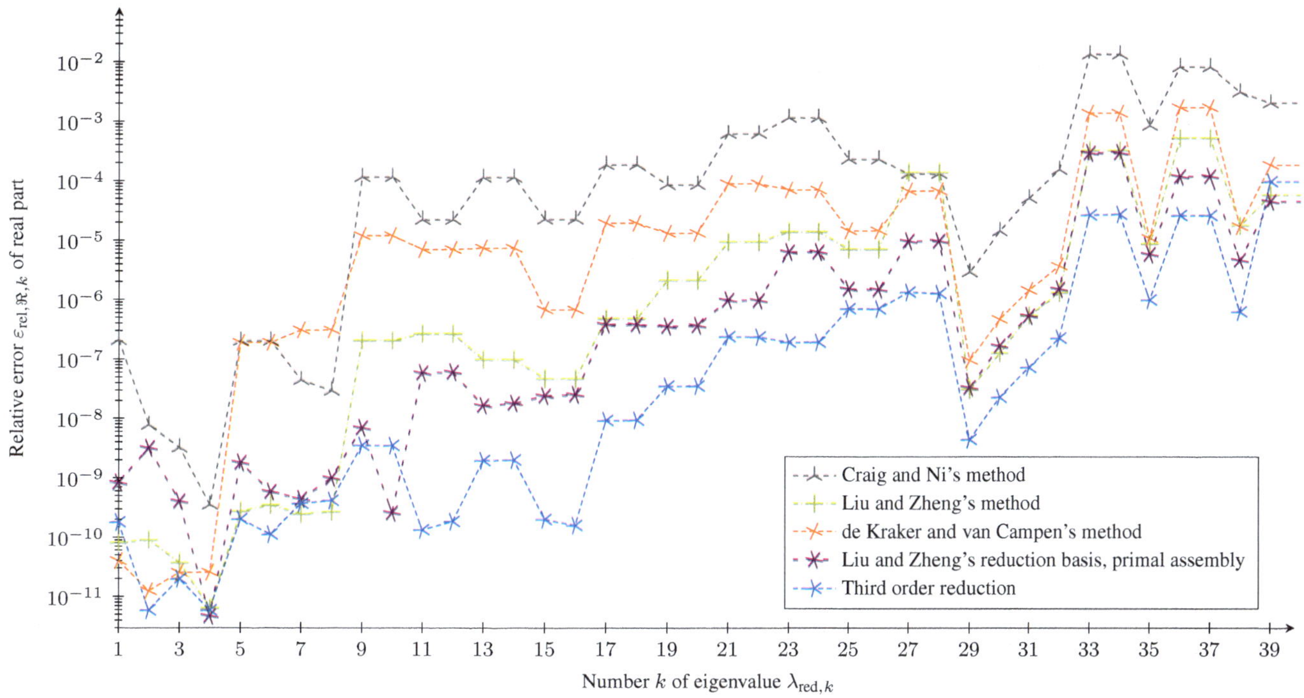

(a)

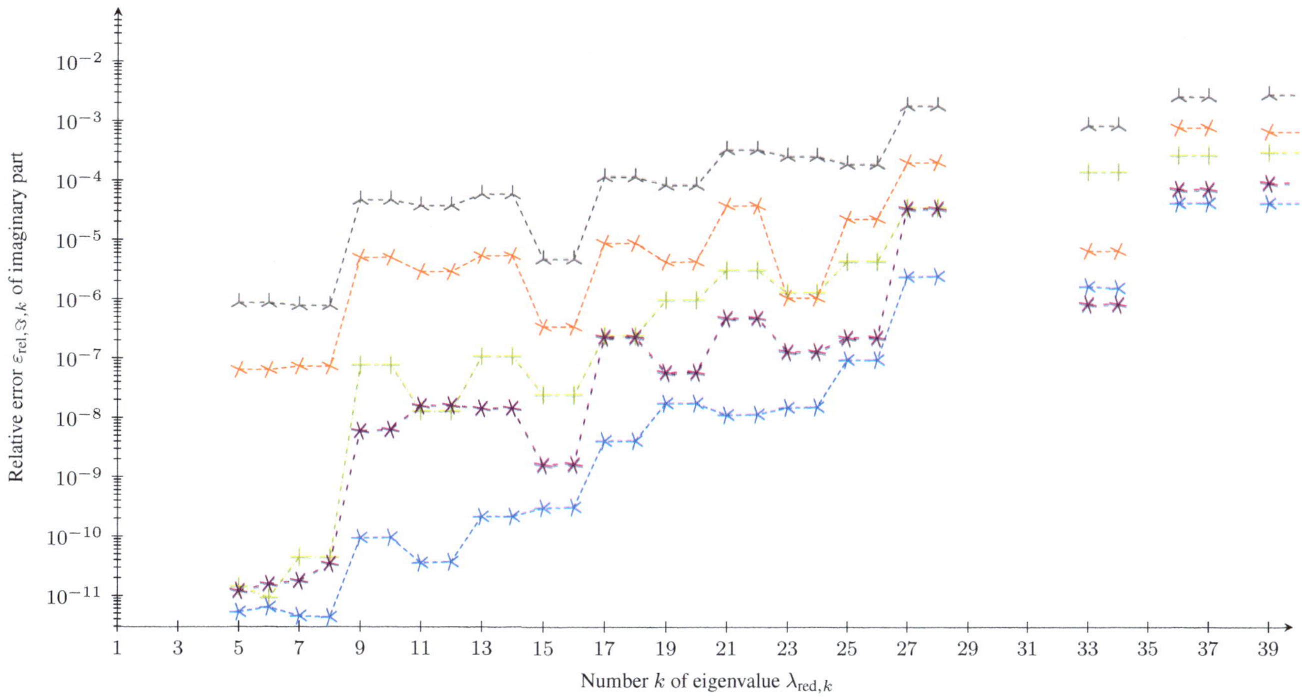

(b)

Fig. 11.15 Relative error of the real and imaginary parts of the approximated eigenvalues λ_{red} of the clamped beam with four substructures of Figure 11.11. The relative errors of the 39 eigenvalues with the lowest absolute value are shown using Craig and Ni's method (CN), Liu and Zheng's method (LZ) and de Kraker and van Campen's method (KC). Additionally, the combination of Liu and Zheng's reduction basis and primal assembly (LZ,KC) as well as the third order reduction (TO) are shown. The number of DOFs of the reduced assembled system is identical for all approaches $n_{\mathrm{red,CN}} = n_{\mathrm{red,LZ}} = n_{\mathrm{red,KC}} = n_{\mathrm{red,LZ,KC}} = n_{\mathrm{red,TO}} = 71$. (**a**) Relative error $\varepsilon_{\mathrm{rel},\Re,k}$ of real part to eigenvalue $\lambda_{\mathrm{red},k}$. (**b**) Relative error $\varepsilon_{\mathrm{rel},\Im,k}$ of imaginary part to eigenvalue $\lambda_{\mathrm{red},k}$

Table 11.4 Modes used for reduction and resulting size of the reduced assembled system

	Craig/Ni's method				Liu/Zheng's method			
Substructure	α	β	γ	δ	α	β	γ	δ
Kept eigenmodes	15 (17)	10 (15)	13 (15)	12 (15)	15 (17)	10 (15)	13 (15)	12 (15)
Attachment modes	2	4	4	2	4	8	8	4
Rigid body modes	0	2	4	3	0	2	4	3
DOFs reduced system	59 (71)				59 (71)			

	de Kraker/van Campen's method				Liu/Zheng's reduction basis with primal assembly				Third order reduction			
Substructure	α	β	γ	δ	α	β	γ	δ	α	β	γ	δ
Kept eigenmodes	15	10	13	12	15	10	13	12	15	10	13	12
Attachment modes	4	8	8	4	4	8	8	4	6	12	12	6
Rigid body modes	0	2	4	3	0	2	4	3	0	2	4	3
DOFs reduced system	71				71				71			

The numbers in brackets are used for a comparison based on the identical size of the reduced system as shown in Fig. 11.15

eigenvalues. This also holds in Fig. 11.15, where the number of kept eigenmodes for Craig and Ni's method and Liu and Zheng's method is increased. Liu and Zheng's reduction basis with primal assembly as well as the third order reduction lead in both Figs. 11.14 and 11.15 to lower relative errors compared to Craig and Ni's method, de Kraker and van Campen's method and Liu and Zheng's method. Overall, the third order reduction achieves the most accurate approximation.

11.6 Conclusions

In this paper, a derivation and comparison of three existing free interface substructuring methods for viscously damped systems were provided and two new approximation approaches were proposed. The three existing investigated methods are Craig and Ni's method, Liu and Zheng's method, and de Kraker and van Campen's method. Liu and Zheng's method can be considered as a second-order extension of Craig and Ni's method. We extended Liu and Zheng's method to a third-order approximation and generalized it further to arbitrarily higher orders. Moreover, a new method combining the reduction basis of Liu and Zheng's approach with the primal assembly procedure applied by de Kraker and van Campen was derived.

The new method combining Liu and Zheng's reduction basis with primal assembly and the third-order method give the best results and are recommended for the approximation of arbitrarily viscous damped substructured systems. The third-order method in particular shows the very best results, but increases the size of the reduced system compared to Craig and Ni's and Liu and Zheng's method. Liu and Zheng's reduction basis with primal assembly outperforms de Kraker and van Campen's method, while both methods generate the same size of the reduced system.

The five methods were applied to three different beam structures. In the future, we want to apply the method to bigger problems with a larger number of DOFs. The examples used in this paper are very illustrative, allow for comparison to results in the literature and demonstrate all critical points for the application of the suggested methodology. However, they are too small for a meaningful comparison in terms of computational time. For this purpose, it is necessary to consider additional numerical examples to further examine the performance of the proposed methods.

References

1. Craig, R.R., Kurdila, A.J.: Fundamentals of Structural Dynamics. Wiley, New York (2006)
2. Craig, R.R., Bampton, M.C.C.: Coupling of substructures for dynamic analyses. AIAA J. **6**(7), 1313–1319 (1968). https://doi.org/10.2514/3.4741
3. MacNeal, R.H.: A hybrid method of component mode synthesis. Comput. Struct. **1**(4), 581–601 (1971). https://doi.org/10.1016/0045-7949(71)90031-9
4. Rubin, S.: Improved component-mode representation for structural dynamic analysis. AIAA J. **13**(8), 995–1006 (1975). https://doi.org/10.2514/3.60497
5. Rixen, D.J.: A dual Craig-Bampton method for dynamic substructuring. J. Comput. Appl. Math. **168**(1–2), 383–391 (2004). https://doi.org/10.1016/j.cam.2003.12.014

6. Craig, R.R., Chang, C.-J.: On the use of attachment modes in substructure coupling for dynamic analysis. In: 18th Structural Dynamics and Materials Conference. Structures, Structural Dynamics, and Materials and Co-located Conferences, pp. 89–99. American Institute of Aeronautics and Astronautics, San Diego (1977). https://doi.org/10.2514/6.1977-405
7. Herting, D.N.: A general purpose, multi-stage, component modal synthesis method. Finite Elem. Anal. Des. **1**(2), 153–164 (1985)
8. Park, K.C., Park, Y.H.: Partitioned component mode synthesis via a flexibility approach. AIAA J. **42**(6), 1236–1245 (2004). https://doi.org/10.2514/1.10423
9. Gruber, F.M., Rixen, D.J.: Evaluation of substructure reduction techniques with fixed and free interfaces. Stroj. Vestn.-J. Mech. **62**(7–8), 452–462 (2016). https://doi.org/10.5545/sv-jme.2016.3735
10. Craig, R.R.: Coupling of substructures for dynamic analyses: an overview. In: Proceedings of AIAA/ASME/ASCE/AHS/ASC Structures, Structural Dynamics, and Materials Conference and Exhibit, Atlanta, GA, USA, pp. 1573–1584 (2000)
11. Gruber, F.M., Rixen, D.J.: Comparison of Craig-Bampton approaches for systems with arbitrary viscous damping in dynamic substructuring. In: Linderholt, A., Allen, M.S., Mayes, R.L., Rixen, D. (eds.) Dynamics of Coupled Structures, vol. 4, pp. 35–49. Springer, Cham (2018). https://doi.org/10.1007/978-3-319-74654-8_3
12. Hasselman, T.K., Kaplan, A.: Dynamic analysis of large systems by complex mode synthesis. J. Dyn. Syst. Meas. Control. **96**(3), 327–333 (1974). https://doi.org/10.1115/1.3426810
13. Beliveau, J.-G., Soucy, Y.: Damping synthesis using complex substructure modes and a Hermitian system representation. AIAA J. **23**(12), 1952–1956 (1985). https://doi.org/10.2514/3.9201
14. de Kraker, A.: Generalization of the Craig-Bampton CMS procedure for general damping. Technical Report, Technische Universiteit Eindhoven, Vakgroep Fundamentele Werktuigkunde, Rapportnr. WFW 93.023 (1993)
15. Craig, R.R., Ni, Z.: Component mode synthesis for model order reduction of nonclassically damped systems. J. Guid. Control. Dyn. **12**(4), 577–584 (1989). https://doi.org/10.2514/3.20446
16. de Kraker, A., van Campen, D.H.: Rubin's CMS reduction method for general state-space models. Comput. Struct. **58**(3), 597–606 (1996). https://doi.org/10.1016/0045-7949(95)00151-6
17. Liu, M.H., Zheng, G.T.: Improved component-mode synthesis for nonclassically damped systems. AIAA J. **46**(5), 1160–1168 (2008) https://doi.org/10.2514/1.32869
18. Caughey, T.K.: Classical normal modes in damped linear dynamic systems. J. Appl. Mech. **27**(2), 269–271 (1960). https://doi.org/10.1115/1.3643949
19. Ma, F., Imam, A., Morzfeld, M.: The decoupling of damped linear systems in oscillatory free vibration. J. Sound Vib. **324**(1), 408–428 (2009). https://doi.org/10.1016/j.jsv.2009.02.005
20. Caughey, T.K., O'Kelly, M.E.J.: Classical normal modes in damped linear dynamic systems. J. Appl. Mech. **32**(3), 583–588 (1965). https://doi.org/10.1115/1.3627262
21. Hasselman, T.K.: Damping synthesis from substructure tests. AIAA J. **14**(10), 1409–1418 (1976). https://doi.org/10.2514/3.61481
22. Rayleigh, J.W.S.B.: The Theory of Sound, vol. 2. Macmillan, London (1896)
23. Frazer, R.A., Duncan, W.J., Collar, A.R.: Elementary Matrices and Some Applications to Dynamics and Differential Equations. Cambridge University Press, Cambridge (1938)
24. Hurty, W.C., Rubinstein, M.F.: Dynamics of Structures. Prentice-Hall, Englewood Cliffs (1964)
25. Lang, G.F.: Understanding modal vectors. In: Mains, M., Blough, J.R. (eds.) Topics in Modal Analysis {&} Testing, Volume 10: Proceedings of the 35th IMAC, A Conference and Exposition on Structural Dynamics 2017, pp. 55–68. Springer, Cham (2017). https://doi.org/10.1007/978-3-319-54810-4_8
26. Craig, R.R., Chung, Y.-T.: A generalized substructure coupling procedure for damped systems. In: 22nd Structures, Structural Dynamics and Materials Conference. Structures, Structural Dynamics, and Materials and Co-located Conferences, pp. 254–266. American Institute of Aeronautics and Astronautics, Atlanta (1981). https://doi.org/10.2514/6.1981-560
27. Balmès, E.: New results on the identification of normal modes from experimental complex modes. Mech. Syst. Signal Process. **11**(2), 229–243 (1997). https://doi.org/10.1006/MSSP.1996.0058
28. Gruber, F.M., Rixen, D.J.: Dual Craig-Bampton component mode synthesis method for model order reduction of nonclassically damped linear systems. Mech. Syst. Signal Process. **111**, 678–698 (2018). https://doi.org/10.1016/J.YMSSP.2018.04.019
29. de Kraker, A.: The RUBIN CMS procedure for general state-space models. Technical Report, Technische Universiteit Eindhoven, Vakgroep Fundamentele Werktuigkunde, Rapportnr. Rapportnr. WFW 94.081 (1994)
30. Zhu, D.C., Shi, G.Q.: A theory of structural dynamics with defective phenomena. In: Computational Mechanics '88, pp. 1146–1147. Springer, Berlin (1988). https://doi.org/10.1007/978-3-642-61381-4_305
31. Muravyov, A.A., Hutton, S.G.: Component mode synthesis for nonclassically damped systems. AIAA J. **34**(8), 1664–1669 (1996). https://doi.org/10.2514/3.13287
32. Martinez, D.R., Carrie, T.G., Gregory, D.L., Miller, A.K.: Combined experimental/analytical modelling using component modes synthesis (1984)

Chapter 12
Model Updating of Fluid-Structure Interaction Effects on Piping System

Srijan Rajbamshi, Qintao Guo, and Ming Zhan

Abstract The main goal of this paper is to propose a simplified model to predict the dynamic behavior; mainly vibration under effects of fluid. Firstly, an experiment is carefully designed incorporating important parameters of the system and experimental modal analysis is performed for solid only and with fluid. Then, modal analysis using FEM is performed in real scale 3D model of the test rig and the obtained results is approximated to experimental frequency of the system. Then, a simplified model is developed, which appends the results of explicitly computed mass and stiffness of supports, to a solid only FEM. The new model is analyzed and updated till it is considerably close to the results obtained previously. Based on sensitivity analysis, set of model parameters is selected for the model updating process. Response surface method is implemented to find values of model parameters yielding results closest to the experimental. The results can provide a basis for further experimental and numerical dynamic analysis and optimization.

Keywords Fluid-structure interaction · Added-mass · Pipelines

12.1 Introduction

Piping systems are designed so as to transfer fuel fluids reliably when subjected to vibration most of the time. Although traditional vibration isolation techniques are invariably used in existing systems, dynamic analysis of such piping systems have to be carried out to optimize the performance and prevent failure. To accurately predict the hydrodynamic loadings in the fluid as well as the stress levels and vibrations in the piping, experimental and theoretical researches have shown that the analysis of the fluid and pipe motions must be carried out simultaneously in a coupled way. Such kind of approach is referred to in the literature as fluid–structure interaction (FSI). These coupled analyses involve solving computationally costly equations in addition to prevalent structure only and fluid only FEM. When the fluid inside the pipe is very slow or almost stationary, the fluid can be modeled just as added mass. This paper intends to provide expressions for explicit added mass of fully filled pipe in free and simply supported conditions in terms of density of water.

12.2 Background

Laithier and Païdoussis described and derived a general formulation of the dynamics of fluid conveying pipes [1]. Inherent and dynamic characteristics of FSI of pipes needs to analyzed carefully so as to optimize fluidic and structural design parameters of pipe systems [2–4]. Zhang, Tisjelling and Vardy worked on calculating the natural frequencies of the liquid filled pipes using method of characteristics (MOC) method [5]. Many researches has been conducted to implement Finite element method for calculation of FSI as well [6, 7]. The lateral vibrations in pipe can be considered as added mass only according to Dubyk and Orynyak [8]. Li, Karney and Liu have provided extensive literature review on fluid structure interaction in their paper [9]. Escaler, Torre and Goggins analyzed mass effects in free-free condition [10], this paper further extends their research for simply supported condition.

S. Rajbamshi (✉) · Q. Guo · M. Zhan
College of Mechanical and Electrical Engineering, Nanjing University of Aeronautics and Astronautics, Nanjing, China
e-mail: srijan@nuaa.edu.cn

A. Linderholt et al., *Dynamic Substructures, Volume 4*, Conference Proceedings of the Society for Experimental Mechanics Series,
https://doi.org/10.1007/978-3-030-12184-6_12

12.3 Analysis

The experiment has been carried out in the Lab at the College of Mechanical and Electrical Engineering in the Nanjing University of Aeronautics and Astronautics. Nominal diameter DN32 and DN50(ISO) of length 1 m steel pipes were used in the experiment. Sponge support were used to simulate free-free conditions for both pipes as shown in Fig. 12.1. For supported pipe test a test rig was built as shown in Fig. 12.2. The test rig is designed with a stiff base plate so as to simulate the fixed base.

12.3.1 Test Rig and Experimentation Setup

Seven uniformly separated locations were identified and marked on each pipe along the length so as to obtain the first three bending mode shapes in the vertical directions. The accelerometer (PIEZOTRONICS PCB 356A16) was magnetically attached on the second point from one end as shown in Fig. 12.3. All seven points were excited sequentially, each point being excited three times individually.

A measuring cylinder was used to fill water into the pipes. The cap was closed after pouring water into the pipe.

An instrumented hammer; Dytran model 5800B4 was used to excite all configurations. The accelerometer was connected to OROS OR363 system.

12.3.2 Numerical Simulation

12.3.2.1 Acoustic Structural FEA

As Escaler, Torre and Goggins [10] pointed out earlier Navier-Stokes (NS) equations are coupled with structural dynamics equation to solve coupled Acoustic-Structural analysis. The structural deformation and pressure waves generated in the fluid act simultaneously on each other. The unsymmetrical coupled FSI matrix system as mentioned in Eq. (12.1) can be obtained from combination of discretized structural and lossy wave equations.

Fig. 12.1 DN32 pipe for free-free simulation

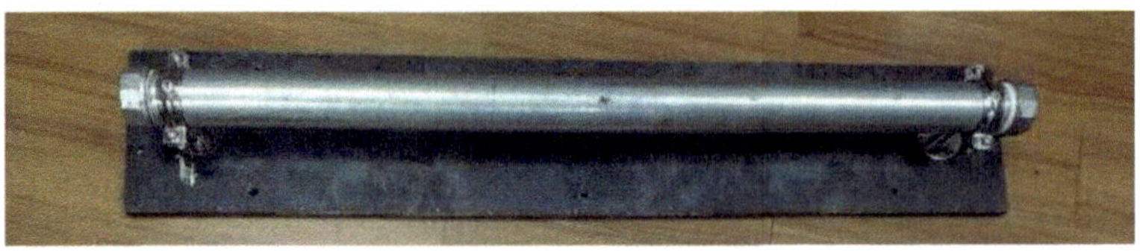

Fig. 12.2 DN50 mounted on support test rig

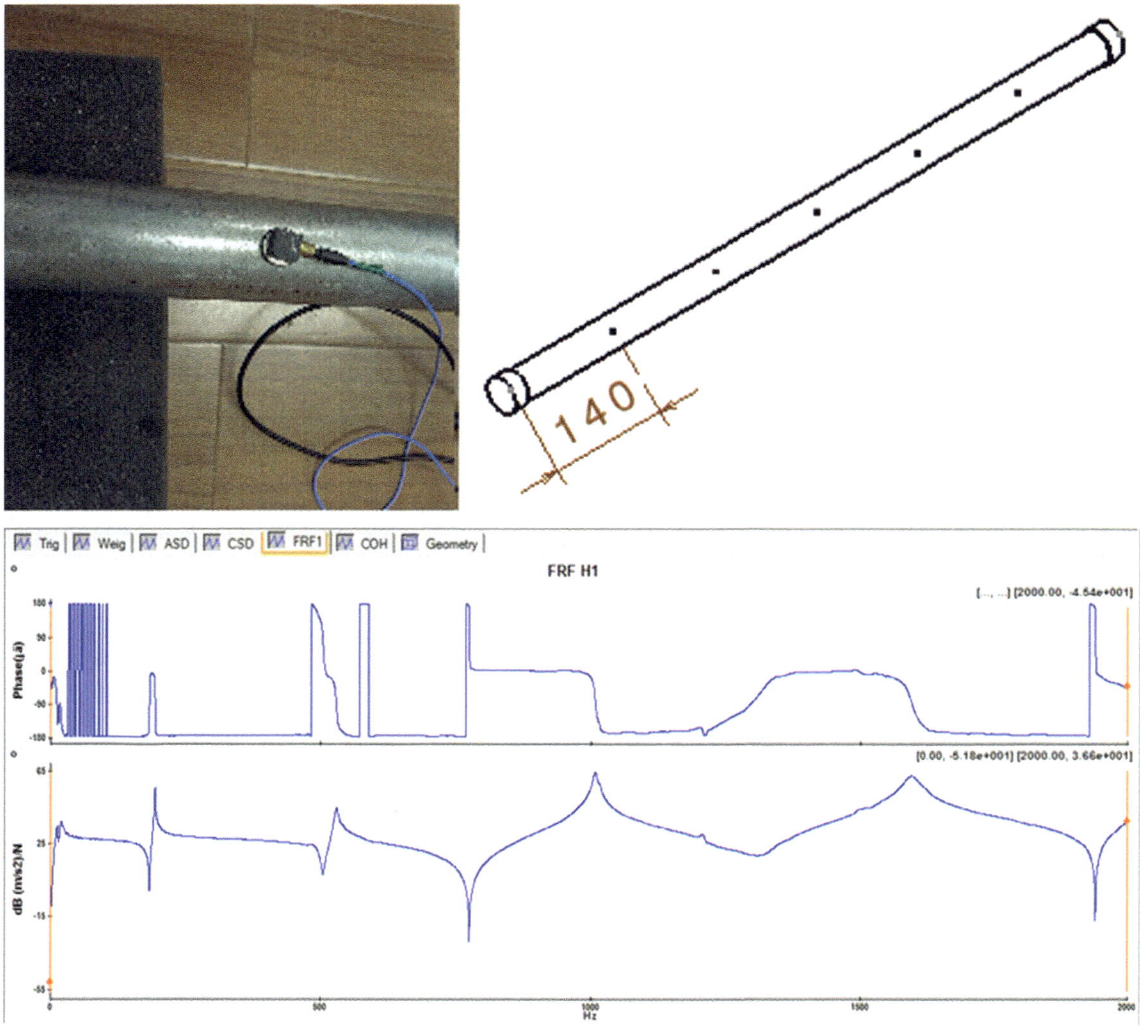

Fig. 12.3 Data acquisition

$$\left(\omega^2\begin{bmatrix} M_s & 0 \\ \rho_0 R^T & M_f \end{bmatrix} + jw\begin{bmatrix} C_s & 0 \\ 0 & C_f \end{bmatrix} + \begin{bmatrix} K_s & 0 \\ 0 & K_f \end{bmatrix}\right)\begin{Bmatrix} u \\ p \end{Bmatrix} = 0 \tag{12.1}$$

where $\mathbf{M_f}$, $\mathbf{C_f}$ and $\mathbf{K_f}$ are the acoustic fluid mass, fluid damping and fluid stiffness matrices, respectively, $\mathbf{R}$ is the acoustic fluid boundary matrix, $\mathbf{f_f}$ is the acoustic fluid load vector, $\mathbf{p}$ is the acoustic pressure vector, $\mathbf{u}$ is the fluid element displacement vector and $\boldsymbol{\rho_o}$ is the mean fluid density, $\boldsymbol{\omega}$ is the angular frequency of the pressure oscillation, and $\mathbf{M_S}$, $\mathbf{C_S}$ and $\mathbf{K_S}$ are the structure solid mass, damping and stiffness matrices, respectively.

12.3.2.2 Model

ANSYS® v. 18 was used to create the model comprising of structure domain for pipe, cap and supports and acoustic fluid domain for the inner liquid filling the pie as shown in Fig. 12.4. The model was built so as to incorporate the additional length and mass due to cap fittings. Effective length and cap mass along with other dimensions are included in Table 12.1. Table 12.2 includes the material properties of Steel, Water and Stainless Steel used for the simulation. Higher order solid elements exhibiting quadratic pressure behavior were used for acoustic domain whereas higher order solid elements exhibiting quadratic displacement were used to model structure domain. Full model analyses were performed with unsymmetric method with acoustic elements and direct method without acoustic elements to extract a mode shapes in the range of 10–2500 Hz;

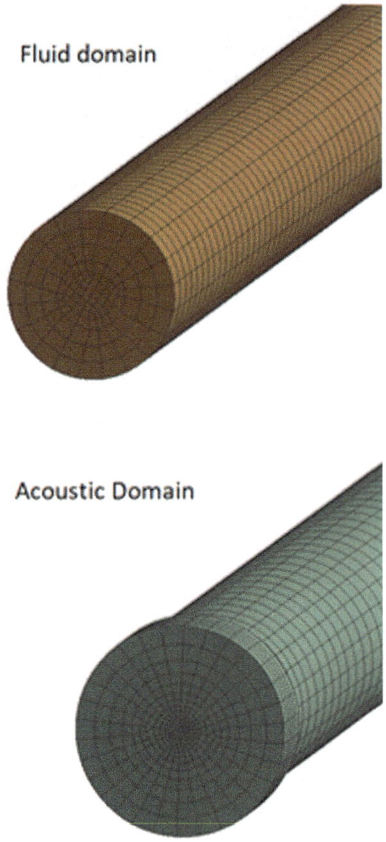

Fig. 12.4 Mesh of acoustic and fluid domain

Table 12.1 Dimensions of pipe

Pipe	Outer diameter (m)	Inner diameter (m)	Thickness (m)	Cap mass (kg)	Effective length (m)	Liquid volume (m^3)
DN32	0.0424	0.0359	0.00325	0.09405	1.04	0.001011
DN50	0.0603	0.053	0.00365	0.18905	1.04	0.002205

Table 12.2 Material properties

Material	Density(kg/m^3)	Young's modulus (Pa)	Sonic speed (m/s)
Steel	7850	2e11	–
Water	998	–	1482
Stainless steel	7750	1.93e11	–

the first three vertical bending mode shapes as shown in Fig. 12.5 always fell inside this range. To simulate the support of pipes three bush elements were used and parameters were tuned to obtain the closest results to experiment.

Then the structure only model was added with distribute mass as shown in Fig. 12.6 to simulate the added mass effect of water. The distribute mass was updated and optimized using Genetic Aggregation RSM and Multi-Objective Genetic Algorithm (MOGA) method built into ANSYS.

12.3.2.3 Results

The results obtained from the experimental and simulation for free-free condition is presented in the Tables 12.3 and 12.4.

The results obtained from the experimental and simulation for the supported condition is presented in Tables 12.5 and 12.6.

We can clearly see that in the free-free condition the maximum deviation from experimental and simulated results is 6%. The initial distributed mass was just used as the dead mass i.e. density times volume of the fluid. This result is also included in Tables 12.3 and 12.4. The added mass parameter was then updated against the experimental and acoustic simulation results

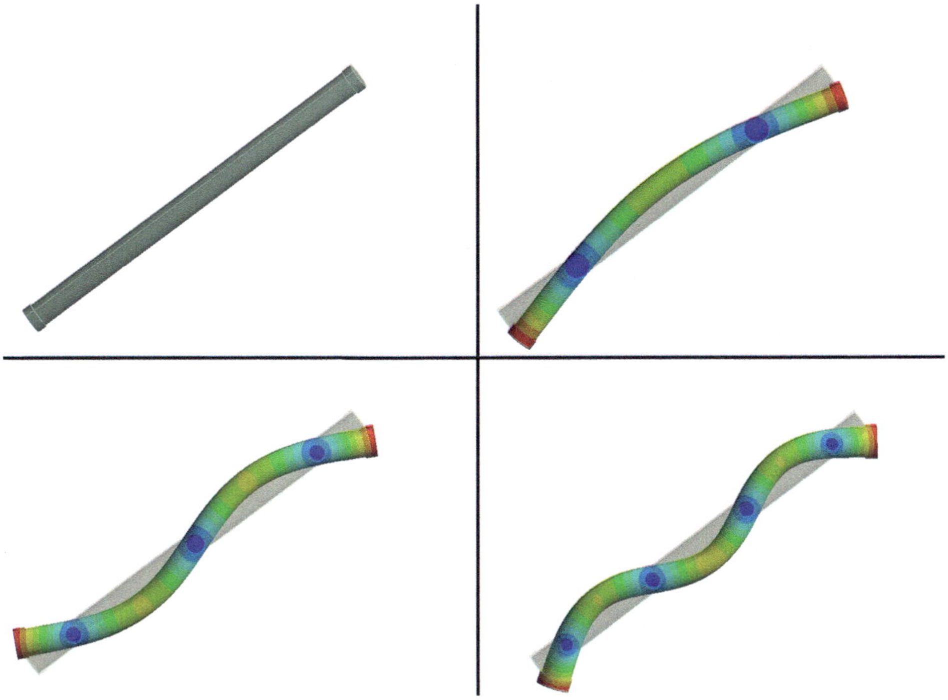

Fig. 12.5 Free mode shapes of the pipe; first mode (*top right*), second mode (*bottom left*) third mode (*bottom right*)

Fig. 12.6 Section view of face selection for distribute mass

Table 12.3 DN32 free-free condition

Condition	Frequency	f1	f2	f3
Without water	EMA (Hz)	224.29	607.83	1160.32
	Sim (Hz)	224.34	613.08	1180
	Dev (%)	0%	−1%	−2%
With water	EMA (Hz)	195.77	528.874	1009.964
	Acoustic sim (Hz)	202.36	550.82	1061.2
	Dev (%)	−3%	−2%	−5%
	Distributed mass (Hz)	201.77	548.24	1054.70
	Dev (%)	−3%	−4%	−4%

as shown in Fig. 12.7. The updated mass parameter was used to simulate the mass effect on the supported condition. In Tables 12.5 and 12.6 the distributed mass used is the function of updated density (ρ') as shown in Eq. (12.2).

$$\rho' = \frac{DM}{\rho_f} \tag{12.2}$$

where DM is the distributed mass used for simulation.

Table 12.4 DN50 free-free condition

Condition	Frequency	f1	f2	f3
Without water	EMA (Hz)	311.78	832.013	1543.37
	Sim (Hz)	315.05	849.01	1601.1
	Dev (%)	−1%	−1%	−2%
With water	EMA (Hz)	257.46	692.429	1287.354
	Acoustic sim (Hz)	269.87	726.15	1368.9
	Dev (%)	−5%	−5%	−6%
	Distributed mass (Hz)	265.96	713.93	1342.80
	Dev (%)	−3%	−3%	−4%

Table 12.5 DN32 with supports

Condition	Frequency	f1	f2	f3
Without water	EMA (Hz)	142.27	608.28	1174.81
	Sim (Hz)	133.73	517.61	1123.8
	Dev (%)	6%	15%	4%
With water	EMA (Hz)	130.93	400.739	1018.853
	Distributed mass (Hz)	115.63	447.76	972.65
	Dev (%)	12%	−12%	5%

Table 12.6 DN50 with supports

Condition	Frequency	f1	f2	f3
Without water	EMA (Hz)	179.63	854.73	1539.925
	Sim (Hz)	187.70	692.82	1455.5
	Dev (%)	−4%	19%	5%
With water	EMA (Hz)	163.34	716.305	1261.704
	Distributed mass (Hz)	152.80	564.30	1187.60
	Dev (%)	6%	21%	6%

As shown in the Fig. 12.7, the three bending frequency are almost linearly dependent on the density coefficient within given range. The optimization algorithm figured out 1.4 and 1.2 as optimal coefficient for DN32 and DN50 respectively for fitting the simulation with experimental results.

For supported condition though the deviation between simulation and experiments were up to 15% for cases without fluid and 21% for cases with fluid. The bushing contacts used for simplification of support may have caused the initial aberrance which cumulated into a massive deviation for fluid enclosed condition.

12.4 Conclusion

In free conditions the additional distributed mass can easily simulate the FSI-mass effect of the still fluid. This method is computationally far less intensive due to low number of elements as a whole and just the use of structural solver. For the supported case the parameters for supports could not be tuned enough to match the experiment without fluid. This may have resulted in overall large deviation for the distribute mass approach in the case with fluid.

Acknowledgements The first author would like to acknowledge the support of China Scholarship Council.

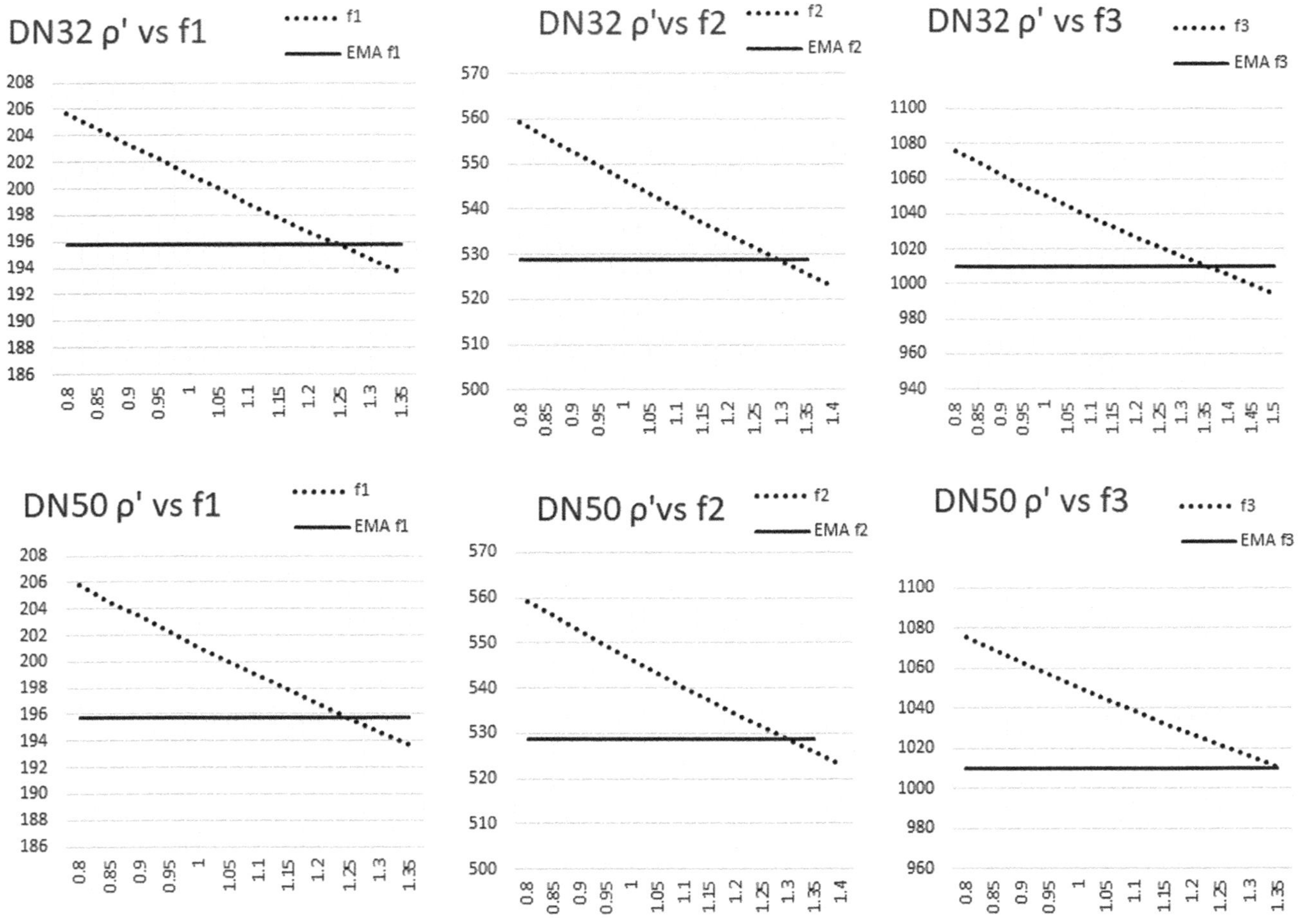

Fig. 12.7 Density coefficient versus frequency

References

1. Laithier, B.E., Païdoussis, M.P.: The equations of motion of initially stressed Timoshenko tubular beams conveying fluid. J. Sound Vib. **79**(2), 175–195 (1981). https://doi.org/10.1016/0022-460X(81)90367-9
2. Hashemi, M.R., Abedini, M.J., Simos, T.E., Psihoyios, G., Tsitouras, C.: Numerical modelling of water hammer using differential quadrature method. AIP Conf. Proc. **936**, 263–266 (2007). https://doi.org/10.1063/1.2790125
3. Ibrahim, R.A.: Mechanics of pipes conveying fluids-part II: applications and fuidelastic problems. J. Press. Vessel Technol. ASME. **133**(2), 1–30 (2011). https://doi.org/10.1115/1.4001270
4. Yi-min, H., Yong-shou, L., Bao-hui, L., Yan-jiang, L., Zhu-feng, Y.: Natural frequency analysis of fluid conveying pipeline with different boundary conditions. Nucl. Eng. Des. **240**(3), 461–467 (2010). https://doi.org/10.1016/j.nucengdes.2009.11.038
5. Zhang, L., Tijsseling, A.S., Vardy, A.E.: FSI analysis of liquid-filled pipes. J. Sound Vib. **224**(1), 69–99 (1999). https://doi.org/10.1006/jsvi.1999.2158
6. Sreejith, B., Jayaraj, K., Ganesan, N., Padmanabhan, C., Chellapandi, P., Selvaraj, P.: Finite element analysis of fluid-structure interaction in pipeline systems. Nucl. Eng. Des. **227**(3), 313–322 (2004). https://doi.org/10.1016/j.nucengdes.2003.11.005
7. Grant, I.: Flow induced vibrations in pipes, a finite element approach. ETD Arch. **633**, 74 (2010)
8. Dubyk, I., Orynyak, I.: Fluid-structure interaction in free vibration analysis of pipelines. Sci. J. Ternopil. Natl. Tech. Univ. **1**(81), 49–58 (2016)
9. Li, S., Karney, B.W., Liu, G.: FSI research in pipeline systems—a review of the literature. J. Fluids Struct. **57**, 277–297 (2015). https://doi.org/10.1016/j.jfluidstructs.2015.06.020
10. Escaler, X., De La Torre, O., Goggins, J.: Experimental and numerical analysis of directional added mass effects in partially liquid-filled horizontal pipes. J. Fluids Struct. **69**, 252–264 (2017). https://doi.org/10.1016/j.jfluidstructs.2017.01.001

Chapter 13
Vehicle Driveline Benchmarking to Support Predictive CAE Modeling Development

J. Furlich, J. Blough, and D. Robinette

Abstract The development of predictive models requires several assumptions along with known system properties and boundary conditions to generate a correlated model. When a prototype product is available, modal analysis can be used to benchmark the current product and extract modal properties. The extracted values are often cross referenced with FEA solutions and utilized to feed forward into CAE models for data replication and future prediction. This study was used to perform modal testing on a full sized pickup truck driveline to build a one-dimensional lumped parameter model. The successful extraction of modal parameters was able to provide benchmark stiffness and damping estimates for use in CAE model updating to achieve better correlation with experimental vehicle data. The resulting lumped parameter changes reduce the number of model assumptions and allow for modification of stiffness design targets for new prototype driveshafts and/or additional driveline components.

Keywords Vehicle benchmarking · CAE · Lumped parameter modeling · Experimental modal analysis

13.1 Introduction

This study performed benchmark testing on a part time four-wheel drive vehicle to build a lumped parameter model for predicting vehicle responses to modified parameters. The test vehicle has a front mounted internal combustion engine, rigidly connected to the automatic transmission and transfer case, all of which are supported by three elastomeric mounts. The transfer case has two outputs; one that supplies torque to the front differential part time and another which supply's torque to the rear differential during all vehicle operation via the propeller shaft. The rear differential splits the torque input to both rear wheels via half shafts and is supported by four elastomeric mounts. The lumped stiffness and damping values for the elastomeric mounts are unknown and thus require testing to benchmark their values for modeling.

The test vehicle is known to experience transient responses during vehicle throttle tip-in and tip-out scenarios and the purpose of this benchmarking is to create a lumped parameter model of the vehicle. This will be used to study parameter modification to reduce the amplitude of transient responses. Transients of concern for the test vehicle include clunk and shuffle which are a result of driver requested torque increases or decreases. The driver makes a torque request by either increasing the throttle pedal position (tip-in), decreasing the throttle pedal position (tip-out), or adjusting the requested cruise control speed of the vehicle. Rapid changes in driver requested torque during tip-in and tip-out scenarios are known to exacerbate transient amplitudes of the vehicle response.

Results of this study support lumped parameter modeling to create a high-fidelity dynamic vehicle model without the high computational cost of a full vehicle FEA model. The lumped parameter model is set up to account for the most important vehicle components as interpreted by the authors. These key vehicle components include the powerplant (combined engine, transmission, and transfer case effective parameters), driveline (propeller shaft, half shaft's and tires), and rear differential. The front differential and half shafts are not an objective to this study because the increase in coupled inertia properties is already known to reduce transient effects as perceived by the vehicle customer.

To support the lumped parameter model, estimates are required for the key vehicle parameters identified in terms of inertia, stiffness, and damping. The key parameter is to estimate damping properties of the vehicle for improved model accuracy. The reason damping estimates are important is in theory mass and stiffness will adjust the frequency of transient oscillations, however damping is the key to reducing peak amplitudes as well as improve decay times and rates.

J. Furlich (✉) · J. Blough · D. Robinette
Mechanical Engineering-Engineering Mechanics, Michigan Technological University, Houghton, MI, USA
e-mail: jefurlic@mtu.edu

A. Linderholt et al., *Dynamic Substructures, Volume 4*, Conference Proceedings of the Society for Experimental Mechanics Series,
https://doi.org/10.1007/978-3-030-12184-6_13

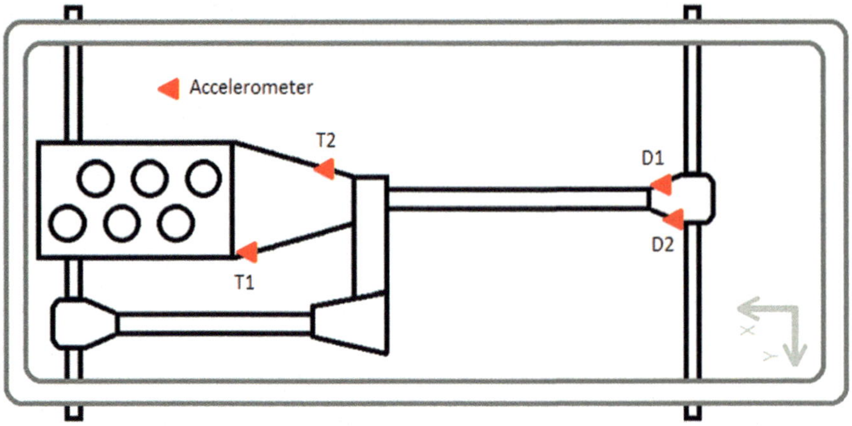

Fig. 13.1 Sketch of test vehicle powertrain including triaxial accelerometer test locations on powerplant and differential

13.2 Background

To perform modal testing and static testing on the vehicle, key locations were identified on the engine and transmission combination to capture rigid body modes to be representative of the overall lumped parameters. The mount locations used for testing are depicted in Fig. 13.1 for SIMO testing as well as static measurements on the powerplant and differential. No mount locations are shown for the propeller shaft to reduce image clutter; however, they were spaced out in even increments along the shaft length and circumference with further clarification in Fig. 13.4. The authors would like to note that the estimated damping values constitute an initial estimate but would like to highlight concerns for the elastomer mounts. This is due to operation of the engine, transmission and differential influenced by temperature effects as the mounts warm up during standard vehicle maneuvers. Thus, their values may be adjusted in the final CAE model to account for temperature discrepancies.

Modal testing was performed on the engine, differential, and propeller shaft using a SIMO testing technique. To perform impact measurements, a modally tuned impact hammer and modal punch were both used when applicable. The modal punch was needed for impacting the engine and transmission due to the large amount of surrounding peripheral components such as exhaust headers, frame cross members, wiring, etc. The impact hammer was used for measurements on the propeller shaft as well as differential because there were less space constraints surrounding these components. Boundary conditions for the modal testing included parking the vehicle on a flat, level surface with all components measured in-situ with no modification to the factory installation conditions. This is not standard practice relative to traditional modal analysis where components are disassembled individually and measured in free-free boundary conditions. The authors wanted to capture relevant boundary conditions while mounted in-vehicle, so the departure was made from standard practice to test as assembled with the knowledge that additional non-linearities and noise may be present and must be interpreted in the final data with care.

Input force and output acceleration from the SIMO testing were captured and processed to compute FRF and Coherence measurements for respective sets of data. The sets of data were separated into three different sets; the powerplant, the propeller shaft, and the differential. The respective measurements for each truncated data set were processed separately with a polyreference modal parameter estimation (MPE) algorithm to estimate natural frequency and damping.

Stationary data was also collected with the vehicle for one test setup in two different vehicle conditions. The first condition was taken after the vehicle had been sitting in place for an appropriate amount of time such that the tire temperature was at equilibrium with the surrounding environment (~20 °C). The second test was performed after driving the test vehicle on a test track to bring the tires and other vehicle components up to an "operating" temperature. In both test conditions the vehicle was parked on a light incline surface with the engine off and transmission shifted into park such that the park pawl impedes rotation of the driveline. A static load was applied to the vehicle frame in tension which loaded the driveline components until the load was released in a step-relaxation manner, like a pluck test. The results from this type of step relaxation test is a decay response which was used to estimate the effective suspension fore-aft stiffness. With the output only of the operational data, a damping value can also be estimated using the log decrement method [1, 2] as shown in Eq. (13.1).

$$\delta = \frac{1}{n} \ln \left(\frac{x_n}{x_{n+T}} \right) = \frac{2\pi\zeta}{\sqrt{1-\zeta^2}} \approx 2\pi\zeta \qquad (13.1)$$

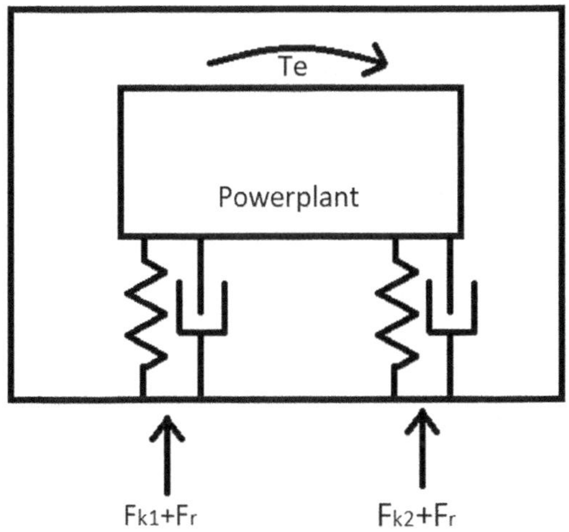

Fig. 13.2 Free body diagram of simplified engine on frame reaction

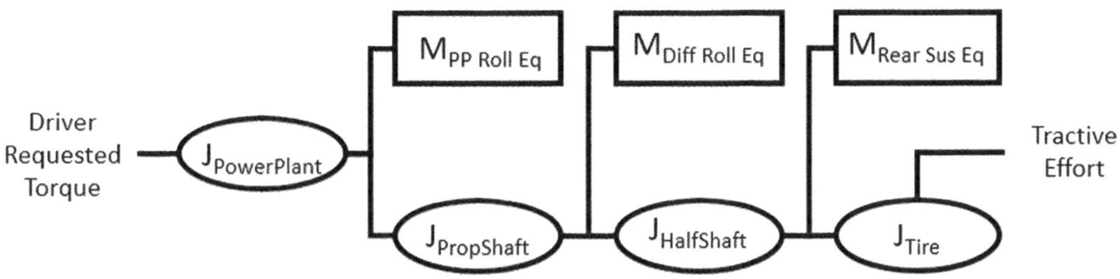

Fig. 13.3 Flow chart of lumped parameter model power flow

The major objective of modal testing on the powerplant, propeller shaft, and differential was to come up with a damping estimate to support a CAE lumped parameter model of the vehicle. The modes of interest include rigid body modes of the powerplant and differential while evaluating the first torsional mode of the propshaft. During tip-in or tip-out events there is a change in driver requested torque which modifies the torque output of the engine. The engine torque output causes a transverse weight to shift due to reaction torques, shown in Fig. 13.2, known as roll which is associated with the rigid body mode about the X-axis.

To model the roll modes of the powerplant and differential, a simple rack and pinion type model can be incorporated in parallel with the power flow that converts rotational degrees of freedom into translational degrees of freedom with a lever arm and spring/damper model to match the equivalent stiffness and damping of the roll mode estimates. This parallel modeling technique is depicted in Fig. 13.3 above. In Fig. 13.3, ellipses represent the torsional elements while rectangles represent the translational elements. Torque is passed back and forth between the elements and whenever the torque splits from one torsional element to a linear element it is incorporating an equivalent mass, stiffness and damping to represent the roll modes of the powerplant, differential and the fore-aft stiffness of the vehicle suspension.

13.3 Analysis

As commonplace with any modal analysis testing method, one of the first things to consider when acquiring data is how much data is appropriate and necessary to capture the relevant mode shapes that are of interest. With respect to the powerplant and differential roll modes, these are low frequency modes and can be found <20 Hz and <200 Hz for the powerplant and differential respectively. However, the test vehicle utilizes an aluminum propeller shaft instead of steel. This requires an analytical estimate to find the shaft natural frequency and set an adequate sample rate for experimental testing.

The torsional natural frequency for the propeller shaft can be calculated by assuming several material properties as well as a constant cross-sectional area for the length of the shaft. The physical shaft has a taper (reduction in diameter) at both the front and rear where the Cardan Joint's connect to the transmission output and differential input respectively. These tapers are ignored when computing the solution to the closed form equation estimate the propshaft natural frequency. The closed form solution may be computed from Eq. (13.2) [3] where f_i is the ith natural frequency, λ_i is the ith dimensionless parameter as a function of the boundary conditions, C is the torsional constant (defined by Blevins as "the moment required to produce a torsional rotation of 1 radian on a unit spanwise length of shaft divided by the shear modulus..." [3]), G is the shear modulus, μ is the volumetric density of the material, and I_p is the polar area moment of inertial about the principal axis.

$$f_i = \frac{\lambda_i}{2\pi L}\sqrt{\frac{CG}{\mu I_p}};\quad i = 1, 2, 3, \ldots \tag{13.2}$$

There were several options for boundary conditions that were assumed for the propeller shaft including; free-free, fixed-free, fixed-fixed, spring-free, fixed-spring, fixed-inertial mass, and free-inertial mass. The propshaft was tested as mounted in the vehicle such that several assumptions could be made as to which boundary conditions were most appropriate. However, to ensure that the sample frequency was high enough to account for improper assumptions, results for the first torsional mode (above rigid body motion) were calculated for all boundary conditions. The results based on the material and geometric properties and assumptions shown in Table 13.1 are shown in Table 13.2.

As shown in Table 13.2, the closed form solution for the propeller shaft itself with no coupled spring or inertia ranges from 639 to 1280 Hz. Based on the assumptions for coupled springs or inertias, the natural frequency ranges from 133 to 1790 Hz which is a larger range than the shaft itself. It would be expected if the shaft was measured in a traditional modal analysis setting with free-free boundary conditions, or close enough as shown in [4, 5], then the natural frequency should lie near 1280 Hz. Because the shaft is measured as installed on the vehicle with the park pawl engaged, it could either be modeled with coupled spring or inertia values. Downstream of the power flow through the propeller shaft is the differential, half shaft, and wheels, which can either be modeled as stiff springs or additional heavy inertia elements (especially the tires). Based on the results in Table 13.2 a bandwidth of 4096 Hz was selected to have adequate overhead to ensure the torsional mode is captured during modal testing.

Results for modal testing on the propeller shaft are shown in Fig. 13.4 for a truncated set of measured FRF's. Due to the nature of the test, it was expected that several non-linearities and noise would be present in the data and should be kept in mind while interpreting results. After combing through the data and resulting mode shapes found through a polyreference MPE algorithm, the first torsional mode was found at 1546 Hz with 2.24% damping. The FRF data shown in Fig. 13.4 depicts points PS6 and PS24 in phase with one another with PS8 and PS24 in opposing phase at resonance. This confirms a torsional mode shape with a nodal line near the center of the shaft. This mode shape is consistent with the closed form solutions for spring-free or free-inertia boundary conditions. When reviewing the results from Table 13.2 it is observed that for the free-inertia boundary condition solution to achieve a natural frequency of ~1550 Hz the coupled inertia value would need to be less than 0.001 kg m^2. Physically this would be likened to the transmission end of the propeller shaft having a very low inertia which is not realistic relative to the computed value. With respect to the spring-free boundary condition, it is shown in

Table 13.1 Propeller shaft material and geometric properties

Material properties	$E = 68.9$ GPa	$\nu = 0.33$	$\mu = 2700\,\text{kg/m}^3$
Physical properties	$C = \pi/2\,(r_o^4 - r_i^4)$	$I_p = \pi/4\,(r_o^4 - r_i^4)$	

Table 13.2 Closed form solution results for propshaft first torsional mode

Boundary condition	Assumption	Fn (Hz)
Free-Free	–	1280
Fixed-Free	–	639
Fixed-Fixed	–	1280
Spring-Free	K = 10 kN m/rad	1310
	K = 100 kN m/rad	1540
	K = 500 kN m/rad	1790
Fixed-Spring	K = 10 kN m/rad	703
	K = 100 kN m/rad	985
	K = 500 kN m/rad	1190
Fixed-Inertia	J = 0.01 kg m^2	363
	J = 0.05 kg m^2	185
	J = 0.10 kg m^2	133
Free-Inertia	J = 0.01 kg m^2	839
	J = 0.05 kg m^2	691
	J = 0.10 kg m^2	666

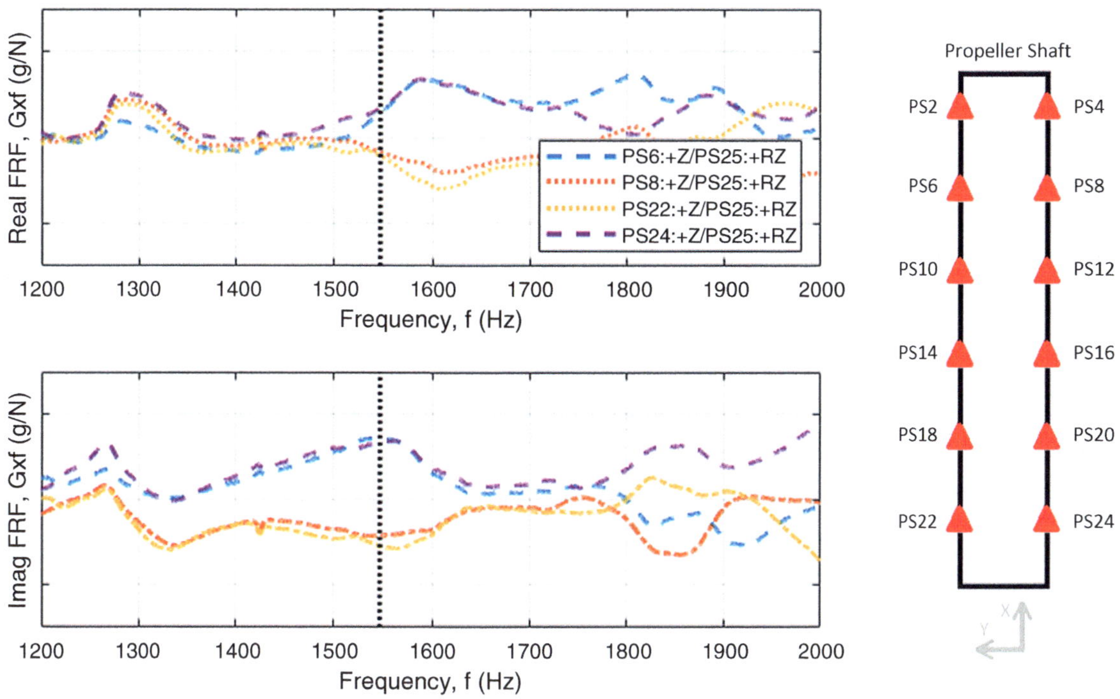

Fig. 13.4 Propshaft torsional mode at 1546 Hz and 2.24% damping with tri-axial response locations

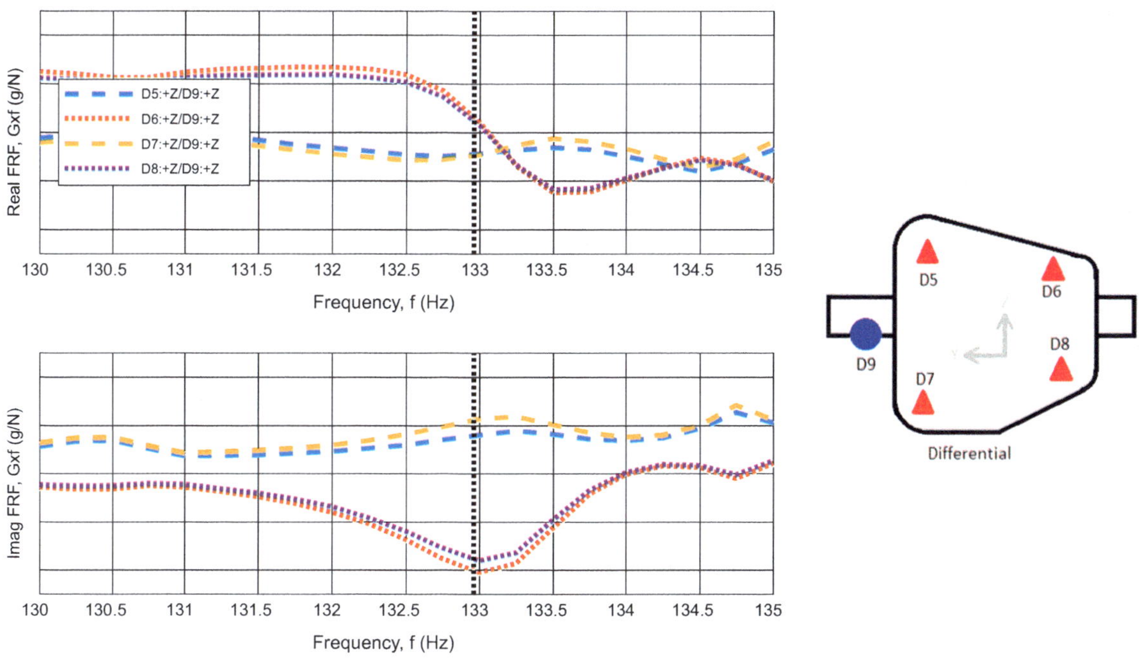

Fig. 13.5 Differential roll mode at 132 Hz and 0.72% damping with uni-axial input and tri-axial response locations

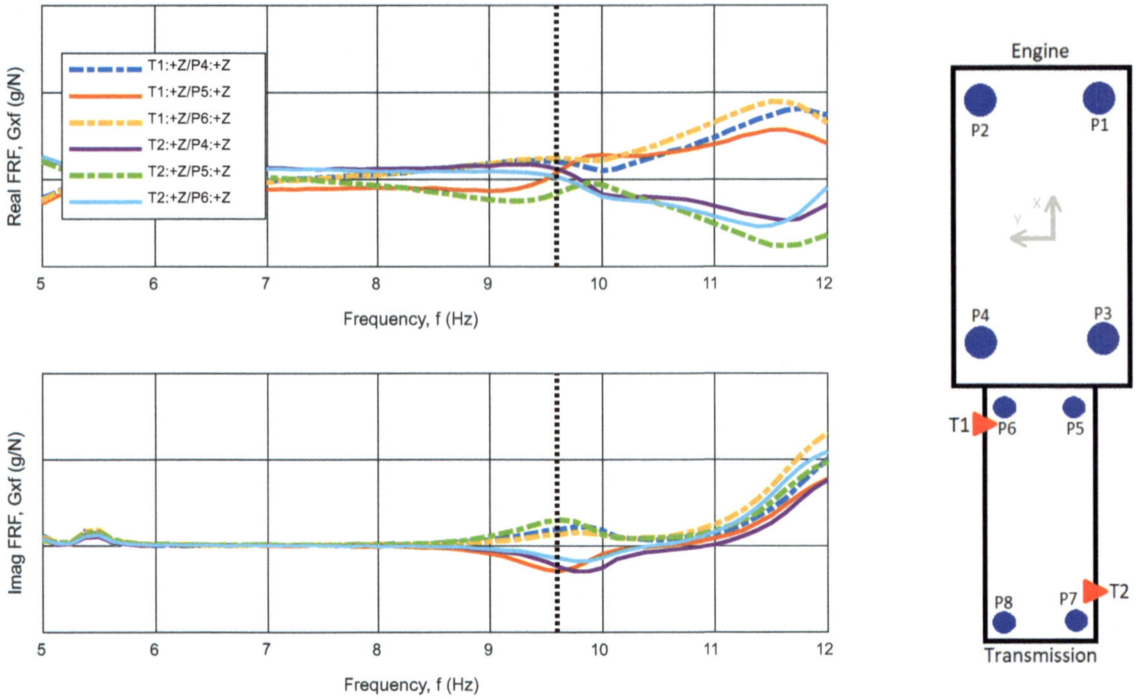

Fig. 13.6 Powerplant roll mode at 9.5 Hz and 2.96% damping with uni-axial input and tri-axial response locations

Table 13.2 that a coupled stiffness of roughly 100 kN m/rad is required to achieve a natural frequency of ~1550 Hz, closer to realistic physical conditions.

The resulting differential roll mode was found using a polyreference MPE algorithm with truncated results shown in Fig. 13.5. The results show the roll mode to occur at 132 Hz with 0.72% damping. The frequency of this mode is as expected but has a lower damping estimate than anticipated. The truncated FRF data shown in Fig. 13.5 shows the two sides of the shaft D5 and D7 out of phase with the other measurements at D6 and D8 as expected for a roll mode. It is possible that the mode found is in fact a torsion mode. Additional testing would be ideal to further confirm this frequency to be the roll mode and not a torsional mode of the differential, however current spatial resolution of the data has a strong indication that this frequency is in fact the differential roll mode.

Like the differential modal analysis, a truncated set of FRF's is shown in Fig. 13.6 for results of modal testing on the powerplant (combined engine and transmission). The results of polyreference MPE indicate the powerplant roll mode to be at 9.5 Hz with 2.96% damping. These values are close to what was anticipated for the powerplant roll mode and animation of the mode shapes show good indication of a roll mode as well. Additional modes such as bounce (near ~5.5 Hz) are observed in the data set and found through MPE but are not deemed necessary for the lumped parameter vehicle model in the CAE simulation.

The stationary step relaxation testing results are shown in Fig. 13.7 for both time data of the cold and hot tire tests as well as frequency content of the data. The response to a step relaxation is like that of a pluck test where it is seen that the fore-aft response of the vehicle in the X direction decays logarithmically as expected for a viscously damped system. The step-relaxation force was not captured during data acquisition, thus only output acceleration was recorded. The output data is processed with Eq. (13.1) and yields a damping value of 2.73% and 3.71% for the hot and cold tire tests respectively. This indicates that when the tire's get warmer they become less effective at damping out fore-aft vibrations and have lower amounts of shear allowing for viscous damping.

The frequency of oscillation can be found in the linear spectrum of the data, like that of an ODS. The fore-aft mode of the vehicle was found to be 1 Hz and 1.13 Hz for the cold and hot tire test's respectively. This indicates that as the tire heat's up either the effective stiffness increases, or the effective inertia decreases. For an elastomer, this is counter intuitive for the stiffness to increase with increasing temperature. Thus, it is understood that as the longitudinal tire stiffness decreases, the tire contact patch increases reducing the amount of coupled inertia. Also, with an increased tire contact patch there is a higher capacity for tractive force in the tire to increase the effective fore-aft stiffness.

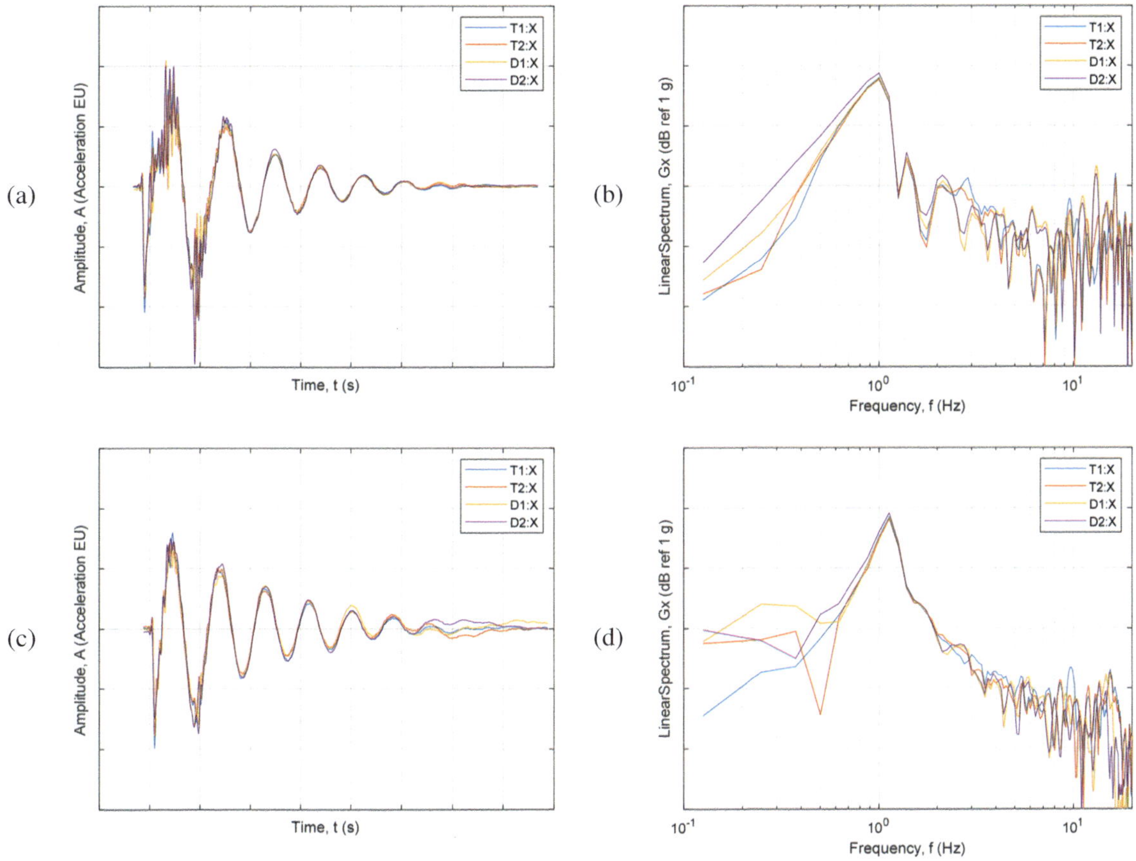

Fig. 13.7 Step relaxation results for cold tire time data (**a**) and linear spectrum (**b**) as well as hot tire time data (**c**) and corresponding linear spectrum (**d**)

13.4 Conclusion

The aim of this study was to use experimental modal analysis to derive stiffness and damping estimates for use in a CAE lumped parameter model for vehicle development. The lumped parameter model is used for reduced computational load compared with FEA methods, however it requires several assumptions to be made for lumped parameter distribution. With the values extracted from MPE, the number of assumptions is reduced in the CAE model and drives towards better correlation with experimental results.

Testing was performed on three different lumped components; the powerplant, differential, and propeller shaft. Each component was evaluated for either the first rigid body roll mode or first torsional mode. An updated material selection for the propeller shaft required a closed form solution to estimate the first torsional mode. The solution identified the mode to be between 133 and 1790 Hz based on estimates for boundary condition and/or coupled stiffness or inertia. This was used to set the bandwidth of the experimental testing to ensure the torsional mode was captured within the dataset. Results from experimental testing identified the roll modes at 9.5 Hz and 132 Hz for the power plant and differential respectively while the propeller shaft torsional mode was found at 1546 Hz.

Static testing was performed on the entire vehicle body to estimate the stiffness and damping of the fore-aft suspension mode. This was done by shifting the transmission into park such that the driveline cannot rotate (other than through the lash zone). The frame was then loaded in tension with a cable and released such that a step relaxation force was input to the body. This resulted in a vehicle oscillation like that of a pluck test and showed the frequency of oscillation to be 1 Hz with cold tires and 1.13 Hz with warm tires. The decay of the output data was also used to estimate damping of 3.71% to 2.73% when the tires were cold and warm respectively. The combination of roll modes, torsional modes, and fore-aft body mode parameters were all used as vehicle parameters for building a lumped parameter CAE model of the test vehicle.

Acknowledgements The authors of this paper would like to acknowledge the project sponsor Ford Motor Company for their support of this research and endeavor.

References

1. Ewins, D.J.: Modal Testing: Theory, Practices, and Application. Research Studeis Press, Philadelphia (2000)
2. Heylen, W., Lammens, S., Sas, P.: Modal Analysis Theory and Testing. Katholieke Universiteit Leuven, Departement Werktuigkunde, Leuven (1997)
3. Blevins, R.D.: Formulas for Natural Frequency and Mode Shape. Van Nostrand Reinhold Company, New York (1979)
4. Avitabile, P.: How free does a test need to be? Does it really matter that much? In: SEM Experimental Techniques (2012)
5. Avitabile, P.: Can the test setup have an effect on the measured modal data? In: SEM Experimental Techniques (2001)

Chapter 14
A Proposal of Dynamic Behaviour Design Based on Mode Shape Tracing: Numerical Application to a Motorbike Frame

Elvio Bonisoli, Domenico Lisitano, Luca Dimauro, and Lorenzo Peroni

Abstract A modal based method to design and to improve the dynamic behaviour of flexible multibody systems is proposed in this paper. The dynamic behaviour of a flexible multibody system depends on the dynamics of its components and on the effects of the connections. When dynamics problems are experienced in particular working conditions, changes to the design are required. Usually connections are standard and there is not space for changes, therefore the components should be improved. Changes at components level improve the global system behaviour, but it is not so easy to identify the most effective component for each specific case. The purpose of the proposed method is to identify the most influent components in specific working frequency ranges of a multibody system. The method is based on a hierarchical procedure from assembly to components which leads to the identification of the components to be modified in order to improve assembly performance.

The method is applied to a whole motorbike frame, a multibody system with comparable components stiffnesses. Numerical modal analysis is performed on the full assembly with connections, front and rear sub-assemblies and on single components: chassis, swingarm, engine and wheel.

The method is applied to the selection of the most influent components mode shapes in the motorbike behaviour during manoeuvres at high velocity. The selected components are the most suitable for structural, geometric and material modifications to effectively improve the global motorbike behaviour, performances and driveability.

Keywords Flexible multibody dynamic · Numerical modal analysis · Motorbike dynamics · Components dynamic design · Weighted mode shape

14.1 Introduction

The structural and dynamic design is the primary step for the manufacturing of a successful and high quality product. In this context, the design takes an even leading role, especially when the design requirements follow advanced and particular standards. Usually, the aim is to reduce and, if it is possible, to eliminate the problematic dynamic effects on the global behaviour of a multibody system. The quick selection of the most critical components of the system on which changes are expected is more and more required by industries. Usually this aim is achieved using Modal Assurance Criterion method (MAC) [1], which compares the mode shapes without weighting their modal participation factor. The method here proposed aims to operate as a filter in the preliminary dynamic design phase or in a re-design phase, to localise the regions suitable for changes. The presented method is an extension of the Design for Modal Assembly (DMA) method proposed by Bonisoli et al. in [2]. In DMA the system global mode shapes are compared with the mode shapes of each component changing a physical parameter of the model. In the actual method the parameter of variation is the excitation frequency and a tracing procedure is developed to follow the dynamic behaviour from system to components, considering a weighted mode shape that includes the modes participation factors.

This work is focus on the frame of a racing motorcycle. In racing sector, very fast design is required to update year by year the motorcycle performance, and often this is done starting from the problems encountered in the release of the previous year.

The study of the handling and dynamic behaviour of a two-wheeler vehicle requires a particular attention with respect to the much more known automobile dynamics, because also the roll and the steer angles must be taken into account, since they

E. Bonisoli (✉) · D. Lisitano · L. Dimauro · L. Peroni
Department of Mechanical and Aerospace Engineering, Politecnico di Torino, Torino, Italy
e-mail: elvio.bonisoli@polito.it

A. Linderholt et al., *Dynamic Substructures, Volume 4*, Conference Proceedings of the Society for Experimental Mechanics Series,
https://doi.org/10.1007/978-3-030-12184-6_14

can generate troublesome consequences for the motorbike driveability and stability. Indeed, the motorbikes are intrinsically unstable and thus the primary goal of a motorcycle dynamic design is avoid that this instability comes off during normal functionality.

In last years many authors developed rigid bodies mathematical models to analyse the dynamic behaviour of motorcycles, considering only the stiffness and damping of connections and neglecting the flexibility of the main components like chassis and rear swingarm [3–7]. In [8] flexible components are implemented to simulate the frame dynamic behaviour and it comes out that the wheels have a sharp effect on the global dynamics. Multibody approach is used in [9] to study the motorbike drivability and the dynamic response of the rear shock absorber under sudden braking and impacts with curbs. The simulation results of a completely rigid model were compared with another model containing a flexible swingarm, obtained using a floating frame of reference (FFR) approach [10], and highlighting in this last case an unstable behaviour of the motorcycle in the case of subsequent impacts.

In [11] Bocciolone et al. characterised the behaviour of the motorbike frame during a high roll angle cornering on high friction passage. It was highlighted that in this condition the torsional stiffness given by the frame flexibility works in series with the front fork: the front fork absorbs only the road irregularity component directed along the shocks axis while the chassis absorbs the other components, working as an equivalent torsion spring. This chassis capability is fundamental to guarantee the adherence in the passages between low and high friction surface. The flexibility of the chassis is also very relevant in comfort studies. Indeed, its high frequencies could be very close to the engine rotation frequencies, causing vibration transmission to the whole structure and a probable fatigue local rupture of a connection between frame and engine. Recently, Lake et al. in [12] give an overview on the effects of flexible components on the global motorcycle dynamics and on the stability properties. They show as the optimal frame stiffness values depend on the purpose of the motorcycle. The low frequency modes are also influenced by the global dynamic properties: the "wobble" mode is stabilised when the frame torsional stiffness is increased, while "weave" mode is stabilised when the bending stiffness is increased. The low frequency wheel dynamics is also important to preserve the tyres integrity; in particular "chattering" phenomena [13] is induced by the violent vibrations of front forks during turning phases. The front wheel can lose contact with the ground because of high variations in the vertical load. The frequency of these vibrations depends on the front unsprung mass and for sports bikes the frequency is about 12–18 Hz. This phenomenon depends on the "stick slip", generated by the static friction of the sliding parts of the front forks. The chattering is increased by the coupling between the structural vibrating modes of the fork with the in-plane mode of the unsprung mass, and thus it is very important to try to uncouple the two systems to preserve the tyres.

In [14] the present authors proposed a mode shape tracing from component-to-system environment using Modal Assurance Criterion. The aim was to follow modifications of components mode shapes in sub-systems of increasing complexity up to the global motorbike frame. In details, four chassis modes, three swingarm modes and four wheel modes were traced from component to assembly, with a particular interest to chassis and swingarm torsion. All the followed component modes had high MAC correlation with respect to the global system, but this does not imply that all of them are influent in the motorbike dynamics and it does not give particular information regarding possible changes in the design to improve system behaviour in working conditions.

The paper is organised as follow: in Sect. 14.2 the modal based method is proposed as an extension of DMA. In Sect. 14.3 the method is applied to the case study of motorbike dynamic behaviour with the aim of suggest the components to improve. Finally, some considerations regarding the most influent modes and their impact on motorbike performances are presented.

14.2 Mode Shapes Tracing via Weighted Mode Shapes

The proposed method aims to select the most influent components for the dynamic behaviour of a flexible multibody system in specific working conditions. Given a working driving frequency ω to the multibody system, the aim is to move the multibody system resonance frequencies as far as possible with respect to the driving frequency. The requirement of shifting assembly natural frequencies means to re-design and thus to update and improve the flexible components of the systems. The main questions in this case are: (1) which components have a high influence on the assembly mode shapes near the driving frequency? (2) Once selected the most influent components, which mode shapes of these components are mainly participating for the system response at that driving frequency in assembly level? To quickly answer to these questions, a method, intended to obtain a very good dynamic behaviour design of a multibody system, is required.

The starting point of the proposed method is the modal analysis of the full assembly, its sub-assemblies, and each single component. Modal analysis is nowadays a standard technique and a very fast tool to perform accurate modal analysis prediction and both Finite Elements Analysis (FEA) commercial software and Experimental Modal Analysis (EMA) tools are widely diffused.

Once the modal properties of the assembly, sub-systems and components are known, the classical answer to the two previous questions is the comparison between modal properties from assembly to component using the well-known Modal Assurance Criterion (MAC) [1]:

$$MAC_{j,k} = \frac{\left[\varphi_j^T \varphi_k\right]^2}{\left[\varphi_j^T \varphi_j\right]\left[\varphi_k^T \varphi_k\right]} \tag{14.1}$$

The MAC index is defined as the normalised scalar product between two modes j and k: hence it results equal to 100% if two modes are identical, while it reaches 0% if they are orthogonal. This is a quite useful tool in comparing mode shapes of two systems, but it does not take into account the participation factor of each mode shape in the global response of the system, therefore it can give not reliable results if it is applied to the proposed application. Indeed, even if a mode shapes is well traced from component to assembly, there is no certainty that the detected mode has a high contribution in the global system dynamics.

The weighted mode shape has been defined in Eq. (14.2) to take into account the participation factors of each mode. The weighted deformation shape of the structure at a certain driving frequency ω is computed as the sum of all the mode shapes $\mathbf{\Phi}_r$, scaled to unitary modal mass, and weighted with respect to resonance and driving frequency. Each mode contributes in a different percentage to the global response and in a different frequency range:

$$\tilde{\mathbf{\Psi}}_{WM}(\omega) = \sum_{r=1}^{n} \frac{1}{\left(\omega_r^2 + 2\mathrm{i}\zeta_r\omega_r\omega - \omega^2\right)}\mathbf{\Phi}_r \tag{14.2}$$

The weighted mode shapes $\tilde{\mathbf{\Psi}}_{WM}$ are complex vectors and thus a further normalisation is suggested to univocally define them. For a complex eigenvector, a useful normalisation is to minimise the imaginary part of the vector or unitary modal mass normalisation can be taken into account, as well [15]. The minimisation of the imaginary part is useful to have almost all the information related to the weighted mode shape in the real part of it. Writing the weighted mode shape $\tilde{\mathbf{\Psi}}_{WM}$ as the sum of a real and an imaginary part:

$$\tilde{\mathbf{\Psi}}_{WM} = \mathrm{Re}\left(\tilde{\mathbf{\Psi}}_{WM}\right) + \mathrm{i}\,\mathrm{Im}\left(\tilde{\mathbf{\Psi}}_{WM}\right) \tag{14.3}$$

the normalised eigenvectors are obtained multiplying for a complex versor $z = \mathrm{e}^{\mathrm{i}\delta} = \cos(\delta) + \mathrm{i}\sin(\delta)$, used to rotate $\tilde{\mathbf{\Psi}}_{WM}$ of an angle δ in the Argand-Gauss plane. The normalised eigenvector is $\mathbf{\Psi}_{WM} = z\tilde{\mathbf{\Psi}}_{WM}$, where the angle δ results:

$$\tan(\delta) = \frac{\mathrm{Im}\left(\tilde{\mathbf{\Psi}}_{WM}^T\right)\ \mathrm{Im}\left(\tilde{\mathbf{\Psi}}_{WM}\right)}{\mathrm{Im}\left(\tilde{\mathbf{\Psi}}_{WM}^T\right)\ \mathrm{Re}\left(\tilde{\mathbf{\Psi}}_{WM}\right)} \tag{14.4}$$

The weighted mode shape $\mathbf{\Psi}_{WM}(\omega)$ of the global system at a driving frequency ω can be compared to the mode shapes of different sub-assemblies using the modal assurance criterion for complex eigenvectors MACX [16]:

$$MACX\left(\mathbf{\Psi}_{WM}^a(\omega), \varphi_k^i\right) = \frac{\left[\left|\mathbf{\Psi}_{WM}^a(\omega)^H \varphi_k^i\right| + \left|\mathbf{\Psi}_{WM}^a(\omega)^T \varphi_k^i\right|\right]^2}{\left[\mathbf{\Psi}_{WM}^a(\omega)^H \mathbf{\Psi}_{WM}^a(\omega) + \left|\mathbf{\Psi}_{WM}^a(\omega)^T \mathbf{\Psi}_{WM}^a(\omega)\right|\right]\left[\varphi_k^{i\,H} \varphi_k + \left|\varphi_k^{i\,T} \varphi_k\right|\right]} \tag{14.5}$$

where $\mathbf{\Psi}_{WM}^a(\omega)$ is the weighted mode shape of the assembly, or more in general of the more complex level compared, φ_k^i is the kth mode shapes of the ith sub-assembly, or more in general simpler level, $(\bullet)^H$ represents the Hermitian and $(\bullet)^T$ is the transposition operator.

From Eq. (14.5) the most influent sub-assemblies can be detected, and for each sub-assembly the most participating mode φ_k, the one with the highest MACX index. The procedure can be repeated, considering the weighted mode shapes of the other sub-assemblies at the natural frequency ω_k related to the selected φ_k and the mode shapes of each component in the sub-assembly. Hence Eq. (14.5), with $\mathbf{\Psi}_{WM}^a(\omega) = \mathbf{\Psi}_{WM}^i(\omega_k)$, i.e. the weighted mode shape of the ith sub-assembly at driving frequency ω_k, and φ_k^i representing the kth mode shapes of the ith component in the considered sub-assembly, is used.

Using this procedure all the components participating to the response of the assembly at the driving frequency ω are selected. For each component the mode shapes on which it is convenient to act to improve the design are highlithed. The re-design can involve localised changes on stiffness or mass in order to shift the frequency range influence of that particular mode shapes (of the selected components) in the global system.

14.3 Motorcycle Chassis Under Engine Excitation

The method presented in the previous paragraph is here applied to the numerical model of a racing motorcycle developed by Mahindra Moto3 Racing Team for the World Championship of 2017. The motorbike analysed includes several components as the chassis, the engine, the rear swingarm, the rear shock absorber and finally the rear wheel. The chassis and the swingarm are two boxes structures in aluminium alloy, made up of several parts connected through welding points. The internal cavities are used for electronic cables and lubricant oil. The wheel is made of magnesium and aluminium alloy, manufactured by multi-directional forging in order to ensure an extremely lightweight component with improved mechanical properties. The wheel includes the transmission pinion and the rear brake disk. It presents seven spokes with curved shape geometry.

The mode shapes tracing is performed from assembly to component in order to understand which of the component mode shapes is influent in the complete motorbike assembly. The work aims to reach different targets: (1) to understand the influence of the component mode shapes in the assembly behaviour, and (2) to understand which components could be improved to avoid resonance frequency of the system in the working condition frequency range.

The first point studies the frequency range of influence of the component mode shapes in the sub-assembly, and the sub-assembly in the global system. It is fundamental to identify the influent mode shapes since for some modes, even if the MAC value is high, its influence on the global behaviour of the assembly could be not relevant. For the second aim an example of working condition is analysed, in particular the vibration reduction in full throttle condition under engine order excitations when the motorcycle travel along a straight or during a traction fast corner.

The two sub-assembly mode shape $\boldsymbol{\varphi}_k^i$ are compared with the relative assembly weighted mode shape $\boldsymbol{\Psi}_{WM}^a(\omega)$ using MACX correlation of Eq. (14.5). Then the sub-assembly weighted mode shapes are compared with the components mode shapes.

The critical frequencies of the motorbike in this case study are evaluated though the rotation speed of the engine. Indeed, it emits a response (acoustic and/or vibrational) at a certain amplitude and this response changes with the variation of the engine speed. The driving frequencies, given by the engine orders, are evaluated using Eq. (14.6):

$$f_{EO} = \frac{N_{EO}\ n_E}{60} \tag{14.6}$$

where f_{EO} is the frequency of the engine order in Hz, N_{EO} is the number of the engine order and n_E is the engine speed in rpm.

The Campbell diagram of the motorbike is reported in Fig. 14.1. The dashed lines represent the first four engine orders; the two dashed-dot green vertical lines indicate the working region of the engine when the motorbike is at the maximum velocity, as it happens in straights or in traction fast rounds. The coloured horizontal lines represent the global flexible modes of the whole motorbike multibody system that intersect the engine orders in the frequency range analysed, while the grey horizontal lines indicate all the other global flexible modes of the motorcycle. The four global flexible modes of the motorbike in the engine speed range of interest are then analysed to better understand the influence of each component on the global dynamic behaviour of the motorbike.

These four modes correspond to: the vertical bending of the chassis coupled with a torsional behaviour of the rear-subsystem (mode 14), the drum mode of the wheel (mode 20), the first global bending with respect to wheel rotation axis (mode 29) and to the second bending with respect to wheel rotation axis (mode 30). Therefore, these four modes return the four driving frequencies ω to be analysed and they should be moved to attenuate vibrations in full throttle working conditions. The natural frequencies should be placed in frequency ranges in which the engine works mostly in transitory.

The mode shape tracing via weighted deformation modes starts from the comparison between full frame/assembly weighted mode deformation and front and rear sub-assembly modes in Figs. 14.2 and 14.3.

In Figs. 14.2 and 14.3, the first six modes of the sub-assemblies are rigid body motion and they are not considered in the comparison. Only the flexible modes below 1000 Hz are taken into account from here after. The two sub-assemblies are quite influent on the global chassis behaviour. The influence of each front assembly mode-shape in the global weighted deformation of the frame is not equally distributed: some modes have an influence widely extend in frequency; other modes instead are very localised or do not have any influence.

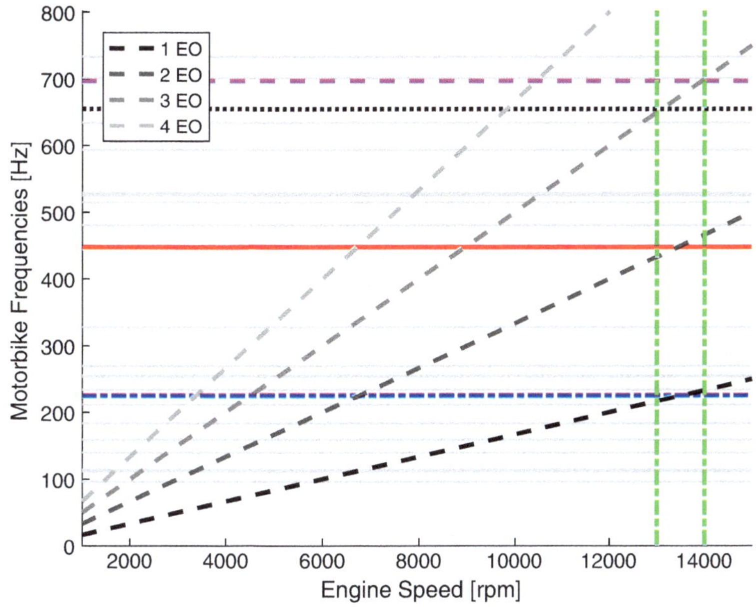

Fig. 14.1 Motorbike Campbell diagram

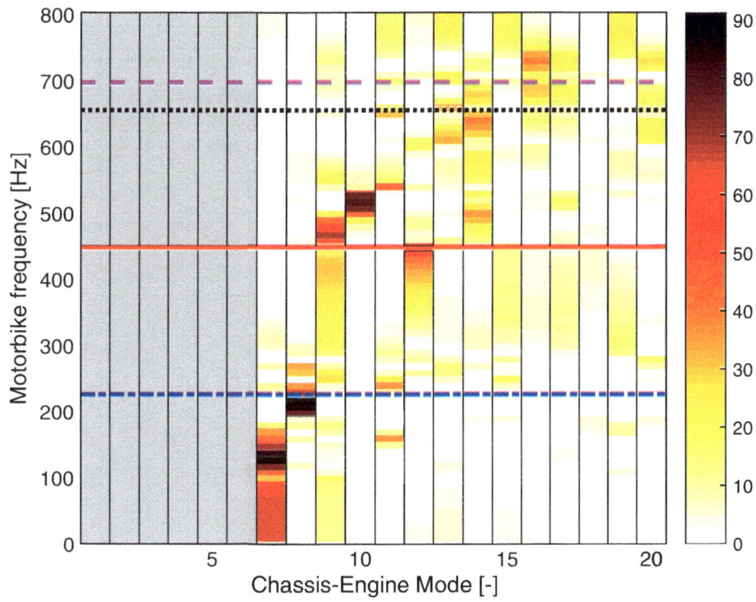

Fig. 14.2 Front sub-assembly vs motorbike

The four horizontal lines correspond to the natural frequencies of the motorbike modes that intersect the engine orders in the working frequency range analysed. In Fig. 14.2 only the first three engine harmonics find out a correspondence in the front sub-assemblies, while the last driving frequency is not influenced by the front sub-assembly. Modes 8, 12 and 13 of the front sub-assembly have a high influence on the motorbike global behaviour at those frequencies, respectively 225 Hz, 455 Hz, and 655 Hz. The influence of the rear sub-assembly is very important for the second and third driving frequency, with high participation of modes 14 (455 Hz), 22 (655 Hz), and a lower influence of mode 16 at 695 Hz as illustrated in Fig. 14.3.

The mode shape tracing should continue with the comparison between sub-assemblies and components, as reported in Figs. 14.4, 14.5, and 14.6. The front sub-assembly can be compared only with respect to the chassis. No comparison with the engine is performed because it is treated as a rigid mass linked to the chassis.

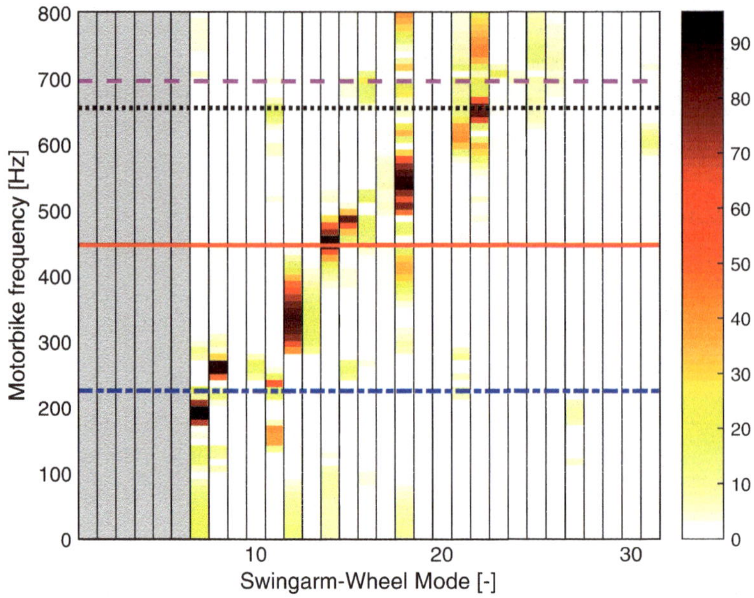

Fig. 14.3 Rear sub-assembly vs motorbike

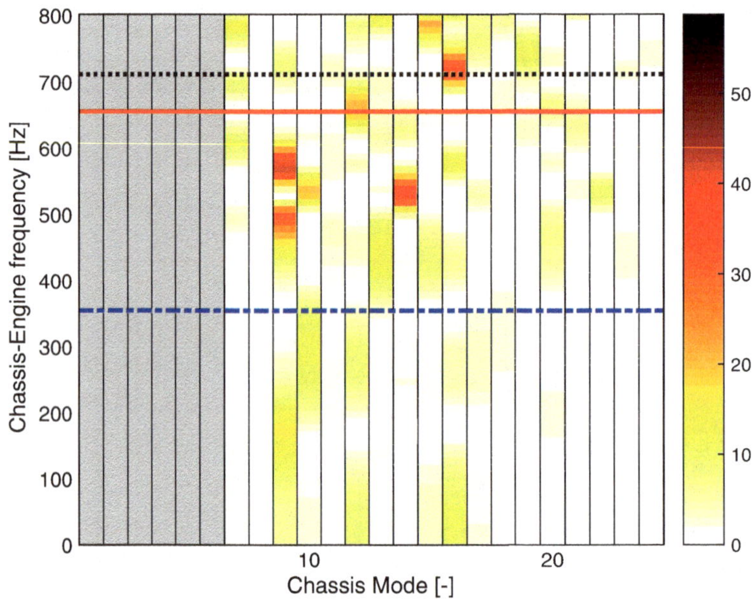

Fig. 14.4 Chassis vs chassis + engine sub-assembly

In Fig. 14.4 the three horizontal lines correspond to the frequencies of the selected modes of the front sub-assembly (8, 12 and 13). It is clear that only the mode 16 of the chassis at 615 Hz has a high influence on the sub-assembly modal behaviour. This is probably due to the effect of the engine and its link that strongly modify the modal properties of the sub-assembly with respect to the chassis component.

The influence of swingarm and wheel in the operative dynamics of the rear sub-assembly is shown in Figs. 14.5 and 14.6. In both figures, blue, red and black lines are relative to the natural frequencies of the three rear sub-assembly modes selected in Fig. 14.4 (modes 14, 22 and 16).

The swingarm modes 8 and more 9 have an effect on the dynamics of the sub-assembly, both with wheel modes 11, 14 and 15. It is very interesting to notice that mode 8 of swingarm at a frequency of 290 Hz has a strong effect on the two modes of the rear-subsystem. Therefore, it is necessary to act on the stiffness or on the mass distribution of the component to change its modal influence and to move the natural frequency of that mode out of the critical working frequency range.

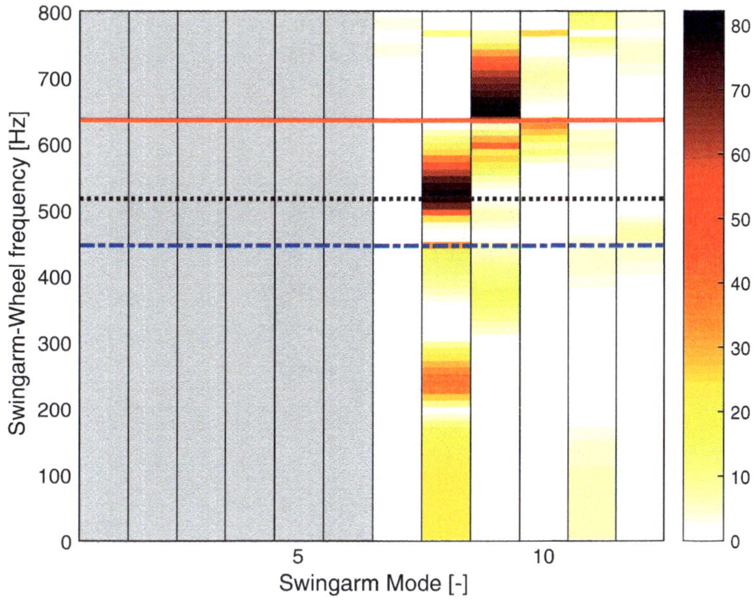

Fig. 14.5 Swingarm vs swingarm + wheel sub-assembly

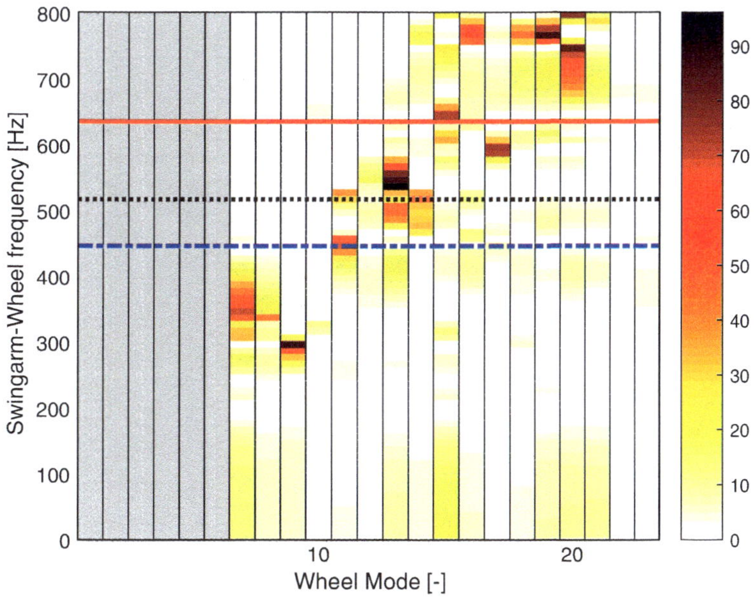

Fig. 14.6 Wheel vs swingarm + wheel sub-assembly

The results of the proposed mode shape tracing method are reported in Figs. 14.7, 14.8, 14.9, and 14.10, from the global assembly and moving to the single components.

Motorbike mode 14 is characterised by a vertical bending of chassis with engine and an almost rigid motion of the wheel with respect to the swingarm, due to the elastic rubber components inside the wheel hub. Mode 8 of front subsystem has high influence and it is consequently weakly influenced by mode 10 of the chassis component.

Mode 20 of motorcycle, exicited by the second engine order, corresponds to a drum mode coupled with a lateral bending of chassis and swingarm. Both the two sub-assemblies participate quite a lot in the global dynamics. In particular mode 14 of the rear sub-assembly, at almost the same frequency, and mode 12 of the front sub-assembly, which frequency is considerably higher. The rear sub-assembly is influenced quite strongly by swingarm mode 8 and wheel mode 11, while mode 12 of the chassis weakly influences the front sub-assembly.

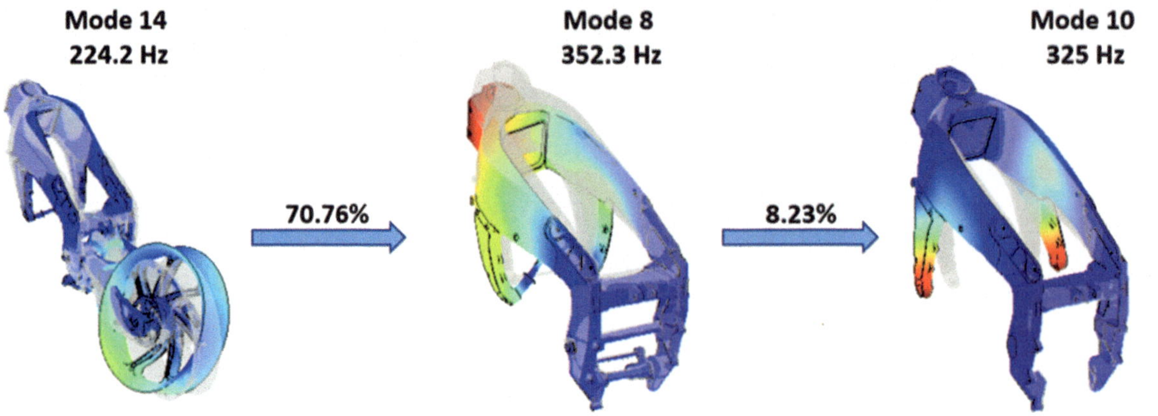

Fig. 14.7 Motorbike mode 14 traced from assembly to components

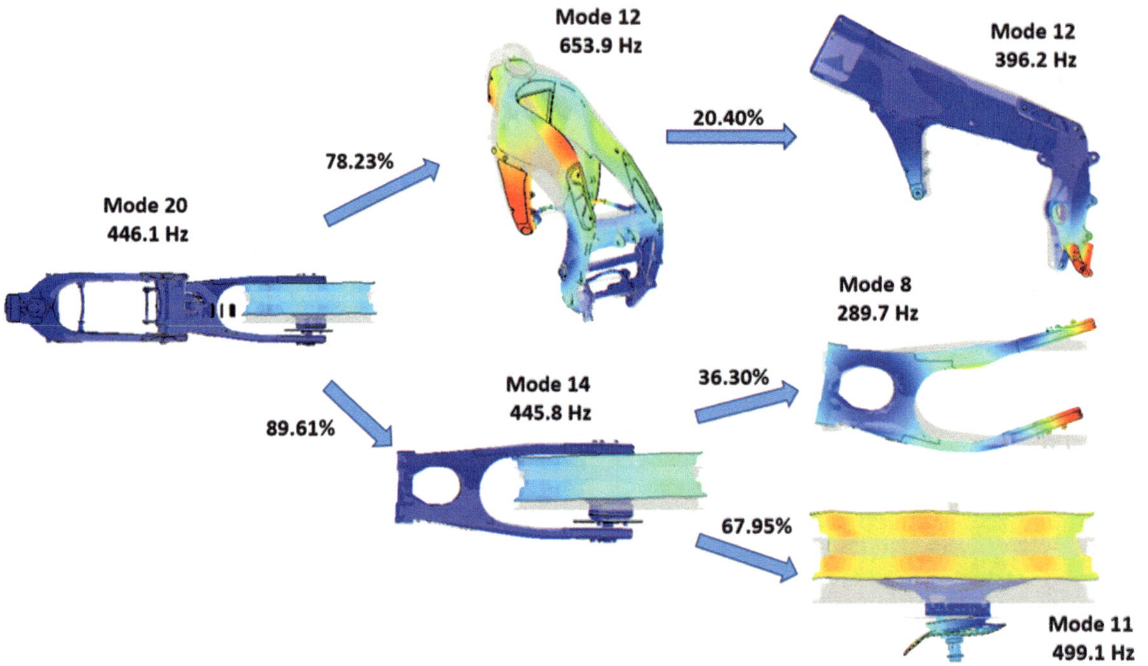

Fig. 14.8 Motorbike mode 20 traced from assembly to components

Motorbike mode 29, excited by the third engine harmonic, is particularly excited during large corners [17] and thus swingarm and wheel components should be improved to avoid that this mode shape be excited at maximum speed range of the engine. The global mode is high correlated with the mode 22 of the rear sub-assembly and mode 13 of the front sub-assembly. The rear sub-assembly mode is strongly influenced by the vertical bending out of phase of the swingarm and again by the deformation of the wheel hub. The front sub-assembly, instead is weakly influenced by the large deformability of the two lateral plates of the chassis.

Motorbike mode 30, also excited by the third engine harmonic, corresponds to the global second bending of the motorbike with respect to the wheel rotation axis. There is a correlation only with mode 16 of the rear sub-assembly, even if the deformability of the swingarm is different. Finally, the sub-assembly is influenced by the bending mode of the swingarm and by mode 14 of the wheel.

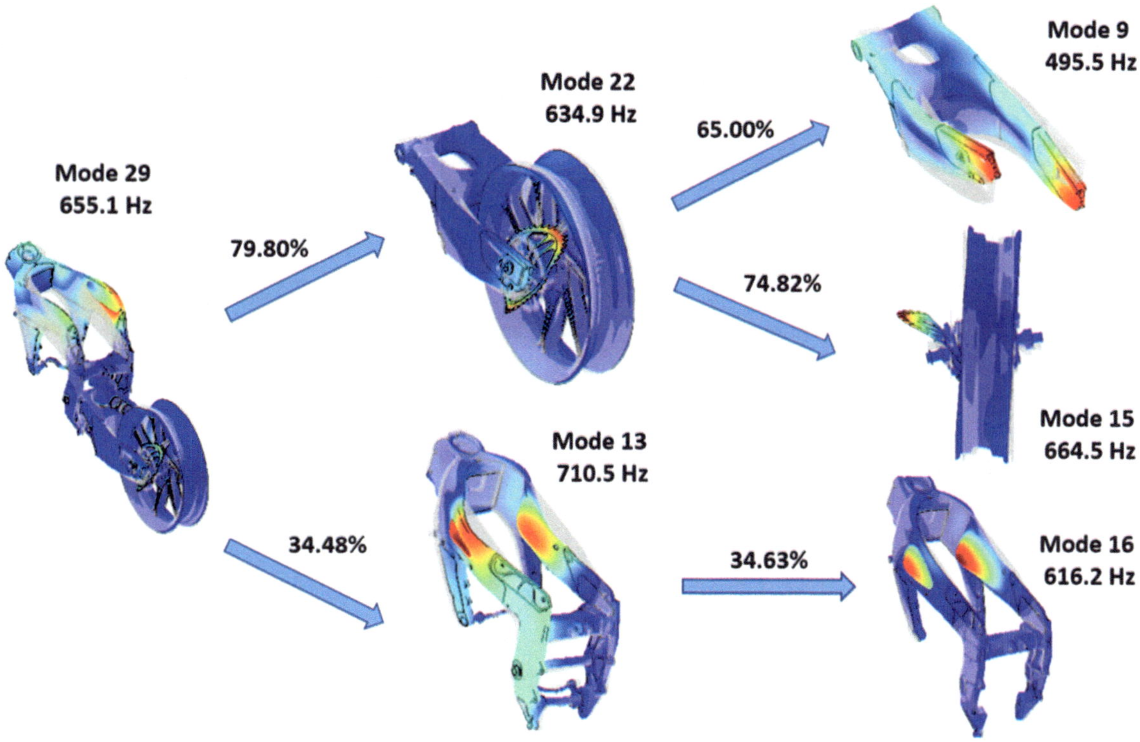

Fig. 14.9 Motorbike mode 29 traced from assembly to components

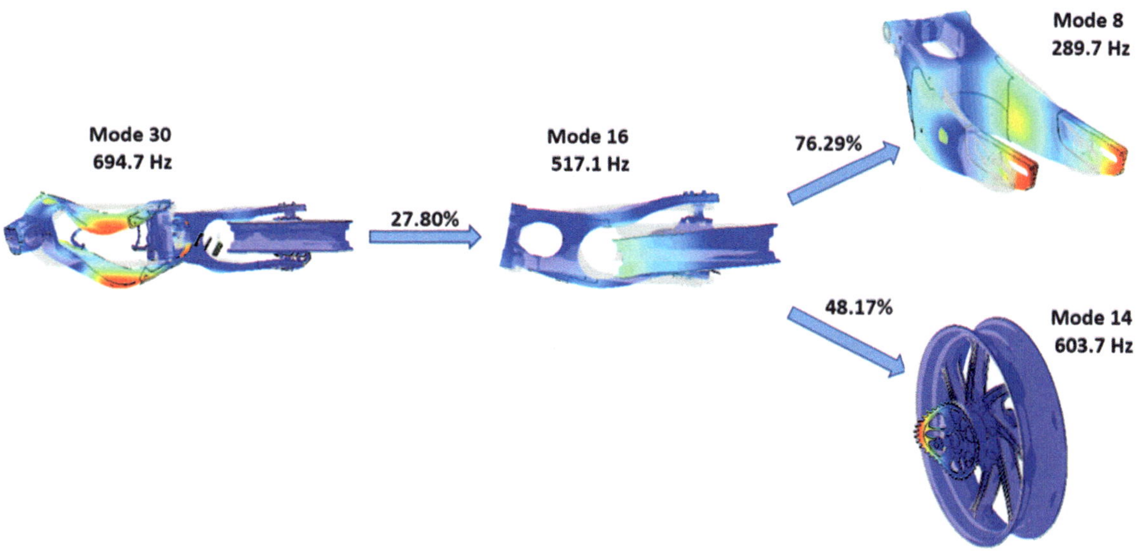

Fig. 14.10 Motorbike mode 30 traced from assembly to components

14.4 Conclusions

The presented approach consists in the comparison of the component mode shapes with respect to the weighted assembly global mode shape. It results to be a very interesting engineering tool for an objective evaluation of which component can be modified to improve the system performance. It is quite fast in highlight components and specific modes involved somehow in specific dynamics behaviour.

The presented case study is an actual engineering problem, since the influence of each component mode shape on the motorbike dynamics is a quite interesting topic to better understand the motorbike dynamic behaviour.

The comparison between weighted mode shapes and traditional mode shapes of components helps to understand the most influent mode shapes in the global motorbike dynamics. Some modes have no influence on the one level up assembly, other modes influence is extended in a large frequency range, while other modes influence is localised only at their natural frequency.

This kind of analysis gives two different information: the former is used to define which mode has a relevant effect on the frequency range of the level-up sub-assembly, while the latter is read to understand how carry-on some structural changes on the components, improving thus the dynamic performance of the global assembly and avoiding component resonances in some frequency ranges. The way forward, in stiffen or give more elasticity to a particular component [18], must take into account also the drivability, handling and reactivity sensations of the rider for a better final comfort both for a racing motorbike and for a touring one.

The method provides encouraging results but should be tested against multi-objective optimisation, considering more than a working condition, to check its capabilities when contradictory aims have to be addressed.

References

1. Allemang, R.J., Brown, D.L.: A correlation coefficient for modal vector analysis. In: Proceedings of 1st IMAC, pp. 110–116. Orlando, FL (1982)
2. Bonisoli, E., Marcuccio, G., Rosso, C.: Crossing and veering phenomena in crank mechanism dynamics. In: Proceedings of 31th IMAC, pp. 175–187. Garden Grove, CA (2013)
3. Sharp, R.S., Limebeer, D.J.N.: A motorcycle model for stability and control analysis. Multibody Syst. Dyn. **6**(2), 123–142 (2001)
4. Cossalter, V., Lot, R.: A motorcycle multi-body model for real time simulations based on the natural coordinates approach. Veh. Syst. Dyn. **37**(6), 423–447 (2002)
5. Sharp, R.S., Evangelou, S., Limebeer, D.J.N.: Advances in the modelling of motorcycle dynamics. Multibody Syst. Dyn. **12**, 251–283 (2004)
6. Leonelli, L., Mancinelli, N.: A multibody motorcycle model with rigid-ring tyres: formulation and validation. Int. J. Veh. Mech. Mobil. **53**(6), 775–797 (2015)
7. Sequenzia, G., Olivieri, S.M., Fatuzzo, G., Calì, M.: An advanced multibody model for evaluating rider's influence on motorcycle dynamics. J. Multibody Dyn. **229**(2), 193–207 (2015)
8. Olivieri, S.M., Calì, M., Catalano, L.: Dynamics of motorcycle using flexible elements. In: International Design Conference, pp. 1227–1236. Dubrovnik, Croatia (2002)
9. Ferretti, G., Scaglioni, B., Rossi, A.: Multibody model of a motorbike with a flexible swingarm. In: Proceedings of the 10th International Modelica Conference, pp. 273–282. Lund, Sweden (2014)
10. Shabana, A.A.: Dynamics of Multibody Systems. Cambridge University Press, Cambridge (1998)
11. Bocciolone, M., Cheli, F., Pezzola, M., Viganò, R.: Static and dynamic properties of a motorcycle frame: experimental and numerical approach. Trans. Model. Simul. **41**, 517–526 (2005)
12. Lake, K., Thomas, R., Williams, O.: The influence of compliant chassis components on motorcycle dynamics: an historical overview and the potential future impact of carbon fibre. Int. J. Veh. Mech. Mobil. **50**(7), 1043–1052 (2012)
13. Cossalter, V.: Motorcycle Dynamics. Lulu, Morrisville (2006)
14. Bonisoli, E., Lisitano, D., Dimauro, L.: Experimental and numerical mode shape tracing from components to whole motorbike chassis. In: International Conference on Noise and Vibration Engineering, ISMA, pp. 1–8. Leuven, Belgium, September 17–19 (2018)
15. Ibrahim, S.R., Sestieri, A.: Existence and normalization of complex modes in post experimental use in modal analysis. In: Proceedings of 13th IMAC, pp. 483–489 (1995)
16. Vacher, P., Jacquier, B., Bucharles, A.: Extension of the MAC criterion to complex modes. In: International Conference on Noise and Vibration Engineering, ISMA, pp. 2713–2725. Leuven, Belgium (2010)
17. Cossalter, V., Doria, A., Massaro, M., Taraborrelli, L.: Experimental and numerical investigation on the motorcycle front frame flexibility and its effect on stability. Mech. Syst. Signal Process. **60–61**, 452–471 (2015)
18. Cocco, G.: Dinamica e tecnica della motocicletta. Giorgio Nada Editore, Milan (1999)

Chapter 15
Rapid Seismic Risk Assessment of Structures with Gaussian Process Regression

Mohamadreza Sheibani, Ge Ou, and Shandian Zhe

Abstract Risk assessment of structures is an important step to predict damages and losses during natural hazards. The main idea is to fit fragility functions from existing earthquake records. Considerable research has been conducted in this area and various fitting approaches have been suggested. However, existing methods for deriving fragility functions require large training data sets and are designed for a single type of structures each time. This study aims to provide the loss estimation data using supervised learning techniques with a limited amount of observations. Observations are made after the strike of the earthquake based on the responses obtained from randomly selected damaged structures. These observations are used to train the model and prediction of the responses are made for the rest of the structures in the area. A case study is conducted based on the Fukushima 2011 earthquake. It is shown that Gaussian Process Regression (GPR), can provide an integrated model prediction for a group of structures with high accuracy and low uncertainties. This way, not only the predictions are made by a single model for different types of structures, but also the number of observations is very small compared to the existing methods.

Keywords Seismic damage prediction · Risk assessment · Supervised learning · Gaussian Process Regression · Fukushima earthquake

15.1 Introduction

One of the main concerns with damage assessment after an earthquake is about the understanding of the extent of damage and destruction caused by the ground motion shortly after the strike. This information can help authorities recognize the most damaged areas and structures expeditiously, and better manage and dispatch reinforcement and rescue teams. One of the main techniques which is commonly used in risk assessment of structures are the fragility functions [1]. In this approach, a number of ground motion records are collected and scaled and then applied to the structure in order to have a wide range of responses for a structure. Fragility curves are then fitted to the cumulative probability distribution of the responses for a desired Intensity Measure (IM) of the ground motions [2]. This method can reliably predict the damage states of structures in cases where there is a plentiful amount of data available [1]. Some of the notable previous works in this field can be found in [1, 3–5]. However, fragility functions do not perform well in cases where provided data is sparse. Most commonly, if a number of different structures are concerned, a separate fragility curve needs to be calculated for each type of structure. If the data is combined for different types of structures, and one curve is fitted, fragility curves will not be distinctive for different damage states [3]. On the other hand, machine learning methods provide much better results in cases of prediction based on a limited amount of observations. In contrast to fragility functions which only consider one property of the ground motion signal, such as Peak Ground Acceleration (PGA), or Spectral Acceleration (Sa), regression learning methods can be fed as many features as desired. Machine learning methods has been implemented to seek solutions in earthquake science for a long time. One of the most common methods in this area, Artificial Neural Networks (ANN), has been widely used in order to predict the responses of structures using huge training sets [6–9]. However, in the current approach, models are being trained with different earthquake motions and are not specific to one single event.

In the post-hazard situation, one may have to visually inspect the whole area to assess the damage level of different buildings. Such data that reflect the observations of the post-seismic building damage conditions have not been integrated

---

M. Sheibani (✉) · G. Ou
Department of Civil and Environmental Engineering, University of Utah, Salt Lake City, UT, USA
e-mail: m.sheibani@utah.edu

S. Zhe
School of Computing, University of Utah, Salt Lake City, UT, USA

A. Linderholt et al., *Dynamic Substructures, Volume 4*, Conference Proceedings of the Society for Experimental Mechanics Series,
https://doi.org/10.1007/978-3-030-12184-6_15

with the predictive models, either fragilities curves or ANN trained models. In this paper, we propose to predict the damage of similar structures in the whole area based on a limited number of visually or instrumentally inspected structures. Visual inspections can be made shortly after the hazard and the information will be sent to the data center for further analysis. Moreover, there may be some structures which are pre-equipped with monitoring equipment, such as accelerometers, and this could be another source of information. Combining these two sources, the data center will train a model to predict the responses for other uninspected structures. In this paper, we aim to train our machine learning model from some limited post-hazard observations, and predict the responses for the larger test set. Data from the 2011 Fukushima earthquake is utilized in the form of 592 ground motions recorded at different locations of the country of Japan. The training set is consisted of 100 randomly selected records and the rest is used for verification of the method. Gaussian Process Regression learning method is used to that end. Four different building types are considered in this study, one of which is randomly selected and assigned to each ground motion record. Finally, the predicted responses are compared to the responses obtained by nonlinear time history analysis.

15.2 Gaussian Process Regression

To implement the Machine Learning (ML) to predict the structure responses, GPR algorithm has been used in this paper. GPR works basically like linear regression but for non-parametric functions. In this case, we have a multi-dimensional vector as the independent variable $\mathbf{x}_i$ and observations y_i are made at different intervals of $\mathbf{x}$. Assuming that all dependent variables, i.e. $y_1, y_2, \ldots, y_n$, are different elements of one point sampled from a multivariate Gaussian distribution, one can relate observations to each other using the covariance matrix $\mathbf{K}$ as follows. Considering the exponential kernel, each element of the matrix $\mathbf{K}$ can be expressed as [10],

$$K_{i,j} = k\left(\mathbf{x}_i, \mathbf{x}_j\right) = \sigma_f^2 \exp\left[-\frac{1}{2}\left(\mathbf{x}_i - \mathbf{x}_j\right)^T \mathrm{diag}\left(\mathbf{m}\right)\left(\mathbf{x}_i - \mathbf{x}_j\right)\right] + \sigma_n^2 \delta_{ij} \tag{15.1}$$

where $\boldsymbol{\theta} = \left[\mathbf{m}, \sigma_f^2, \sigma_n^2\right]$ is the vector of the hyperparameters. The maximum allowable covariance is represented by σ_f^2, the variance of the Gaussian distributed noise affecting the observations is denoted by σ_n^2, and δ_{ij} is the Kronecker delta function. Different choices for the matrix $\mathbf{m}$ can be found in [10]. Using the above-mentioned n observations, the prediction of y_* given x_* is achieved by GPR. In order to predict, two more Matrices need to be defined,

$$\mathbf{k}_* = \left[k\left(x_*, x_1\right) k\left(x_*, x_2\right) \ldots k\left(x_*, x_n\right)\right] \tag{15.2}$$

$$\mathbf{k}_{**} = k\left(x_*, x_*\right) \tag{15.3}$$

As mentioned before, the observations are considered as a sample from a multivariate Gaussian distribution. Based on this assumption, and considering the conditional probability, we have,

$$y_* \mid \mathbf{y} \sim N\left(\mathbf{k}_*\mathbf{K}^{-1}\mathbf{y}, \mathbf{k}_{**} - \mathbf{k}_*\mathbf{K}^{-1}\mathbf{k}_*^T\right) \tag{15.4}$$

Therefore, the mean and variance of the predicted instance can be expressed as

$$\mathrm{E}\left(y_*\right) = \mathbf{k}_*\mathbf{K}^{-1}\mathbf{y} \tag{15.5}$$

$$\mathrm{var}\left(y_*\right) = \mathbf{k}_{**} - \mathbf{k}_*\mathbf{K}^{-1}\mathbf{k}_*^T \tag{15.6}$$

It is clear that in order to have predictions with lower variances, the testing instances should be well covered with the training data.

15.3 Case Study Results and Discussion

15.3.1 Model Description

The structure model which has been used in this paper is consisted to be a multi-story steel frame building. A five story shear frame with the same steel columns for all stories was tested experimentally before at [11] and a numerical model was fitted to its components using the Bouc-Wen nonlinear model. Based on the calibrated nonlinear model, four different structures with the same components but a different number of stories such as a 3-story, a 5-story, a 7-story, and a 9 story model, are considered here. The height of stories are about 0.17 m and the weight of each story is 23.2 kg. Newmark Beta has been used for nonlinear time history analysis of the structures and the time step is considered to be 0.001 s for all time history analyses.

Damage state of a structure is best reflected in the maximum experienced drift ratio. Since the drift ratio can be obtained from relative displacement of each story normalized to the story height, damage states can be defined directly on a hysteresis loop of a typical component. In this case, four different damage states are defined, namely none, slight, moderate, and collapse damage. The hysteresis behavior of the nonlinear model subjected to a cyclic pushover test and the abovementioned damage states are shown in Fig. 15.1. The limit state of slight damage Δs, is considered as the yielding point of the component which is about 1.2 cm. The collapse damage state Δc is considered as 5 cm, and the moderate damage state Δm is taken as half of the displacement of the collapse damage. The considered limit states are shown in Fig. 15.1 and described in Table 15.1.

15.3.2 Prediction Objective

Since in real life we get a limited chance to rapidly identify the extent of damage in different areas, the training set size is considered to be a small fraction of the whole data. In this case, out of 592 ground motion records available, 100 randomly selected records (~17%) are used for training purposes and the rest remain untouched to test the model. We might not have

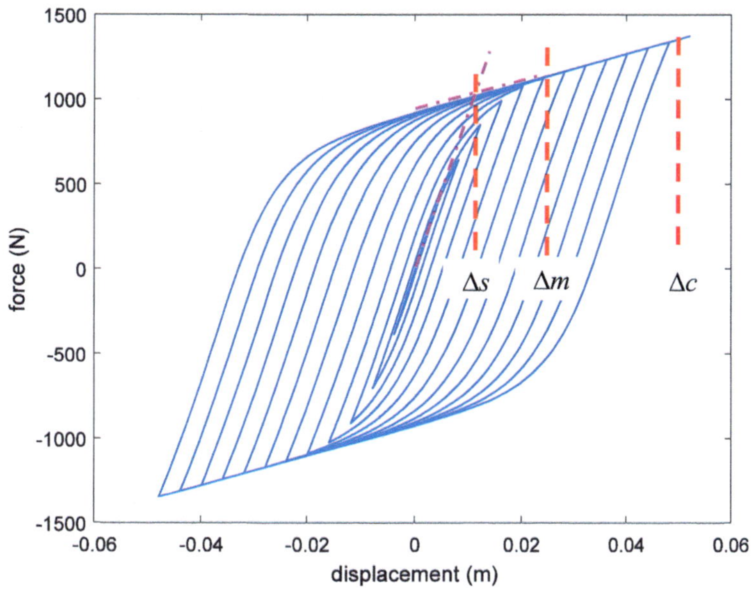

Fig. 15.1 The hysteresis behavior of the component subjected to a cyclic pushover test

Table 15.1 Definition of damage states

Inter-story displacement (m)	Drift ratio	Damage state
ID < 0.012	DR < 0.07	None
0.012 < ID < 0.025	0.07 < DR < 0.15	Slight
0.025 < ID < 0.05	0.15 < DR < 0.29	Moderate
0.05 < ID	0.29 < DR	Collapse

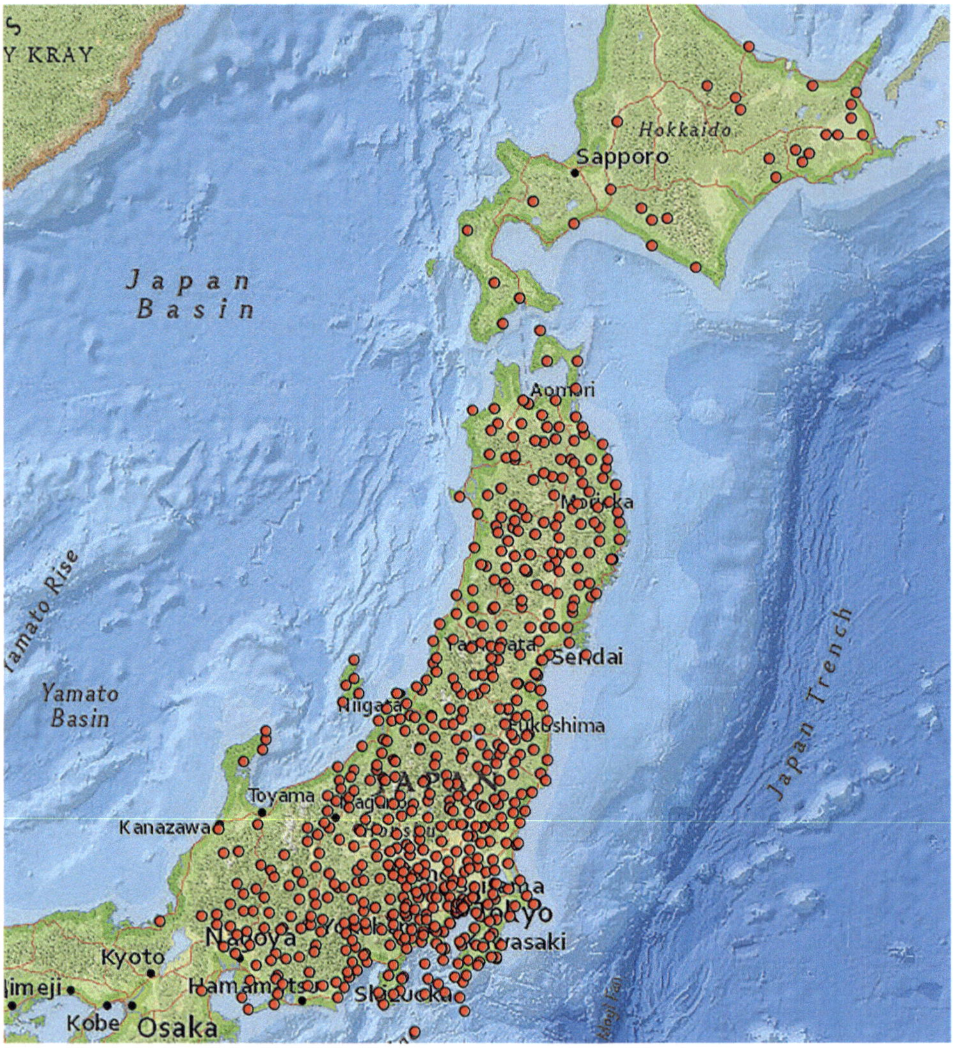

Fig. 15.2 Locations of the recorded ground motions on geographical map of Japan

a chance to record the ground motion at every location that a building is found at in real situations, and thus the attenuation of the ground motion should be considered for other spots. But, in this case, it is assumed that at the location where each ground motion is recorded, one of the four building types is located there and we aim to predict the response of that building. The location of each recorded ground motion with respect to the map of Japan can be seen in Fig. 15.2. The true responses of the structures subjected to the ground motion records are numerically calculated using the nonlinear Newmark Beta method, based on the Bouc-Wen components explained in Sect. 15.3.1. These simulations represent the inspections which are made in real situations.

15.3.3 Results

Supervised machine learning techniques are used in order to help us predict the responses using a trained model. Various supervised regression techniques are available in this field, however, one of the most robust techniques is GPR, which has been used in this paper. One of the main advantageous of GPR is that it provides us with a statistical representation of the prediction, and thus one can see both the mean value and the confidence interval of the predicted value.

In order to train the model, the first stage is defining the features which are going to represent our instances. Two major aspects need consideration in deriving the response of a structure subjected to an earthquake. First, the ground motion signal which carries a significant amount of information with it. Second, the properties of the structure itself, such as the number

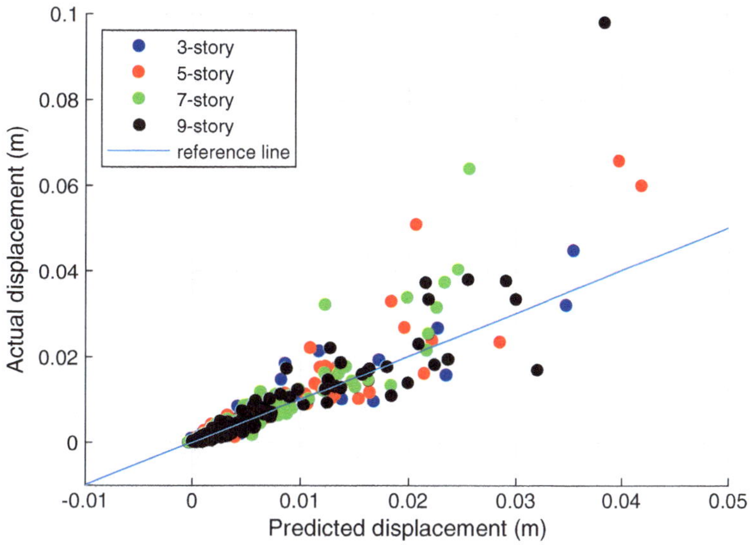

Fig. 15.3 Predicted versus actual maximum displacements for the test set

Table 15.2 Prediction accuracy for each damage state

Damage state	Number of training set	Number of test set	Prediction accuracy %
None	84	423	98.1
Slight	11	48	75.0
Moderate	3	16	31.3
Collapse	2	5	0.0
Total	100	492	92.7

of stories, construction materials, lateral resisting system, etc. Therefore, our model should be trained with a combination of ground motion features and structural features. In this case, nine features are extracted from each ground motion signal, and since the structures are assumed to be identical with only a different number of stories, only one feature is considered for the structures. More information concerning the ground motion features can be found in [12]. Consequently, the model is trained with the above-mentioned features as the feature space and the maximum experienced drift ratio as the label space.

Matlab's GPR modeling command is used in order to train the model. Optimized hyperparameters are obtained and the values of each feature are normalized to unity. The predicted versus actual labels are plotted in Fig. 15.3 for the testing set.

It can be seen in Fig. 15.3 that the trained model has predicted the responses with acceptable accuracy. However, there are some deviations where the components go into the nonlinear phase, and the reason is that the model has been trained with few instances where nonlinear behavior happens. This statement can be illustrated if the output is discretized and the accuracy of each damage state is calculated separately, as in Table 15.2. In order to better clarify this issue, the number of instances in the training set as well as the number of the instances in the test set and their prediction are compared based on their labels in Fig. 15.4a. Moreover, the prediction results are shown on a geographical map of Japan in Fig. 15.4b.

As mentioned earlier, GPR provides a 95% confidence interval for each prediction. Therefore, evaluating the results should not be merely based on the mean values. Confidence intervals as well as the damage limit states are shown in Fig. 15.5 for each prediction. Based on this information, classification of the predicted values is performed from another aspect in Table 15.3. Table 15.2 shows the coverage area of the 95% confidence interval for each damage state. For instance, if the lower bound and upper bound of the confidence interval lies in the same zone as the true response, it is considered to be an exact prediction. On the other hand, if the lower bound is in one zone and the upper bound lies one zone away, and the true response is in either of these zones, it is stated as ± 1 level of damage state and so on.

Considering Table 15.3 results, it can be said that the model delivers an accurate result, such that for none damage state, 99.5% of the times the 95% confidence interval includes the true response's zone or, at most, 1 level beside that. This percentage decreases to 93.8% for the slight damage state and to 75% for the moderate damage state. Finally, it can be stated that for the collapse damage state, the 95% confidence interval covers at most two zones beside the true response, 60% of the time. As mentioned earlier, the reduction of accuracy for higher damage states was expected since the GPR model is trained with a limited number of instances at these damage states. Looking at the cumulative results, 75% of the times we are 95%

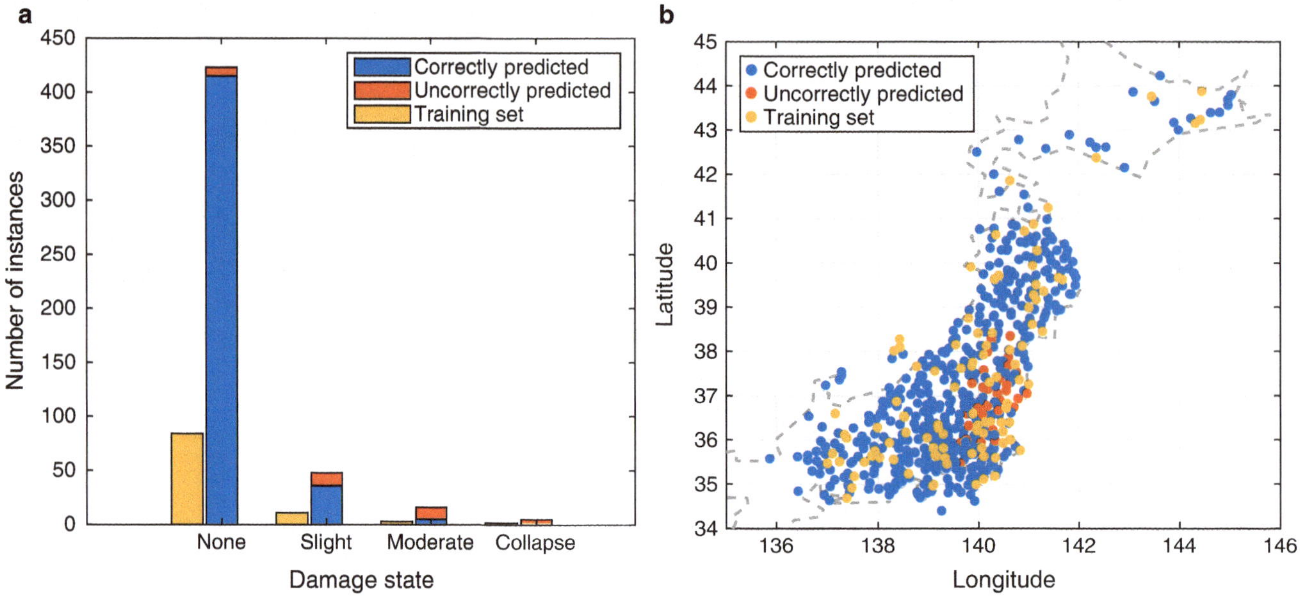

Fig. 15.4 (**a**) Prediction accuracy with respect to number of instances for each damage state, (**b**) correctness of the predicted labels on the geographical map of Japan

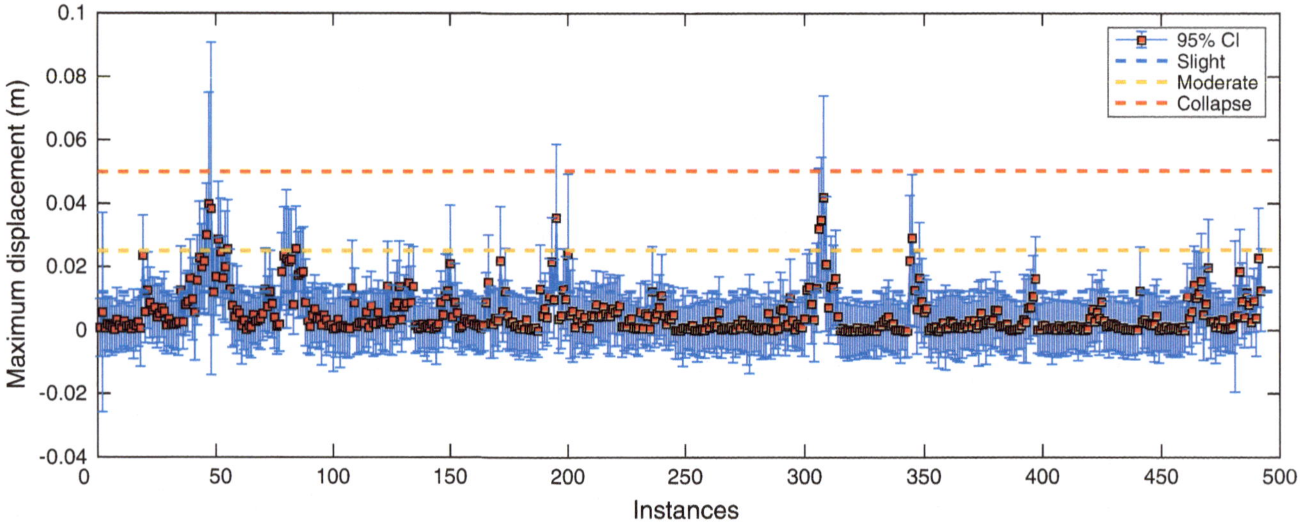

Fig. 15.5 Distribution of the predicted labels along with the 95% confidence intervals

Table 15.3 Predictions based on the 95% confidence interval

		Exact		±1 level		±2 level	
Damage state	Total no.	No.	%	No.	%	No.	%
None	423	369	87.2	52	12.3	2	0.5
Slight	48	1	2.1	44	91.7	0	0.0
Moderate	16	0	0.0	12	75.0	2	12.5
Collapse	5	0	0.0	0	0.0	3	60.0
Cumulative	492	370	75.2	478	97.1	485	98.5

confident that we have the exact prediction of our responses and, for 97.1% of the time, this confidence predicts the exact damage state level or at most one damage state level apart.

15.4 Conclusion

We propose to develop a model which utilizes limited inspection observations after earthquakes to predict building damage conditions based on Gaussian Process Regression. GPR model was able to predict responses with a satisfying accuracy of 92.7%. Moreover, considering the 95% confidence intervals for the predicted data, it was demonstrated that the model is 95% certain to predict the responses within at most one damage state apart, 97.1% of the time. Using this method, predictions can be made within a short period of time after the natural hazard for all the buildings in a vast area.

References

1. Noh, H.Y., Lignos, D.G., Nair, K.K., Kiremidjian, A.S.: Development of fragility functions as a damage classification/prediction method for steel moment-resisting frames using a wavelet-based damage sensitive feature. Earthq. Eng. Struct. Dyn. **41**(4), 681–696 (2012)
2. Kafali, C., Grigoriu, M.: Seismic fragility analysis: application to simple linear and nonlinear systems. Earthq. Eng. Struct. Dyn. **36**(13), 1885–1900 (2007)
3. King, S., Kiremidjian, A., Pachakis, D., Sarabandi, P.: Application of empirical fragility functions from recent earthquakes. In: Proceedings of the 13th World Conference on Earthquake Engineering, Vancouver, BC, Canada. Paper (No. 2829) (2004)
4. Barbat, A.H., Pujades, L.G., Lantada, N.: Performance of buildings under earthquakes in Barcelona, Spain. Comput. Aided Civ. Inf. Eng. **21**(8), 573–593 (2006)
5. Shibata, A.: Estimation of earthquake damage to urban systems. Struct. Control Health Monit. **13**(1), 454–471 (2006)
6. Molas, G.L., Yamazaki, F.: Neural networks for quick earthquake damage estimation. Earthq. Eng. Struct. Dyn. **24**(4), 505–516 (1995)
7. De Lautour, O.R., Omenzetter, P.: Prediction of seismic-induced structural damage using artificial neural networks. Eng. Struct. **31**(2), 600–606 (2009)
8. Morfidis, K., Kostinakis, K.: Approaches to the rapid seismic damage prediction of r/c buildings using artificial neural networks. Eng. Struct. **165**, 120–141 (2018)
9. Lagaros, N.D., Papadrakakis, M.: Neural network based prediction schemes of the non-linear seismic response of 3D buildings. Adv. Eng. Softw. **44**(1), 92–115 (2012)
10. Seeger, M.: Gaussian processes for machine learning. Int. J. Neural Syst. **14**(02), 69–106 (2004)
11. Ou, G.: Robust hybrid simulation with improved fidelity: theory, methodology, and implementation. Doctoral dissertation, Purdue University, West Lafayette (2016)
12. Ye, L., Ma, Q., Miao, Z., Guan, H., Zhuge, Y.: Numerical and comparative study of earthquake intensity indices in seismic analysis. Struct. Design Tall Spec. Build. **22**(4), 362–381 (2013)

Chapter 16
Modeling Rail-Vehicle Coupled Dynamics by a Time-Varying Substructuring Scheme

Luigi Carassale, Paolo Silvestri, Roald Lengu, and Paolo Mazzaron

Abstract The dynamic interaction between a flexible rail line and a traveling wheel can be treated in the framework of dynamic substructuring, assembling line and wheel through a moving contact point. The rail line is idealized as a continuous beam supported by the sleepers. It is discretized by the finite element method assuming the displacement and rotation of the support points as degrees of freedom. Besides, the particular solution of the static equation is included for the beam portion that is directly loaded by the moving wheel. The wheel is modelled as a dynamical system comprising two masses connected by visco-elastic devices. Following a dual assemblage approach, the equation of motion is obtained in the form of a set of linear time variant ordinary differential equations, whose unknowns are the rail line degrees of freedom, the displacement of the sprung and unsprung masses of the moving wheel and the contact rail-wheel force.

Keywords Rail-wheel interaction · Substructuring · Time-variant systems

16.1 Introduction

Trains transfer their weight to ground through their suspension systems and wheels, which roll on rails that are supported by sleepers. This dynamical system may be reviewed as a series combination of simple mechanical components (car body, bogie, wheelset, rail, sleepers), but has two peculiarities: (1) the rail-wheel contact point move with the time; (2) wheels may be affected by geometrical defects (ovalization, flat) so that the axle-rail distance changes with time. Due to these fundamental features, the problem is usually treated in the framework of the multibody method, including the continuous behavior of rails, which is usually retained through a modal representation. This approach has two drawbacks: (1) the time integration of the dynamical system is computationally intensive as the rail-wheel contact condition needs to be imposed by an iterative method; (2) when the target of the analysis is the deformation of the rail, the number of modes to be retained in the model needs to be very high.

In this paper we propose a rail-wheel modeling approach formulated in the context of the dynamic substructuring. The rail is represented as a continuous beam; sleepers are connected to the ground and the rail through springs representing, respectively, the terrain and railpad flexibility [1]. The rail-wheel compatibility condition is enforced in the wheel instantaneous position, giving rise to a time-variant assemblage relationship. A further time-variation of the compatibility condition is produced by the presence of wheel defects. It is demonstrated that, adopting a dual assemblage scheme, the equations of motion governing the system are obtained in the form of a set of linear equations with time-variant coefficients.

16.2 Description of the Mechanical System

For the sake of simplicity, we consider a 2D idealization of the problem in which a single rail is traveled by a single wheel supporting a sprung mass (Fig. 16.1).

Let $X(t)$ be the position of the wheel at the time t. In this point of the line, the wheel applies the force g_l on the rail and, vice versa, the rail exerts the force g_a on the wheel (Fig. 16.2).

L. Carassale (✉) · P. Silvestri
Department of Mechanical, Energy, Management and Transportation Engineering, University of Genova, Genova, Italy
e-mail: luigi.carassale@unige.it

R. Lengu · P. Mazzaron
Ansaldo STS, Genova, Italy

A. Linderholt et al., *Dynamic Substructures, Volume 4*, Conference Proceedings of the Society for Experimental Mechanics Series,
https://doi.org/10.1007/978-3-030-12184-6_16

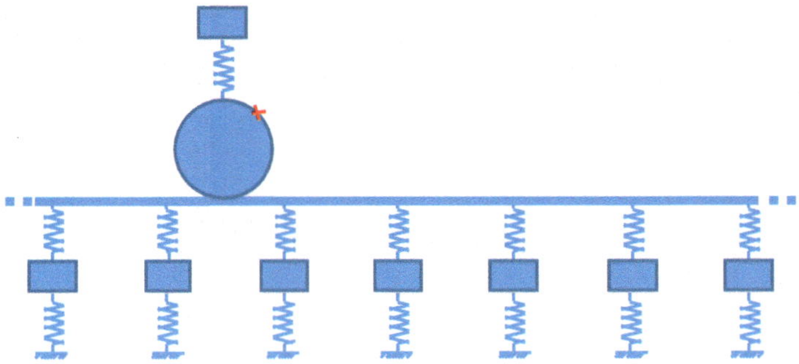

Fig. 16.1 Rail line and traveling wheel idealization

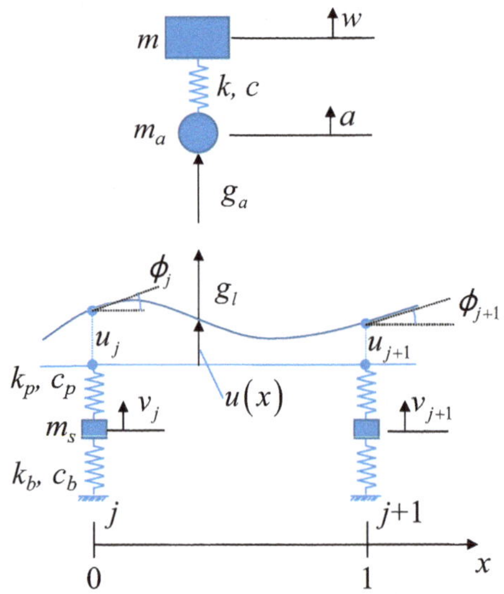

Fig. 16.2 Subsystems

16.2.1 Rail Line Model

The rail line is idealized by a continuous Euler-Bernoulli beam supported by the sleepers having a constant spacing l. The sleepers are connected to the ground and to the rail through visco-elastic devices representing, respectively the behavior of the terrain and the flexibility of the railpads [1] (Fig. 16.1). Let k_b and c_b be the stiffness and damping coefficient of the ground-sleeper connection, while k_p and c_p are the stiffness and damping coefficient of the rail-sleeper connection; m_s is the mass of the sleepers.

The motion of the rail is governed by a partial differential equation, however the system can be discretized according to the Finite Element (FE) approach obtaining the equation of motion in the form:

$$\mathbf{M}_0\ddot{\mathbf{u}} + \mathbf{C}_0\dot{\mathbf{u}} + \mathbf{K}_0\mathbf{u} = \mathbf{f} + \mathbf{g} \tag{16.1}$$

where $\mathbf{u}$ is the displacement vector organized as

$$\mathbf{u} = \begin{bmatrix} \mathbf{u}_1 \\ \vdots \\ \mathbf{u}_N \end{bmatrix}; \quad \mathbf{u}_j = \begin{bmatrix} u_j \\ \phi_j \\ v_j \end{bmatrix} \tag{16.2}$$

where N is the number of sleepers included in the model, u_j and ϕ_j are, respectively, the vertical displacement and rotation of the rail at the support point j (Fig. 16.2), and v_j is the vertical displacement of the jth sleeper. The vector $\mathbf{f}$ contains the external force applied to the FE nodes, while the vector **g** contains the rail-wheel contact force reported to the FE nodes. $\mathbf{M}_0$, $\mathbf{C}_0$ and $\mathbf{K}_0$ are, respectively, the mass, damping and stiffness matrices of the system obtained through the FE method.

The vertical displacement of the line in the position X is provided by the relationship:

$$u(X) = \mathbf{\Psi}(x)\mathbf{D}(X)\mathbf{u} + u_0(x)g_l \tag{16.3}$$

where x is the non-dimensional local abscissa that parametrize the beam element between the sleepers j and $j+1$ in which the rail-wheel contact appears (Fig. 16.2); $\mathbf{\Psi}(x)$ is a vector of shape functions and $u_0(x)$ is the displacement in the point x of a clamped-clamped beam, produced by a unitary force applied at x [2]

$$\mathbf{\Psi}(x) = \begin{bmatrix} 1 - 3x^2 + 2x^3 \\ \left(x - 2x^2 + x^3\right) l \\ 3x^2 - 2x^3 \\ \left(-x^2 + x^3\right) l \end{bmatrix}^{\mathrm{T}}; \quad u_0(x) = \frac{1}{3EJl^3} x^3 (1-x)^3 \tag{16.4}$$

E and J are, respectively, the Young modulus and the moment of inertia of the rail and $\mathbf{D}(X)$ is a Boolean matrix such that:

$$\begin{bmatrix} u_j \\ \phi_j \\ u_{j+1} \\ \phi_{j+1} \end{bmatrix} = \mathbf{D}(X)\mathbf{u} \tag{16.5}$$

The vector **g** can be obtained by projecting the rail-wheel contact force into the space of the FE degrees of freedom (dofs)

$$\mathbf{g} = \mathbf{D}^T \mathbf{\Psi}^T g_l \tag{16.6}$$

16.2.2 Moving Wheel Model

The moving wheel is modelled by an unsprung mass m_a representing the wheel and a sprung mass m connected to the wheel by a visco-elastic suspension with stiffness k and dumping c. The equation of motion of this system can be written in the form:

$$\mathbf{M}_1 \begin{bmatrix} \ddot{a} \\ \ddot{w} \end{bmatrix} + \mathbf{C}_1 \begin{bmatrix} \dot{a} \\ \dot{w} \end{bmatrix} + \mathbf{K}_1 \begin{bmatrix} a \\ w \end{bmatrix} = \begin{bmatrix} f_a \\ f_w \end{bmatrix} + \begin{bmatrix} g_a \\ 0 \end{bmatrix} \tag{16.7}$$

where a and w are the vertical displacement of the unspung and spung masses, respectively; f_a and f_w are the external forces applied to those masses; g_a is the force transferred from the rail to the wheel. The system matrices $\mathbf{M}_1$, $\mathbf{C}_1$ and $\mathbf{K}_1$ are given as:

$$\mathbf{M}_1 = \begin{bmatrix} m_a & 0 \\ 0 & m \end{bmatrix}; \quad \mathbf{C}_1 = \begin{bmatrix} c & -c \\ -c & c \end{bmatrix}; \quad \mathbf{K}_1 = \begin{bmatrix} k & -k \\ -k & k \end{bmatrix} \tag{16.8}$$

16.2.3 Substructure Assemblage

The assemblage of the rail and axle models is carried out adopting a dual schema [3]. Accordingly, all the dofs of both the models are retained and the equation of motion is obtained as:

$$\begin{bmatrix} \mathbf{M}_0 & \\ & \mathbf{M}_1 \end{bmatrix} \begin{bmatrix} \ddot{\mathbf{u}} \\ \ddot{a} \\ \ddot{w} \end{bmatrix} + \begin{bmatrix} \mathbf{C}_0 & \\ & \mathbf{C}_1 \end{bmatrix} \begin{bmatrix} \dot{\mathbf{u}} \\ \dot{a} \\ \dot{w} \end{bmatrix} + \begin{bmatrix} \mathbf{K}_0 & \\ & \mathbf{K}_1 \end{bmatrix} \begin{bmatrix} \mathbf{u} \\ a \\ w \end{bmatrix} = \begin{bmatrix} \mathbf{f} \\ f_a \\ f_w \end{bmatrix} + \begin{bmatrix} \mathbf{g} \\ g_a \\ 0 \end{bmatrix} \tag{16.9}$$

The equilibrium condition $g_a + g_l = 0$ is fulfilled by imposing that the vector of the internal (rail-wheel) forces is written in the form

$$\begin{bmatrix} \mathbf{g} \\ g_a \\ 0 \end{bmatrix} = - \begin{bmatrix} \mathbf{D}^{\mathrm{T}}\boldsymbol{\Psi}^{\mathrm{T}} \\ -1 \\ 0 \end{bmatrix} \lambda \tag{16.10}$$

where λ is a Lagrange multiplier constituting a further unknown for the problem and having the meaning of rail-wheel contact force. The compatibility condition $a = u(X)$ is enforced explicitly through the equation:

$$\mathbf{B}\widehat{\mathbf{u}} = u_0\lambda; \quad \widehat{\mathbf{u}} = \begin{bmatrix} \mathbf{u} \\ a \\ w \end{bmatrix}; \quad \mathbf{B} = [\boldsymbol{\Psi}\mathbf{D} \quad -1 \quad 0] \tag{16.11}$$

Combining Eqs. (16.9)–(16.11), the equation of motion of the full system is given as:

$$\begin{bmatrix} \mathbf{M}_0 & & \\ & \mathbf{M}_1 & \\ & & 0 \end{bmatrix} \begin{bmatrix} \ddot{\widehat{\mathbf{u}}} \\ \ddot{\lambda} \end{bmatrix} + \begin{bmatrix} \mathbf{C}_0 & & \\ & \mathbf{C}_1 & \\ & & 0 \end{bmatrix} \begin{bmatrix} \dot{\widehat{\mathbf{u}}} \\ \dot{\lambda} \end{bmatrix} + \begin{bmatrix} \mathbf{K}_0 & & \mathbf{B}^{\mathrm{T}} \\ & \mathbf{K}_1 & \\ \mathbf{B} & & -u_0 \end{bmatrix} \begin{bmatrix} \widehat{\mathbf{u}} \\ \lambda \end{bmatrix} = \begin{bmatrix} \mathbf{f} \\ f_a \\ f_w \\ 0 \end{bmatrix} \tag{16.12}$$

It should be noted that the matrix **B** and the coefficient u_0 depend on the wheel position X and therefore make the Eq. (16.12) time-variant.

16.2.4 *Modeling Wheel Defects*

The introduction of wheel defects in the numerical scheme here presented is straightforward. To this purpose, it is sufficient to update the compatibility condition as:

$$a = u(X) + (R(\theta) - R_0) \tag{16.13}$$

where $R(\theta)$ is the wheel radius at the contact point as a function of the wheel position θ and R_0 is the nominal radius. This modification can be incorporated into Eq. (16.12) as:

$$\begin{bmatrix} \mathbf{M}_0 & & \\ & \mathbf{M}_1 & \\ & & 0 \end{bmatrix} \begin{bmatrix} \ddot{\widehat{\mathbf{u}}} \\ \ddot{\lambda} \end{bmatrix} + \begin{bmatrix} \mathbf{C}_0 & & \\ & \mathbf{C}_1 & \\ & & 0 \end{bmatrix} \begin{bmatrix} \dot{\widehat{\mathbf{u}}} \\ \dot{\lambda} \end{bmatrix} + \begin{bmatrix} \mathbf{K}_0 & & \mathbf{B}^{\mathrm{T}} \\ & \mathbf{K}_1 & \\ \mathbf{B} & & -u_0 \end{bmatrix} \begin{bmatrix} \widehat{\mathbf{u}} \\ \lambda \end{bmatrix} = \begin{bmatrix} \mathbf{f} \\ f_a \\ f_w \\ R(\theta) - R_0 \end{bmatrix} \tag{16.14}$$

16.3 Numerical Example

As an example of application we consider a portion of rail line composed by 30 sleepers. Figure 16.3 shows the deflection of the rail calculated at the mid point between two sleepers. The blue line represents the deflection calculated neglecting the term u_0 in Eq. (16.3); the red line represents the solution obtained including this term. The figure shows that this term has a significant impact on the estimation of the rail deflection.

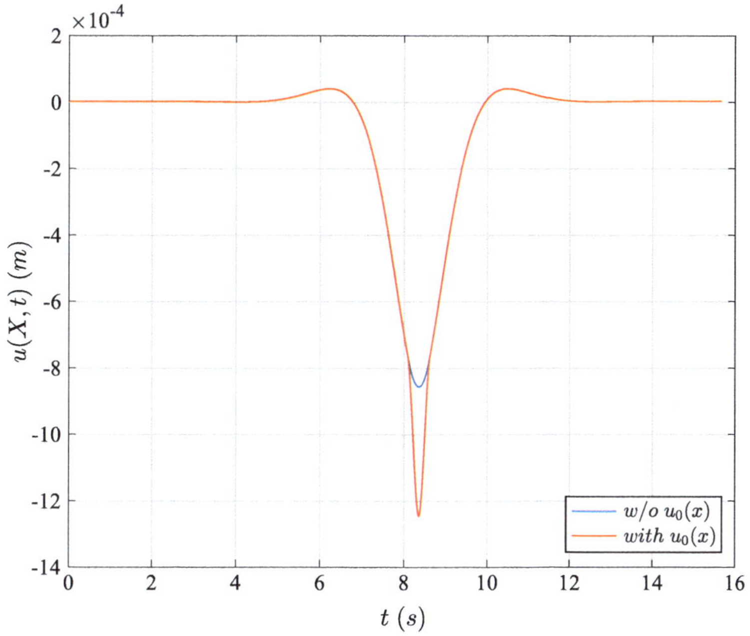

Fig. 16.3 Deflection in the mid point between sleepers # 15 and 16. Solution with and without the term $u_0(x)$ in Eq. (16.3)

16.4 Conclusion

The coupled dynamics of a rail with a traveling wheel can be formulated as a set of time-variant linear ODEs, whose unknowns are the displacement and rotation of the rail at its support points, the displacement of the sleepers, the displacement of the sprung and unsprung masses of the wheel and the contact force.

The linearity of the system, though with time variant coefficient, allows numerical solutions that are very efficient if compared with models based on nonlinear contact elements.

The dual assemblage provides very simple equations of motion. In contrast, the primal approach leads to very complicated equations containing the derivatives, up to the second order, of the shape functions $\boldsymbol{\Psi}$, as well as of the wheel defect $R(\theta)$.

The modelling of multiple wheels is relatively simple as the equation of motion can be obtained from Eq. (16.14) by replacing the unknowns a, w and λ with vector-valued quantities.

The inertia of the rail is approximated since the mass matrix $\mathbf{M}_0$ is obtained by projecting the inertial forces on the shape functions, i.e. approximating the beam displacement by a cubic function.

Acknowledgements This research was partially funded, in the IN2SMART Project, reference n. 730569, by the SHIFT2RAIL Joint Undertaking and the European Commission, within the SHIFT2RAIL Research Programme, call for members S2R-CFM-IP3-02-2016 Intelligent maintenance systems and strategies.

References

1. Esveld, C.: Modern Railway Track. MRT-Productions, Duisburg (1989)
2. Przemieniecki, J.S.: Theory of Matrix Structural Analysis. Dover, New York (1985)
3. de Klerk, D., Rixen, D.J., Voormeeren, S.N.: General framework for dynamic substructuring: history, review, and classification of techniques. AIAA J. **46**(5), 1170–1181 (2008)

Chapter 17
Planning of a Black-Box Benchmark Structure for Dynamic Substructuring

D. Roettgen and A. Linderholt

Abstract Experimental Analytical substructuring has been a popular field of research for several years and has seen many great advances for both frequency based substructuring (FBS) and component mode synthesis (CMS) techniques. To examine these technical advances, a new benchmark structure is being designed through the SEM Dynamic Substructuring technical division to act as a black-box study for anyone researching in the field. This black-box challenge will include experimental subcomponents and a finite element model to which these experimental results are to be connected. We often know the truth target when completing substructuring research; however, this challenge allows researchers to test their techniques without the aid of a truth measurement. IMAC XXXVII attendees are encouraged to contact the authors to join the team designing the black-box hardware and/or eventually have their institution participate in this challenge. The results from the assembled truth measurement will be presented at IMAC XXXIX.

Keywords Dynamic substructuring · Experimental analytical substructuring · Component mode synthesis · Frequency based substructuring · Structural modification

17.1 Research Strategy

Over the past decade there has been a flurry of research surrounding experimental-analytical substructuring. In 2011, the experimental substructures focus group developed a testbed structure that could be used for international collaboration. The structure selected was the Ampair 600 Wind Turbine. The origin of this test bed is discussed in detail in [1]. Since this initial testbed creation several universities have studied this structure. Sandia National Labs and University of Wisconsin studied the structure [2, 3] with applications of the component mode synthesis technique the Transmission Simulator method [4]. Simultaneously, Rahimi [5] and Brunetti [6] studied this structure with applications of Frequency Based Substructuring. The testbed became widely used by several other universities across the globe including University of Massachusetts Lowell [7], University of Stuttgart [8], Linnaeus University [9], and many more. During these studies, several new methods and technologies were developed, but the Ampair 600 Wind Turbine turned out to be quite a challenging structure.

First, as discussed by Harvey in [7], the blades of the wind turbine appeared to have large variance in interface tolerances which could have large impacts on the testbed research as depending on what blades a researcher obtained they may find very different assembled results. Second, as has been the study of much research, the bolted joints in the Ampair Wind Turbine create a significant source of nonlinearity which can have adverse effects when exploring the world of experimental analytical substructuring. A detailed procedure for assembly of the testbed, including potting of the hub, was written and is contained on the Dynamic Substructuring Wiki. Despite these difficulties the testbed proved to be a useful tool that has propelled

Sandia National Laboratories is a multimission laboratory managed and operated by National Technology and Engineering Solutions of Sandia, LLC., a wholly owned subsidiary of Honeywell International, Inc., for the U.S. Department of Energy's National Nuclear Security Administration under contract DE-NA-0003525. This paper describes objective technical results and analysis. Any subjective views or opinions that might be expressed in the paper do not necessarily represent the views of the U.S. Department of Energy or the United States Government.

D. Roettgen (✉)
Sandia National Laboratories, Albuquerque, NM, USA
e-mail: drroett@sandia.gov

A. Linderholt
Department of Mechanical Engineering, Linnaeus University, Växjö, Sweden
e-mail: andreas.linderholt@lnu.se

A. Linderholt et al., *Dynamic Substructures, Volume 4*, Conference Proceedings of the Society for Experimental Mechanics Series,
https://doi.org/10.1007/978-3-030-12184-6_17

Table 17.1 Proposed challenge timeline

Task	Timing
Initiate black-box challenge design work	June 2018–December 2018
Kick-off structure's detailed design by committee	January 2019–April 2019
Manufacture Final Design	May 2019
Test Assembly and Piece Part Variability	June 2019–September 2019
Initiate black-box challenge for participants	December 2019
Participants to present predicted results	January 2021

the dynamic substructuring research community down a positive, successful path with continuously growing interest from industry and researchers!

At IMAC XXXVI the focus group on Dynamic Substructuring officially transitioned into a Technical Division of SEM titled Dynamic Substructures. During this inaugural meeting, the Technical Division discussed the creation of a new benchmark structure to kick-off a new era of research in experimental analytical substructuring. At IMAC XXXIV Roettgen et al. [10] presented an experimental-analytical substructuring paper where the authors made blind predictions at the response of an assembled structure. A handful of team members performed the experimental work, then the substructuring team made predictions for a modified structure to assess the predictive capability of the Transmission Simulator method. This blind prediction was a foray at taking dynamic substructuring from being a research technology to an applicable resource in the world of structural dynamics. It was determined that a blind black-box benchmark structure would be an appropriate test of the current capabilities of the Dynamic Substructures Technical Division. This would provide a litmus test of current capabilities and provide an avenue for future collaboration and research.

17.2 Challenge Schedule

A proposed timeline for this black-box challenge is presented above in Table 17.1. A small team has begun detailing designs that will be presented during IMAC XXXVII where attendees will have a chance to see some of the designs in progress.

At IMAC XXXVII a large committee will be formed to finalize the design choices and complete a detailed design of the black-box structure. The new black-box testbed will then be fabricated and tested for variability to understand the variance introduced into the challenge. The results from this variability study will be available on the Dynamic Substructures Technical Division website and will be presented at IMAC XXXVIII. At that time, universities and industry groups will be able to sign up to use the experimental hardware to test their latest substructuring methodologies, where results shall be presented at IMAC XXXIV (2021). Upon completion of this initial black-box challenge, the testing results from several different research groups will be made available on the Dynamic Substructures Technical Division website and the results from truth testing will be presented. The data from these experiments will then be available for several future research tasks as the community continues to integrate analytical and experimental subcomponent models.

17.3 Design Considerations

The structure will consist of at least two experimental subcomponents and a finite element (FE) model, which these parts must be connected to. The experimental hardware will be available for checkout and the finite element model will be available through the Dynamic Substructures Technical Division website. The interfaces between the FE model and the experimental components shall consist of one continuous interface (to facilitate interesting component mode synthesis trials) and one interface consisting of 2–3 single point connection locations (to facilitate interesting frequency based substructuring trials); thus, the assembled structure will contain both a multi-point and continuous interface.

17.4 Closing Remarks

The objective of this proposed work is to continue to foster collaborative and cutting-edge research in the field of experimental-analytical substructuring. If successful, this black-box challenge will be useful in completing two main tasks. First, it will provide research institutions a useful means to test their substructuring capabilities on a structure with an unknown assembled response. Second, after the challenge is complete, it will provide several experimental data sets to test out future dynamic substructuring tool development as the field continues to grow. Attendees of IMAC XXXVII are encouraged to contact the authors of this abstract to show interest in both participating in this challenge and/or being a member of the design team to select a final testbed design.

Disclaimer This manuscript has been authored by National Technology and Engineering Solutions of Sandia, LLC. under Contract No. DE-NA0003525 with the U.S. Department of Energy/National Nuclear Security Administration. The United States Government retains and the publisher, by accepting the article for publication, acknowledges that the United States Government retains a non-exclusive, paid-up, irrevocable, world-wide license to publish or reproduce the published form of this manuscript, or allow others to do so, for United States Government purposes.

References

1. Mayes, R.L.: An introduction to the SEM substructures focus group test bed – the ampair 600 wind turbine. In: Topics in Experimental Dynamics Substructuring & Wind Turbine Dynamics, Jacksonville, FL (2012)
2. Rohe, D.P., Mayes, R.L.: Coupling of a bladed hub to the tower of the ampair 600 wind turbine using the transmission simulator method. In: Topics in Experimental Dynamics Substructuring, vol. 2, Orlando, FL (2014)
3. Roettgen, D.R., Mayes, R.L.: Ampair 600 wind turbine 3-bladed assembly substructuring using the transmission simulator method. In: Proceedings from the XXXIII IMAC Conference, Orlando, FL (2015)
4. Allen, M.S., Gindlin, H.M., Mayes, R.L.: Experimental modal substructuring to estimate fixed-base modes from a test on a flexible fixture. J. Sound Vib. **330**(18–19), 4413–4428 (2011)
5. Rahimi, S., deKlerk, D., Rixen, D.J.: The Ampair 600 Wind Turbine Benchmark: results from the frequency based substructuring applied to the rotor assembly. In: Proceedings of the 31st International Modal Analysis Conference, Orlando, FL (2013)
6. Brunetti, J., Culla, A., D'Ambrogio, W., Fregolent, A.: Experimental dynamic substructuring of the ampair wind turbine test bed. In: Proceedings of the International Modal Analysis Conference, Orlando, FL (2014)
7. Harvey, J., Avitable, P.: Comparison of some wind turbine blade tests in various configurations. In: Proceedings from the 30th International Modal Analysis Conference, Orlando, FL (2012)
8. Gross, J., Seeger, B., Peter, S., Reuss, P.: Applying the transmission simulator techniques to the ampair 600 wind turbine testbed. In: Proceedings of the 34th IMAC Dynamics of Coupled Structures, Vol. 4, Orlando, FL (2016)
9. Cwenarkiewicz, M., Johansson, T. Experimental dynamic substructirng of an ampair 600 wind turbine hub together with two blades. Linnaeus University Thesis Under Advisor: Andreas Linderholt (2016)
10. Roettgen, D.R., Seeger, B., Tai, W., Baek, S., Dossogne, T., Allen, M.S., Kuether, R.J., Brake, M.R.W., Mayes, R.L.: A comparison of reduced order modeling techniques used in dynamic substructuring. In: Proceedings of the International Modal Analysis Conference XXXIV, Orlando, FL (2016).

Chapter 18
Study on the Technology of Reliable Life Prediction of Plate Heat Exchanger for Ship

Longbo Liu, Na Han, Lingli Fu, and Jun Yao

Abstract Reliability prediction for new type plate heat exchanger, taking into accounts similar quantitative information obtained during use of the product. This paper presents a reliability prediction method using the similar product information. Similarity ratio is usually obtained by the experts' experience. Because the usual method exits some subjectivity, we use the concept of fuzzy closeness degree to analyze the performance parameter of plate heat exchanger. Calculate the field testing time based on the actual using time of similarity product. To estimate the reliability use the trouble-free truncated life evaluation method. Above method offer a new method to calculate the similarity ratio, optimized the expert assessment in part on the similarity of the subjective factor.

Keywords Reliability prediction · Similar product information · Similarity ratio · Fuzzy closeness degree · Trouble-free truncated

18.1 Introduction

The reliability prediction is a work in order to estimate the reliability of products in a given working conditions. It is not only a kind of technical means of reliability analysis, but also one of the six requirements of the reliability program of equipment development. Among them, the similar product technology is widely used in the process of reliability prediction, but the estimation of similarity ratio of this technology needs the support of expert's experience and manufacturer's evaluation on the similarity of the prototype. However, this technology has certain subjectivity, lack of objective evaluation on the overall situation of the products. Therefore, at present, the focus of study is on how to objectively determine similarity between two products and improve the accuracy of reliability prediction.

Aiming at the inseparability of various components of the system in the mission reliability prediction technology, the author of a paper [1] had proposed a mission prediction technology based on the information of similar products. The technology obtains the similarity ratio by considering the similarity between each subsystem and calculating the average weight coefficient. The authors [2, 3] had used the information of similar products as a sample, and adopted the similarity system analysis to determine the similarity between the historical sample and the current sample, and then summarized the inheritance factor. This technology makes full use of the information of similar products. The author of a paper [4] had proposed using non equal distance coefficient method to determine the inheritance factor. This technology, to a certain extent, overcomes the subjectivity that experts [5, 6] assess the similarity between two products based on their own experience, but it does not consider the performance characteristics of products. The authors [7–10] had took two similar equipments which had the same local parts as model, established an index model of reliability evaluation based on the case of the relevant

L. Liu (✉)
Naval Academy of Armament, Beijing, China
e-mail: 993618630@qq.com

N. Han · L. Fu · J. Yao
School of Reliability and System Engineering, Beihang University, Beijing, China

A. Linderholt et al., *Dynamic Substructures, Volume 4*, Conference Proceedings of the Society for Experimental Mechanics Series,
https://doi.org/10.1007/978-3-030-12184-6_18

information, and then proposed a technology of reliability evaluation under the condition of similar information of products by fusion algorithm. The author [11] had combined fuzzy number with the technology of reliability prediction of similar products, and then proposed a technology of reliability prediction of engine by combining a fuzzy similar product and fuzzy comprehensive evaluation. By the summary on above literatures, as to the reliability problem of similar products, many researchers [12–14] had focused on considering multiple inheritance factors, and then determined the similarity ratio in the system. But because the problem determining the similarity ratio between the two products is mainly through the experts' evaluation, it's lack of the objective calculation of similarity ratio.

On the basis of above researches, this paper researches on the model of reliability evaluation based on the information of similar products. In this case that similarity ratio is uncertain. In order to overcome the problem determining subjectively the similarity ratio, the degree of similarity between the two products is assessed objectively from the stand point of the failure mechanism of products, by quantitative estimation of the performance parameters and the calculation of conversion factor. This paper combines the data of practical application with the conversion factor to transform it into the outfield experimental data of the products and establishes the technology of reliability prediction based on similar products.

18.2 Establish the Reliable Life Prediction Model with Information of Similar Products

18.2.1 Basic Hypothesis

According to the reliability prediction technology of similar products, this paper made following hypotheses:

Hypothesis 1: Both of working environment and the use environment of products are similar, and the difference is minor.
Hypothesis 2: the failure is similar, the failure mechanism is the same and failure pattern is similar.
Hypothesis 3: the consistency of the products was good, inheritance is obvious. In other words, both of manufacturing process and design flow are similar.

18.2.2 The Principles and Approaches of the Assessment on Reliable Life

In response to these hypotheses, this model combines the method of reliability prediction of similar products. The key problem with this approach is to select the appropriate sensitivity stress that is the stress causing the main failure mode. The reliable life of new products is predicted in following steps and Fig. 18.1:

1. The analysis on the similar factor of products. In the respects like structure, function, design principle, working principle, material composition and the working environment of products, this paper comprehensively considers the similar factors of products;
2. Research the main failure mode of the products from the performance index of products. Through the analysis on the failure mechanism of products, this paper determines the main sensitive stress;
3. Calculate the value of unit sensitive stress. Because the two heterogeneous physical conditions, such as volume, quality of products are different, this paper unitizes the sensitive stress, and then compares and analyzes it;
4. For other factors that remove the stress-sensitive, this paper cannot compare and analyze it through assessing quantitatively. At this time, consult the opinions of relevant experts to get the conversion coefficient between products;
5. Analyze comprehensively multiple conversion coefficients of products. From the function and the failure mechanism of products, the technology taking the intersection or weighted is used to calculate the conversion coefficients of products;
6. Assess the reliability of new products by combining conversion coefficient products with historical information.

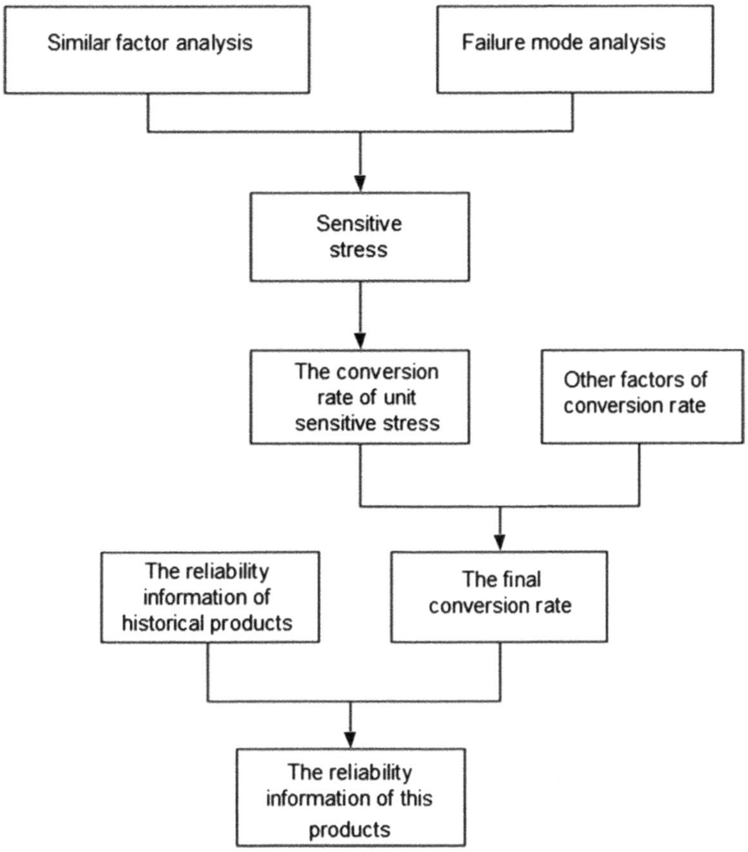

Fig. 18.1 The flow chart of the reliability prediction of new products based on the information of historical products

18.2.3 Calculation of Conversion Ratio of Similar Products Based on the Fuzzy Degree of Nearness

Reliability prediction of new products with similar products is based on many similarities between these products and the historical products.

Assuming that the new products and the historical products had a number of similarities, N characteristics were selected, and then, this paper considered from M aspects. If A1, A2 ... Am was used respectively as the characteristic factors, for the M characteristics, then, under conditions of uncertainty, the paper used the fuzzy membership function to represent what extent the new and existing products had these characteristics and how had the new product similarity degree with the existing products.

According to the concept of fuzzy sets close degree of fuzzy mathematics, suppose there were two objects A1 and A2. The feature set is T = {t1, t2 ... tm}, and t1, t2... tm is Eigen elements. Feature vector is $\left\{\mu_{A_i}\left(t_j\right) | i = 1, 2; j = 1, 2, \cdots, m\right\}$. Then the close degree of A1 and A2 is:

$$k = \frac{\sum_{R=1}^{m} \left(\mu A_1 (\mu R) \cup \mu A_1 (\mu R)\right)}{\sum_{R=1}^{m} \left(\mu A_2 (\mu R) \cup \mu A_2 (\mu R)\right)} \tag{18.1}$$

Normally a characteristic element has different impact on the reliability. The weight of the impact on the size is ω_1, ω_2, ... , ω_m. Then when the importance of various factors is considered, the measure of similarity between both is:

$$k = \frac{\sum_{R=1}^{m} \left(\omega R \left(\mu A_1 (\mu R) \cup \mu A_1 (\mu R)\right)\right.}{\sum_{R=1}^{m} \left(\omega R \left(\mu A_2 (\mu R) \cup \mu A_2 (\mu R)\right)\right.} \quad (18.2)$$

Obviously, when feature weight is not taken into consideration or the importance of attribute is identical, Eq. (18.2) translates into Eq. (18.1), which also means Eq. (18.1) is peculiar case of Eq. (18.2).

18.3 Reliability Lifetime Prediction for Plate Heat Exchanger

Taking plate heat exchanger [15, 16] for instance, reliability prediction technology based on similar products is proposed. By comparing plate materials, plate structures, ocean water flow rate, heat transfer temperature, sealing rubber and so on, conversion factor can be got. With application data of history products and the calculated conversion factor, this paper transfers application data of history products into outfield experimental data of this product to evaluate its life.

18.3.1 Analysis on Similar Factors of Plate Heat Exchanger

The cooler has high reliability when equipped in ships. As it is the first time that the cooler is equipped in certain naval vessel, before conducting reliability test, the data from civilian can be only used as references, and based on similarity analysis, its reliable life is analyzed and evaluated after calculating its life index.

In the view of working mechanism similarity and structure similarity, both of heat exchangers adopt plate structure, so they share the same working mechanism; they are both designed in the standard of "plate heat exchanger" BN/T 47004-2009, and they both meet the requirement.

According to the analysis on the similarity of structure and appearance, the compared sample exchange more heat and has a lager mass; however, the surface area, punching, groove and wave shape are different.

Judged from performance index, the two products are obviously different in lube, water flow, temperature difference between import and export, quantity of heat exchanged and so on. The comparing product works in a more severe circumstance.

18.3.2 Usual Failure Mode and Sensitive Stresses on Plate Heat Exchanger

Failure as leakage, fouling blocking stress corrosion damage and so on may come up to plate heat exchanger when it is designed or produced improperly, or the sealing structure is not reasonable, or the operation of installation and compaction is not proper, or the operating conditions of hot and cold media (pressure, temperature, velocity of flow and component) does not meet the requirements, or inspection, repair, maintenance are not timely. There are two main failure modes, i.e. deformation failure and corrosion failure. The main influencing factors for the deformation failure are the thickness of the plate and the fluid medium. Corrosion on plate heat exchanger occurs mainly on the cold side of the plate. Current velocity of the cold side has important influence on the corrosion rate. From the analysis above, the main sensitive stresses of plate heat exchanger are scouring water and heat transfer of cold side.

18.3.3 Calculating Conversion Ratio by Severity of Working Environment and Own Factor

(a) *Flow velocity comparison*

Though parameters of the two products are not all the same, flow velocity of medium is adjacent, calculated below: after calculation, flow velocity of the compared product is 0.4 m/s, less than 0.8 m/s; while velocity of the comparing product is 0.35m/s, also less than 0.8m/s. So the two products both meet the requirements. Only the flow velocity of seawater side, i.e. the cola side needs to be taken into consideration judging from the perspective of erosion corrosion. The conversion coefficient from consideration of the current of the cold side is K1 = 0.4/0.35 = 1.14.

(b) *Comparison of heat transfer coefficient*

Considering product performance parameters, heat exchange is a key indicator of product performance parameters. The heat transfer coefficient per unit mass is

$$\alpha = \frac{\text{heat transfer}}{\text{flow of cold side}} \tag{18.3}$$

By Eq. (18.3), heat transfer coefficient of compared product and heat transfer coefficient of this product are figured out: $\alpha 1 = 3.34$;$\alpha 2 = 11.6$; mass coefficient due to different mass is $\beta = 1410$ kg/1458 kg = 0.97; considering the quality and heat transfer, by formula (18.1), the conversion coefficient is figured out: $K2 = (\alpha 1/\alpha 2)*\beta = 0.279$.

(c) *Conversion of other similarity*

According to the similarity analysis above, this paper conducts comparison on principle similarity, structure and shape similarity, material similarity and working environment. By consulting experts, the conversion coefficient is obtained: K3 = 0.9.

In summary, by formula (18.2), the conversion coefficient between this product and compared product is:

$$K1^*K2^*K3 = 0.286.$$

Therefore, the data of the compared product, and approximately reckon 78 thousand hours, the service life of the compared product are converted as time of field test. According to the conversion coefficient above, test time after conversion is 78,000 h*0.286, 22,308 h.

18.4 Reliability Life Prediction of Plate Heat Exchanger Based on Conversion Factor

Considering the similarity between the two products, the conversion of field test time of plate heat exchanger is conduct with conversion factor.

By using no failure censored evaluating technology, the lubricating oil cooler research report points out that the product obeys the exponential [17] distribution. The field test time is 22,308 h and the no failure censored evaluating technology is used to evaluate the product's life.

References indicate lower confidence limit of average life of products θ under confidence level $1 - \alpha$ is:

$$\theta_L = \frac{\sum_{i=1}^{n} l_i}{-\ln \alpha} \tag{18.4}$$

Substituting experiment data into formula (18.4), the lower confidence limit of average life under confidence level 0.9, 0.8, 0.7 is figured out as Table 18.1:

Table 18.1 Relationship between reliability and lower confidence limit of reliable life

Reliability	0.9	0.8	0.7
Lower confidence limit	9688 h	13,861 h	18,529 h

18.5 Result and Discussion

This paper presents a reliability prediction technology based on similar products. The technology developed in this paper involves selecting the key performance parameters starting from the failure mechanism of products, comprehensively considering various performance factors, determining the degree of similarity between the two samples by means of quantitative conversion factor calculation, and then quantifying the inheritance factor. In the paper, a ship plate heat exchanger is taken as an example. By contrast analysis on plate material, plate structure, and water flow rate, heat transfer, temperature difference, and sealing rubber, conversion factor is figured out, and with application data of historical product, it is transferred into application data of this product. Evaluation on reliability life of the product by no failure censored evaluating technology indicates the product meets the design requirements. And that indicates this method has a certain practical significance for engineering application.

This technology provides a new idea about determination of ratio of similarity in reliability prediction of similar products. The technology, based on the failure mechanism of the product, to a certain extent, optimizes the subjective factors of expert assessment of two similar products, thus the results of the assessment accuracy is improved. The evaluating method may be also applied in a number of similar products and similar systems.

References

1. Zhuo, H.Y.: Method of mission reliability prediction based on similar product information. Ordnance Ind Autom. **33**(6), 8–15 (2014)
2. Chen, Z.Z., Xu, L.T., Yu, W.B., et al.: Similarity analysis of storage reliability of ammunition-borne photoelectric system. Equipment Environ Eng. **10**(6), 136–139 (2013)
3. Yang, J., Shen, L., Huang, J., et al.: Bayes comprehensive assessment of reliability for electronic products by using test information of similar products. Acta Aeronaut Astronaut Sinica. **29**(6), 1150–1153 (2008)
4. Yang, J., Huang, J., Shen, L.J., et al.: Bayes comprehensive assessment of reliability for binomial products by using test information of similar product. J Beijing Univ Aeronaut Astronaut. **7**, 786–788 (2009)
5. Zhang, L.B., Wang, H.L., Chen, C., et al.: The method of reliability assessment for electronic products based on multiple similar samples. Aerosp Control. **30**(10), 84–87 (2012)
6. Zhang, Z., Yang, Z.: Fuzzy mathematics assessment in the evaluation of air quality. In: International Conference on IEE Multimedia Technology, pp. 6378–6381 (2011)
7. Guo, Z., Tian, J.: Discussion about similarity measures in pattern recognition of fuzzy information. In: International Conference on IEEE Computational Intelligence and Security Workshops, CISW, vol. 2007, pp. 299–303 (2007)
8. Fang, M., Li, X.J., Liu, J.G.: An assessment method of the reliability for the similar equipment which has similar product information. Math Pract Theory. **38**(6), 107–112 (2008)
9. Zhang, H.Q., Chen, C.L., Cao, Y.K., et al.: Evaluation study on service status of armored vehicle engine based on the similarity system theory. Equipment Environ Eng. **10**(6), 37–40 (2013)
10. Du, L., Zhang, H.L., Huang, H.Z., et al.: Combining fuzzy similar product and fuzzy comprehensive evaluation method for reliability prediction of diesel engine. J Univ Electron Sci Technol China. **39**(4), 629–633 (2010)
11. Guo, H.K., Zhao, X.W., Duan, M.Q., et al.: Reliability prediction model based on minimal cut sets. J Sichuan Ordnance. **35**(3), 148–152 (2014)
12. Xu, Y., Shao, W.H.: Groundwater quality evaluation in yang village by fuzzy mathematics method. In: International Conference on. IEEE Digital Manufacturing and Automation (ICDMA), vol. 1, pp. 790–794 (2010)
13. Luo, M.Z., Kang, R., Liu, F.W.: A review of reliability prediction methods for electronic products. Electron Sci Technol. **1**(2), 246–256 (2014)
14. Li, R.Y., Kang, R., Dang, W.: Comparison and selection of reliability prediction method for mechanical products. Constr Machinery Equipment. **40**, 53–57 (2009)
15. Oronel, C., Amaris, C., Valles, M., et al.: Experiments on the characteristics of saturated boiling heat transfer in a plate heat exchanger for ammoniac/lithium nitrate and ammonia/(lithium nitrate + water). In: 3rd International Conference on IEEE, pp. 217–225 (2010)
16. Yu, L.J., Zhang, Y.F., Zhou, J.X.: Application of domestic super large scale plate shell heat exchanger in petrochemical units. Petro Chem Equipment. **39**(5), 69–72 (2010)
17. Li, B., Liu, J., Liu, D.: Reliability analysis on zero-failure data of missile power system. In: AIMSEC, 2nd International Conference on IEEE, pp. 5722–5725 (2011)

Chapter 19
Real-Time Hybrid Substructuring Results of the Mars Pathfinder Parachute Deployment

Michael J. Harris and Richard E. Christenson

Abstract Spacecraft are subjected to a variety of extreme loads during the course of a mission. One period during which these demanding loads are observed occurs during the parachute deployment stage of reentry. The deployment process utilizes a mortar in order to deploy the parachute and the corresponding reaction forces from deployment generate large impulsive loads on the spacecraft and cause vibrations throughout the spacecraft. Accurate prediction of the forces exerted on the spacecraft during deployment is paramount to the design and safety of the spacecraft. Typically the time history of the reaction forces exerted during parachute deployment are measured experimentally using a tripod-mounted mortar assembly. This approach introduces coupling between the force generated by the mortar and the dynamics of the tripod legs, and does not account for the structural compliance of the actual spacecraft system. This will lead to erroneous results if compared to the true reaction forces observed during the in-service deployment of the parachute during reentry.

In this paper, a cyber-physical test procedure called real-time hybrid substructuring (RTHS) is utilized to test examine the reaction forces which occurred during the parachute deployment of the Mars Pathfinder spacecraft. The RTHS test couples, in real-time, a numerical substructure consisting of a frequency-domain model of the Mars Pathfinder with a physical substructure consisting of a parachute being fired from a mortar in the Shock and Vibration Laboratory at the University of Connecticut. Reaction forces during parachute deployment was tested for both a fixed-base scenario as well as in the case where the dynamics of the spacecraft hull were incorporated into the test. The Mars Pathfinder RTHS test demonstrates a new approach in aerospace testing that can allow for component testing during the design phase to provide more realistic load profiles and dynamic responses of critical components of the spacecraft.

Keywords Real-time hybrid substructuring · Parachute deployment

19.1 Introduction

Throughout the aerospace industry, the use of devices such as exploding bolts and pyrotechnic actuators have become commonplace due to their versatility and reliability [1, 2]. These devices can be used for a number of applications including deployments and stage separations. One such application, parachute deployment, is pivotal to the safety and success of a spacecraft's mission. Parachute deployment is typically accomplished with a mortar explosion that is used to eject the parachute out of the wake of the spacecraft during re-entry. Typically the time history of the reaction force of the mortar is experimentally measured during the test program using a rigid mounting system. This approach neglects the structural compliance of the spacecraft and thus neglects the dynamic interaction between the deployment forces and spacecraft. This may lead to differences between the force profile observed during laboratory testing and those observed during the course of a mission.

Response to pyroshock can be divided into two general regimes; near-field and far-field. The near-field response occurs close to the site of the pyrotechnic detonation. The excitation force of the detonation contains significant high-frequency energy (upwards of 10 kHz) which induces high-frequency oscillations in the spacecraft. These oscillations typically dampen

M. J. Harris · R. E. Christenson (✉)
Department of Civil and Environmental Engineering, University of Connecticut, Storrs, CT, USA
e-mail: Michael.harris@uconn.edu; rchriste@engr.uconn.edu

A. Linderholt et al., *Dynamic Substructures, Volume 4*, Conference Proceedings of the Society for Experimental Mechanics Series,
https://doi.org/10.1007/978-3-030-12184-6_19

out rapidly (within 20 ms) and the low-frequency modal responses of the structure begin to dominate the structural vibrations [1]. The energy contained within far-field shock spectra is generally contained within lower frequencies than in the case of near-field shock. Specifically, the response of the structure at far-field excitation points will be concentrated at the frequencies corresponding to structural modes. These relatively low-frequency modes often exhibit more compliance and larger displacements than do the local, high-frequency modes. As such, these low-frequency modes may be the greatest cause for uncertainty in the force history profile.

A cyber-physical testing technique called real-time hybrid substructuring (RTHS) was utilized evaluation of the forces acting on the Mars Pathfinder spacecraft during parachute deployment. The RTHS test coupled a numerical substructure consisting of a frequency-domain dynamic model of the Mars Pathfinder in real-time with a physical substructure consisting of a mortar being fired in the Shock and Vibration Laboratory at the University of Connecticut. A more detailed discussion of the numerical and physical substructures is given in previous work [3].

Real-Time Hybrid Substructuring came about as an extension of Pseudo-dynamic testing (PDT) that allows for the testing of nonlinear, complex, or rate-dependent components such as viscoelastic dampers [4–6]. In PDT, the boundary conditions and excitations are imposed on the structure while the restoring forces of the structure are measured [7]. The equilibrium equation is then solved for the structure and the process continues at an elongated time scale. In contrast, RTHS utilizes increased computation speeds to solve, and enforce the boundary conditions between numerical and physical substructures in "real-time." This enables RTHS to analyze rate-dependent and nonlinear by localizing these effects in the physical substructure.

19.2 RTHS Test Equipment

To conduct the RTHS test in 'real-time,' specialized equipment is utilized to perform numerical calculations at high speeds and to impose boundary conditions on the physical substructure at high rates. This research project calculates the response of the numerical substructure using dSPACE 1103 controller running at 4096 Hz. The reaction force from the parachute canister will be converted from an analog to a digital signal and used as the input at the canister mounting point within the numerical model of the spacecraft hull. The numerical model of the Mars Pathfinder spacecraft is being provided by NASA/JPL-Caltech. The output of the numerical substructure is the displacement of the spacecraft hull at the canister mounting point. This target displacement will be sent to a MTS Flex Test 40 digital controller and used to control a hydraulic actuator. A high-speed MTS 244.41S, 110 kip hydraulic actuator will be used to impose the boundary constraint on the physical substructure of the parachute mortar assembly. This actuator is capable of effecting high-speed displacement control of the parachute canister. The actuator is also equipped with a load cell which will be used to sense the reaction force of the parachute deployment. The load cell will provide the means of measuring the reaction forces and inputting them into the numerical substructure for the next numerical integration time step. However, in order to interface the actuator piston with the mortar assembly, a steel fixture was added to mount the tube. As this additional inertial mass would not be present during in-service deployment of the parachute, the force on the actuator load cell caused by this fixture needed to be removed from the total force sensed by the actuator load cell. The contribution of the fixture to the force sensed by the load cell was calculated through the use of a PCB 353B33 accelerometer. The acceleration of the load cell and fixture were then multiplied by the inertial mass of the fixture to obtain the force contribution of the fixture to the load cell force. The contribution and cancelation of these experimental inertial forces was first described by Stehman and Nakata [8]. A depiction of the cycle undergone in RTHS is depicted in Fig. 19.1. In this figure, P is the force from the mortar, R_p is the reaction force at the base of the parachute canister to be imposed on the numerical substructure, and x_n is the numerically calculated displacement at the base of the parachute canister. This point at the base of the canister corresponds to the interface between the physical and numerical substructures. Data collection equipment during testing included a DataPhysics Spectrum Analyzer sampling at 4096 Hz as well as a Photron Fastcam high-speed camera running at 4000 frames per second. This high-speed camera was utilized to calculate the ejection velocity of the parachute packs for each test.

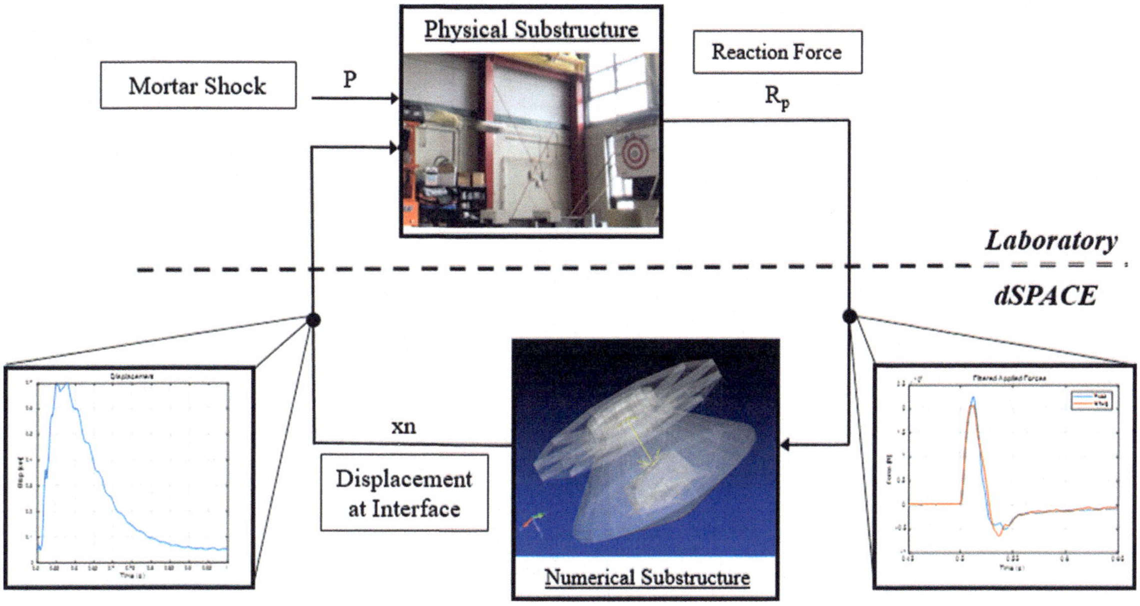

Fig. 19.1 Depiction of RTHS testing procedure. Image of numerical substructure courtesy: NASA/JPL-Caltech

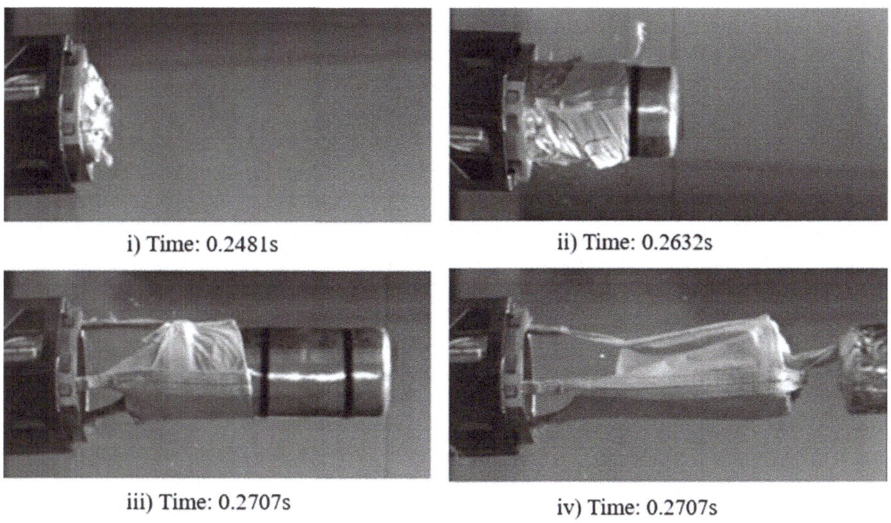

Fig. 19.2 Image captures from the high-speed footage recorded during testing

19.3 Results

The reaction forces during parachute deployment were tested via RTHS testing for two scenarios. In the first "Fixed" scenario, a null displacement was commanded of the actuator. This scenario was similar to a true fixed-base test. In the second "RTHS" scenario, the displacement command signal of the actuator was the output of the reaction force to the parachute mounting point displacement. Images from high-speed footage as well as the reaction forces obtained for the two test scenarios can be seen in Figs. 19.2 and 19.3 respectively. In order to isolate dynamics within the testing region analyzed, the reaction forces have been sent through a 2-pole, 40 Hz, digital low-pass filter.

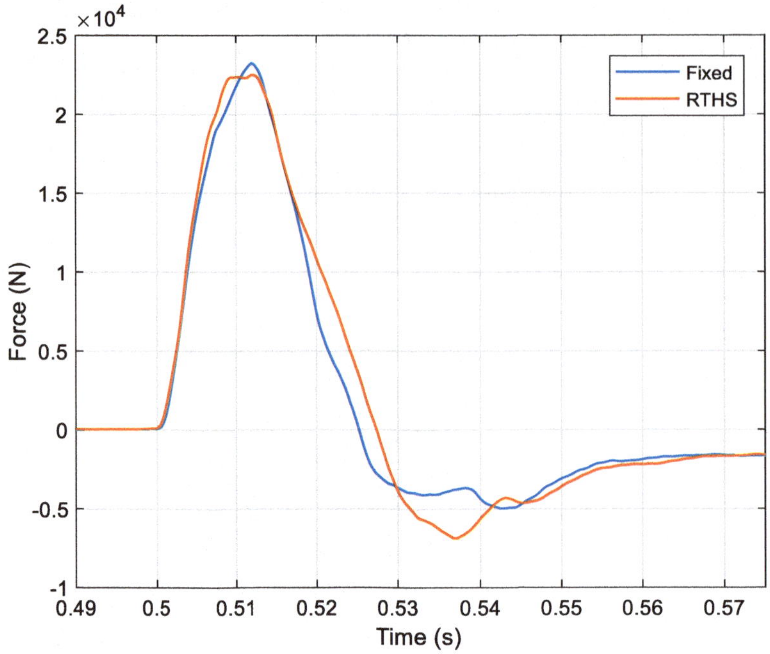

Fig. 19.3 A comparison of reaction force profiles for the Fixed and RTHS test scenarios

19.4 Conclusions

Experimental testing of parachute deployment with the Real-Time Hybrid Substructuring technique shows clear differences between the fixed-base (Fixed) and coupled (RTHS) scenarios. In particular, test results show a decrease in the peak reaction force observed during the RTHS case when compared to the fixed case. In addition, the RTHS scenario shows an increase in the duration of the impulse compared to the force profile observed in the fixed-base scenario. Further processing and analysis is underway to understand and justify these results.

References

1. Harris, C.M., Piersol, A.G.: Harris' Shock and Vibration Handbook, 5th edn. McGraw-Hill, New York (2002)
2. Ryschkewitsch, M.G.: National Aeronautics and Space Administration. Pyroshock test criteria: NASA-STD-7003A, Washington, DC (2011)
3. Harris, M.J., Christenson, R.E.: Experimental test of spacecraft parachute deployment using real-time hybrid substructuring. In: Conference Proceedings of the Society for Experimental Mechanics Series, pp. 67–70 (2018)
4. Blakeborough, A., Williams, M.S., Darby, A.P., Williams, D.M.: The development of real-time substructure testing. Philos. Transact. A Math. Phys. Eng. Sci. **359**(1786), 1869–1891 (2001)
5. Botelho, R.M.: Real-time hybrid substructuring for marine applications of vibration control and structural acoustics. Ph.D. Thesis, Department of Civil, and Environmental Engineering, University of Connecticut (2015)
6. Nakashima, M., Kato, H., Takaoka, E.: Development of real-time pseudo dynamic testing. Earthquake Eng Struct Dyn. **21**, 79–92 (1992)
7. Mahin, S.A., Shing, P.B.: Pseudodynamic method for seismic testing. J Struct Eng. **111**(7), 1482–1503 (1985)
8. Stehman, N., Nakata, N.: Experimental investigation of inertial force control for substructure shake table tests. In: World Conference on Earthquake Engineering, Lisbon, Portugal (2012)